TISSUE CULTURE IN FORESTRY AND AGRICULTURE

BASIC LIFE SCIENCES

Alexander Hollaender, General Editor

Council for Research Planning in Biological Sciences, Inc., Washington, D.C.

Recent volumes in the series:

Volume 18 TRENDS IN THE BIOLOGY OF FERMENTATIONS FOR FUELS AND CHEMICALS
Edited by Alexander Hollaender, Robert Rabson, Palmer Rogers, Anthony San Pietro, Raymond Valentine, and Ralph Wolfe

Volume 19 GENETIC ENGINEERING OF MICROORGANISMS FOR CHEMICALS
Edited by Alexander Hollaender, Ralph D. DeMoss, Samuel Kaplan, Jordan Konisky, Dwayne Savage, and Ralph S. Wolfe

Volume 20 MOLECULAR AND CELLULAR MECHANISMS OF MUTAGENESIS
Edited by J. F. Lemontt and W. M. Generoso

Volume 21 GENETIC TOXICOLOGY: An Agricultural Perspective
Edited by Raymond A. Fleck and Alexander Hollaender

Volume 22 PLANT CELL CULTURE IN CROP IMPROVEMENT
Edited by S. K. Sen and Kenneth L. Giles

Volume 23 INDUCED MUTAGENESIS: Molecular Mechanisms and Their Implications for Environmental Protection
Edited by Christopher W. Lawrence

Volume 24 ORGAN AND SPECIES SPECIFICITY IN CHEMICAL CARCINOGENESIS
Edited by Robert Langenbach, Stephen Nesnow, and Jerry M. Rice

Volume 25 BASIC BIOLOGY OF NEW DEVELOPMENTS IN BIOTECHNOLOGY
Edited by Alexander Hollaender, Allen I. Laskin, and Palmer Rogers

Volume 26 GENETIC ENGINEERING OF PLANTS: An Agricultural Perspective
Edited by Tsune Kosuge, Carole P. Meredith, and Alexander Hollaender

Volume 27 NEUTRONS IN BIOLOGY
Edited by Benno P. Schoenborn

Volume 28 GENETIC CONTROL OF ENVIRONMENTAL POLLUTANTS
Edited by Gilbert S. Omenn and Alexander Hollaender

Volume 29 SISTER CHROMATID EXCHANGES: 25 Years of Experimental Research
Part A: The Nature of SCEs
Part B: Genetic Toxicology and Human Studies
Edited by Raymond R. Tice and Alexander Hollaender

Volume 30 PLASMIDS IN BACTERIA
Edited by Donald R. Helinski, Stanley N. Cohen, Don B. Clewell, David A. Jackson, and Alexander Hollaender

Volume 31 GENETIC CONSEQUENCES OF NUCLEOTIDE POOL IMBALANCE
Edited by Frederick J. de Serres

Volume 32 TISSUE CULTURE IN FORESTRY AND AGRICULTURE
Edited by Randolph R. Henke, Karen W. Hughes, Milton J. Constantin, and Alexander Hollaender

---

TISSUE CULTURE IN FORESTRY AND AGRICULTURE

Edited by

Randolph R. Henke
Karen W. Hughes
University of Tennessee
Knoxville, Tennessee

Milton J. Constantin
Phyton Technologies, Inc.
Knoxville, Tennessee

and

Alexander Hollaender
Council for Research Planning in Biological Sciences, Inc.
Washington, D.C.

Technical Editor

Claire M. Wilson
Council for Research Planning in Biological Sciences, Inc.
Washington, D.C.

Springer Science+Business Media, LLC

Library of Congress Cataloging in Publication Data

Main entry under title:

Tissue culture in forestry and agriculture.

(Basic life sciences; v. 32)
"Proceedings of the Third Tennessee Symposium on Plant Cell and Tissue Culture . . . held September 9–13, 1984, at the University of Tennessee, Knoxville"—P.
Bibliography: p.
Includes index.
1. Plant tissue culture—Congresses. 2. Plant propagation—Congresses. 3. Agriculture—Congresses. 4. Forests and forestry—Congresses. I. Henke, Randolph. II. Tennessee Symposium on Plant Cell and Tissue Culture (3rd: 1984: University of Tennessee, Knoxville) III. Series.
SB123.6.T57 1985 582'.007'24 85-585
ISBN 978-1-4899-0380-8

ISBN 978-1-4899-0380-8 ISBN 978-1-4899-0378-5 (eBook)
DOI 10.1007/978-1-4899-0378-5

Proceedings of the Third Tennessee Symposium on Plant Cell and Tissue Culture, entitled Propagation of Higher Plants Through Plant Tissue Culture III, held September 9–13, 1984, at the University of Tennessee, Knoxville, Tennessee

This symposium was supported by Grant No. 40-3187-4-1561 from the U.S. Department of Agriculture Forest Service (USDA); Grant No. DE-FG05-84ER13271 from the U.S. Department of Energy (DOE); and Grant No. PCM 83182-38 from the National Science Foundation (NSF). Any opinions, findings, and conclusions or recommendations expressed in this publication are those of the authors and do not necessarily reflect the views of the USDA, DOE, or NSF.

Originally published by Plenum Press, New York in 1985
Softcover reprint of the hardcover 1st edition 1985

DEDICATION

L. Evans Roth

This publication of the University of Tennessee's third symposium on "The Propagation of Higher Plants Through Tissue Culture" is dedicated to Dr. L. Evans Roth, Professor of Zoology, for his leadership, encouragement, and contributions to the biological sciences at the University. As Vice Chancellor for Research and Graduate Studies, Dr. Roth recognized the new opportunities and bright future developing in the plant sciences, particularly in the areas of tissue culture and molecular biology. Through this close involvement with the plant science faculty, Dr. Roth has increased the mutual cooperation and exchange of ideas with our local and national colleagues through seminars, workshops, and symposia. Dr. Roth provided the initial support that led to the first symposium that was held at the University of Tennessee in 1978, he played a role in the 1980 symposium, and he was involved in this symposium, held in September, 1984.

We feel particularly fortunate to have Dr. L. Evans Roth as an advocate for biology and as a fellow biologist at the University of Tennessee, and we are grateful for his service to the profession and for his friendship.

-- The University of Tennessee Symposium Organizing Committee

ACKNOWLEDGMENTS

We extend our appreciation and gratitude to the University of Tennessee and to several government agencies, commercial firms, and numerous individuals who contributed funding and services in support of the symposium. Specifically, we would like to thank the University of Tennessee's Office of Research, the Agricultural Experiment Station, and the College of Liberal Arts for their generous and continuing support of this symposium series. Local support was also provided by the Tennessee Technology Foundation, Oak Ridge Chamber of Commerce, and Phyton Technologies, Inc. We are also grateful for the continuing industrial support of this series.

The industry sponsors were: Advanced Genetic Sciences, Agrigenetics, ARCO Plant Cell Research Institute, Ciba-Geigy, DNA Plant Technologies, E.I. DuPont de Nemours, FMC Corporation, Hershey Foods Corp., General Foods, Martin Marietta, Lilly Research Laboratories, Pioneer Hi-Bred, Monsanto, Proctor and Gamble, Plant Genetics, Inc., and Zoecon.

We would also like to thank the National Science Foundation, the U.S. Department of Agriculture Forest Service, and the U.S. Department of Energy for their supporting grants.

We are greatly indebted to our symposium organizing committee, especially to Peter Carlson, Robert Lawrence, and Richard Zimmerman, for their extensive donations of time and effort in developing the symposium program. We would also like to thank the local organizing committee, Jim Caponetti, Bob Conger, Milton Constantin, Donald Dougall, Dennis Gray, Randolph Henke, Otto Schwarz, James M. Stuart, and Russell Weigel, for their help with the details of the symposium management.

A special recognition is due to Dr. Karen Hughes, who chaired the organizing committee, for her dedication and commitment to the details that led to a successful symposium.

-- The Editors

SPECIAL ACKNOWLEDGMENT

We are especially indebted to Dr. Alexander Hollaender for his guidance and encouragement in the development of this symposium. Dr. Hollaender and the Council for Research Planning in Biological Sciences have been invaluable in the planning and execution of both the symposium and the symposium volume. We appreciate the very fine job of Claire Wilson and her associate, Gregory Kuny, in assembling the manuscripts and preparing them for publication.

We would also like to thank Roberta Schwarz for the cover design of the brochures.

Randolph R. Henke
Karen Hughes
Milton J. Constantin

PREFACE

This symposium is the third in a series featuring the propagation of higher plants through tissue culture. The first of these symposia, entitled "A Bridge Between Research and Application," was held at the University in 1978 and was published by the Technical Information Center, Department of Energy. The second symposium, on "Emerging Technologies and Strategies," was held in 1980 and published as a special issue of Environmental and Experimental Botany. One of the aims of these symposia was to examine the current state-of-the-art in tissue culture technology and to relate this state of technology to practical, applied, and commercial interests. Thus, the third of this series on development and variation focused on embryogenesis in culture: how to recognize it, factors which affect embryogenesis, use of embryogenic systems, etc.; and variability from culture. A special session on woody species again emphasized somatic embryogenesis as a means of rapid propagation. This volume emphasizes tissue culture of forest trees. All of these areas, we feel, are breakthrough areas in which significant progress is expected in the next few years.

There were approximately 300 attendees, equally representing academia, government, and industry. This excellent mix of basic and applied researchers provided a unique forum for interaction and exchange of knowledge. The symposium planners developed a schedule of workshops to exploit this professional mix and to facilitate an open and free exchange of information, ideas, and needs. The workshops ranged from practical topics in management and design of commercial tissue culture laboratories and the micropropagation of trees and ornamentals, to pure science workshops on somatic cell strategies of sexually produced crops.

Manuscripts from most of the principal speakers, along with abstracts of the posters presented, constitute the Proceedings of this third symposium. We are hopeful that this book will serve as a useful source of the details and concepts that were presented during the meeting.

CONTENTS

KEYNOTE ADDRESS

REGENERATION PHENOMENA

DEVELOPMENTAL MECHANISMS

APPLICATIONS OF DEVELOPMENTAL SYSTEMS

GENETIC VARIABILITY AND STABILITY

PROPAGATION OF WOODY SPECIES

TISSUE CULTURE FOR FOREST IMPROVEMENT

POSTER ABSTRACTS

A RANDOM WALK THROUGH PLANT BIOLOGY

Leon Dure, III

Biochemistry Department
University of Georgia
Athens, Georgia 30602 USA

INTRODUCTION

The fact that our Symposium organizer, Dr. Karen Hughes, asked me to give this address made it obvious to me that she did not want an introductory talk that dealt specifically with plant tissue culture. This was obvious because what I know about the subject comes from reading and listening rather than from participating in its practice.

From this I have assumed that I have been given the license to talk from this outside perspective on most anything I want. People rarely have such an opportunity; I intend to take advantage of it.

As a practicing biochemist and a peripheral participant in the alleged biotechnological revolution based on genetic engineering, I would like to present some random impressions of what we are living through in the plant sciences in these exciting, highly charged days.

PLANT TRANSFORMATIONS: OVERSOLD OR ON SCHEDULE?

When it became apparent that we had the ability to cut and paste genes into vectors, to deliver these vectors to cultured cells, and to regenerate these cells into mature, reproducing plants that express the newly introduced genes, visions of a new era in agriculture were widely publicized. After all, the only thing we needed was for someone to give us a list of the appropriate genes to introduce to get better plants. Have these visions proven to be naive and bankrupt, or are they on the verge of realization?

Neglect of Plant Biochemistry

Some of the early commentary was obviously naive. Many plant scientists were aware early on that we were abysmally ignorant of the nuances of plant metabolism and of its various levels of regulation, and that this ignorance would present a formidable obstacle to a reasoned application of the new technology. No one had a list of the appropriate genes whose products would enhance plant productivity or efficiency. In fact, we could not define succinctly what was meant by a "better plant." No one had the slightest idea of what would happen should wheat plants be given the ability to synthesize nitrogenase and the other proteins of the *nif* complex. Could the transformed plant support the very high-energy requirement for endogenous nitrogen fixation? Could this process coexist in the same cell with nitrate reduction? Could it be kept anaerobic?

Most of the clearly desirable characteristics of plants are the products of batteries of genes, and, knowing little about the regulation and integration of metabolism, we knew very little about what genes are involved. Nor could we predict the effect on the overall metabolic equilibrium of a cell by juggling the steady-state concentration of a single component.

Why had our knowledge of plant metabolism and its regulation not kept pace with microbial and animal biochemical research? The answer to this has always been clear. There have never been very many scientists worldwide doing this type of research. There never developed a critical mass of researchers to establish model systems with the massive information base as exemplified by *Escherichia coli* or rat liver. This in turn can be traced to the reactionary provincialism and lack of imagination and inspiration of botany departments, and to the fact that food has not been a problem in the developed, research-active world, whereas people do get sick.

Perhaps of paramount significance was the fact that the basic plant sciences were significantly underfunded by the U.S. granting agencies. It appeared that some plant sciences supporters lost sight of the dependence of applied science on basic research, on the necessity to fund and promote research based purely on intellectual curiosity and finding out how nature works for its own sake.

As we (but not everyone) know, answers in science rarely result from a direct assault on a given question. Since, in conducting fundamental research, we deal with the unknown, we really cannot even formulate the question properly. As experiments are performed, the questions change; the old questions are found to be foolish and beside the point. Instead, answers are often found to questions that heretofore we did not know to ask.

In short, the institutions that could have fostered sophisticated explorations into the molecular bases of plant processes did a poor job of it over the past 35 years.

Oversimplification of the Problem

A second aspect of the naiveté prevalent at the launching of the genetic engineering era was the failure to realize that we had very little information on what is involved in the tissue-specific expression of genes; we still don't. Furthermore, we have only rudimentary ideas on how the level of expression is regulated. What has been achieved in most instances in the expression of foreign genes in plants is the low-level constitutive expression of marker genes. The instances where high levels of expression or near-tissue specific expression that have been achieved have not been by design.

This aspect of naiveté probably resulted from the fact that many of the early boosters of the new era were brought up as microbiologists. Most of the problems mentioned above do not exist for procaryotes. The quantum jump in complexity between modulating gene expression in procaryotes and eucaryotes was not allowed to dampen our prospects.

New Synergisms Appear

Is the ignorance of basic plant biochemistry proving to be a big impediment to putting the technology of molecular biology to work in agriculture? Are we having to wait until this work is done? In my opinion, the answer to these questions is a resounding no.

First of all, while we await the development of an integrated view of plant metabolism so as to identify limiting factors in plant productivity and, thus, be in a position to engineer superplants, it has turned out that there is plenty to do in the meanwhile that is instructive, fruitful, and exciting. This has been the result of a synergistic coming together of areas of research that heretofore appeared to have little commonality. The collective front produced by this merging of interest and efforts can be labelled "the study of plant stress."

A second factor that has had profound influence on the progress-despite-ignorance status of bioengineering in plants is the enormous financial investment in people and facilities for plant molecular research by the private sector, e.g., risk-taking individuals. In the space of 2 to 3 years, the strong resistance to change at the university level and lack of adequate funding by granting agencies at a national level was bypassed. The fallout is just beginning to be felt as these latter institutions are being forced to be citadels of modern plant biology.

STRESS SERENDIPITY

Stress as a unifying theme in plant biology was unheard of several years ago. These various stresses could be seen as individual phenomena that could be studied in isolation. Today it is clear that a great deal is being learned about plant biology as the result of studying plant reactions to different types of stress, both biological and environmental.

One of the obvious results of these studies is the recognition of the multiplicity of responses that plants make use of to survive. Table 1 lists a few of these. Many of these responses are in place before the stress is experienced. Much of what is in a seed, we now realize, is there to make seed eating unpleasant or deleterious to the potential seed eater, e.g., surface lectins to agglutinate fungi or bacteria, proteinase inhibitors to give insects a bellyache, etc. The other responses are on-call, elicitable by the stress itself. Plant metabolism obviously must be able to change in response to the stress, and consequently our views of the regulation of metabolism must accommodate this ability. This is not a requirement shared with animal intermediary metabolism, I should point out.

There are a number of examples of how the study of stress phenomena on a molecular level has had a synergistic effect; how work in one area has opened up opportunities in another.

A few years ago very few individuals studying the movement of electrons from photosystem II to photosystem I in photosynthesis would have thought that their findings would be instrumental in the creation of plants that are resistant to some of the commonly used herbicides of the agricultural world. Yet this is precisely what

Tab. 1. A collation of some of the biosynthetic responses to various stresses. The response involves an elevation in concentration of the molecules listed.

PLASTICITY OF RESPONSE

STRESS	RESPONSE (PRE-PROGRAMMED OR ON-CALL)
Biological	
Insects	-Proteinase Inhibitors
Fungi	-Flavanoids, Other Phytoalexins
Bacteria	-Chitinase, Other Glucanases
	-OH-Proline Cell Wall Protein
	-Cyanogenic Substrates
	-Agglutinating Lectins
Environmental	
Heat ------------------------------	Heat Shock Proteins
Chilling --------------------------	?
Water Stress, Salts ---------------	Proline, Betaines, Polyamines
Anaerobiosis ----------------------	Alcohol Dehydrogenase + Others
UV Light --------------------------	Carotenoids
Toxic Metals ----------------------	Metalothionine
Toxic Chemicals (Herbicides) -------	Detoxifying Enzymes

has happened as the mode of action of the triazine herbicides has been uncovered. Once the molecular basis of herbicide action is known, the tools of genetic engineering can be (and are being) brought to bear on making more commercial species resistant to them. Table 2 indicates some of the progress in this single area. The growing season of 1986 should find genetically engineered species resistant to many of these compounds being used commercially.

Another instance of the opening up of new vistas as heretofore unrelated pursuits suddenly find a common ground is the story of oligosaccharides as potential informational molecules in plant life. The development of this idea is an outgrowth of studies of plant reactions to stress.

Oligosaccharides as Plant Messengers

In the middle 1970s Clarence Ryan (Washington State University) purified and characterized biochemically the proteinase inhibitors from a number of plant species and tissues. He noted that certain species (potatoes and tomatoes in particular) would dramatically increase their levels of several trypsin-like proteinase inhibitors in their leaves if they had been wounded (mechanically abraded, in this case). Since abrasion is analogous to tissue destruction by insect feeding, the increase in proteinase inhibitors could be considered a stress response to insects. The fact that leaves on parts of the plant distal to the wounded leaf would also respond indicated that information about the wounding was transmitted through the plant. This, in turn, meant that molecules, generated in response to the wound, were moving. Ryan named this moving informational molecule "PIIF" (Proteinase Inhibitor Inducing Factor). Through prodigious

Tab. 2. Herbicide resistance of various species. The institution principally responsible for determining the mechanism of resistance is given in parentheses beneath the herbicide.

HERBICIDE	TARGET	RESISTANT SPECIES	MECHANISM
ATRAZINE (DOE/MSU)	"32 KD PROTEIN"	PIGWEED CHLAMYDOMONAS MAIZE	SER → GLY SER → ALA DETOXIFICATION BY OXIDATION
GLYPHOSATE (Calgene)	ENOYL PYRUVYL SHIKIMATE PHOSPHATE SYNTHETASE	SALMONELLA	PRO → SER
ALACHLOR (Monsanto)	?	MAIZE SOYBEANS	DETOXIFICATION BY GLUTATHIONE S-TRANSFERASE
SULFONYL UREAS (Dupont)	ACETOLACTATE SYNTHETASE	YEAST	PRO → SER

effort he purified the active molecular fraction. Preliminary analysis indicated that it was a polysaccharide fraction composed of galacturonic acid residues. When this finding was announced by Ryan, Peter Albersheim (University of Colorado) realized that among the natural products of the enzymatic breakdown of sycamore cell walls was a fraction comprised of linearly linked galacturonic acid residues. Ryan and Albersheim exchanged preparations of their materials and Albersheim's degradation fraction proved to be a potent PIIF. Ryan's PIIF, whose structure was determined by Albersheim, proved to be linearly linked galacturonates of about the same average chain length (8-15 residues) as the sycamore degradation fraction. Thus, PIIF could be seen as a degradation product of the cell walls of the wounded leaf cells, probably generated by the cells' own hydrolyases that are released upon cell death. What was exciting about these observations was that nonwounded cells have receptors for small oligosaccharides that, upon interacting with the oligosaccharide effector, cause the elevation in level of specific proteins that, in this case, can be seen as toxic to invading organisms. (Plant cell endoproteinases are generally not of the trypsin type mechanistically.) Ryan subsequently demonstrated that the elevated levels of the proteinase inhibitors are the result of elevated levels of their mRNAs. Thus, the ultimate effect of PIIF was on gene expression.

Meanwhile, Charles West (UCLA) was investigating the "suicide" synthesis of the phytoalexin casbene by plant leaves which is induced by fungal invasion. This phytoalexin, like many others, is toxic to both the plant cell and to the invading fungal hyphae. West found that he could mimic the fungus-induced casbene synthesis with Ryan's PIIF. In this case the oligosaccharide presumably induces the expression of a different set of genes than it does in potatoes or tomatoes: those for casbene synthesis. The overall plant strategy, however, is the same: protection from pathogens in response to a cell wall oligosaccharide fragment.

Subsequently, it has been found in many instances of plant/pathogen interactions that small oligosaccharides, emanating from the plant cell as the result of pathogen hydrolyases, or emanating from the pathogen cell wall as the result of plant hydrolyases, can trigger the various plant defense systems. The overall scheme is shown as Fig. 1.

The fact that cell wall fragments modulate gene expression in plant defense responses has led Albersheim to explore the idea that cell wall fragments may play roles in normal, nonpathogenic plant development, again by influencing gene expression. The idea here is that the sensing tissues of plants (leaves in the case of light) may transmit their information to the responding tissues (meristems) through the release of specific cell wall oligosaccharides. Such an idea requires that the sensing tissue respond itself by synthesizing stimulus-specific cell wall hydrolyases.

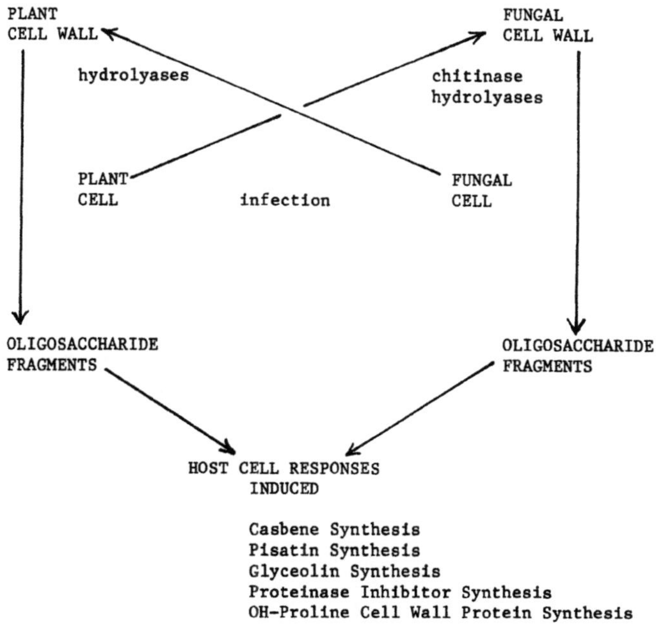

Fig. 1. A composite diagram of some of the events triggered by plant/fungus interactions. The likely involvement of ethylene is not shown since an obligatory role of this regulator in any of these processes has not been demonstrated as yet.

It is premature to present detailed results of the experiments of Albersheim and his colleagues on this idea, since they are still in press. However, it has been reported that specific morphogenic effects (both positive and negative) have been obtained when plant tissue pieces are exposed to minute levels of specific cell wall breakdown preparations.

The full import of these observations is difficult to gauge at this point. Clearly they could herald an enormous breakthrough in comprehending how plants work. No one set out specifically to examine the effects of oligosaccharides on plant morphogenesis. There was no reason to. Stress serendipity played a role. Surely, the world of plant biotechnology will have to accommodate these observations in its plans.

(Experimental documentation of the stress phenomena alluded to in this presentation can be found in The Cellular and Molecular Biology of Plant Stress, (1985) T. Kosuge and J.L. Key, eds. Alan R. Liss, Inc., New York.)

PATTERNS OF DEVELOPMENT IN CULTURE

Philip V. Ammirato*

Department of Biological Sciences
Barnard College
Columbia University
New York, New York 10027

INTRODUCTION

Although organized structures were seen in cell and tissue cultures from the earliest days of research (35,61), investigations of development in plant cultures are usually traced from the classic work of Skoog and Miller (40) on the hormonal manipulation of shoot and root formation in tobacco pith cultures, and that of Steward et al. (41) and Reinert (38,39) on the appearance of adventive embryos from somatic cells of carrot taproot. These pioneering studies established the basic patterns of treatment and response that have dominated the field since that time: masses of cells are moved through a sequence of media changes; the subsequent morphogenesis follows either the organogenic or the embryogenic pathway of development.

However, as research has continued and investigators have looked more closely at these responses and at more plant species, it has become apparent that the relatively simple ways plant development in culture has been viewed do not adequately reflect the complexities that undoubtedly exist in those processes. For example, the pathways of development in culture have a higher degree of plasticity and result in a greater variability of patterns than previously appreciated. This is not to say that the original or subsequent investigators did not realize these issues. To the contrary, much of what concerns us today was duly noted early on. However, in

* Concurrently at DNA Plant Technology Corporation, Cinnaminson, New Jersey 08077

the overall enthusiasm of fostering morphogenesis in culture and in response to the remarkable similarities between in vivo and in vitro organs and embryos, the complexities were often lost to the generalities. An appreciation of this plasticity and variability in development is important if we are to better understand underlying mechanisms. It is particularly important today if we wish to apply plant cell cultures to practical problems (6).

This paper will address a number of the topics that are currently of interest and importance in the general realm of pattern and variation. By design, the discussion will be limited and selective. It will not, for example, address cytodifferentiation in culture, including tracheary element formation (18), flower formation (54,55), or the development from shoot meristems of various organs of perennation such as tubers and bulbs (5). As a brief, introductory essay, it aims to provide a prelude to the papers that follow.

BASIC MORPHOGENETIC RESPONSES

For the past 25 years, it has been generally thought that cells in culture progressed through an organogenic pathway leading either to root or shoot development, or through an embryogenic pathway leading to somatic or gametic embryos. Neither of these responses in culture is particularly surprising. One of the basic patterns of plant development is the continuous production of axillary shoots on shoot meristems and lateral roots on roots. In addition, many plants can regenerate these organs adventitiously from other regions. Adventive embryony is a well-known phenomenon (52); embryos have been observed to arise naturally from unreduced megasporocytes, the embryo itself, cells in the embryo sac, and, in the largest number of cases, from the nucellus. These are all ovular tissue, and embryos have been grown from these cells in plant tissue cultures (4). The unusual aspect of adventive embryogeny in culture is the appearance of embryos from somatic or gametic cells that typically do not produce embryos in vivo.

Although shoots and roots, or shoots, roots, and somatic embryos, have been observed on the same tissue, it is generally considered that the processes leading to their appearance are mutually exclusive, i.e., cells either become committed to the organogenic or the embryogenic pathway, and if the former, to either shoot or root formation (22). Practically, a particular treatment usually produces only shoots or roots, or embryos, and conditions must be changed if other organs are desired (e.g., rooted shoots). Similarly, in many systems such as thin cell layers, not all morphogenetic patterns have been initiated from the same material (56).

Distinguishing Features

Organogenesis. Developing meristems in cultures are first apparent as masses of highly cytoplasmic, deeply staining cells located either at the periphery of the explant or callus tissue in the case of shoots, and usually deeper in the tissue in the case of roots (49,50). In a number of cases, meristems arise from groups of cells that have been called "meristemoids" (11) with the suggestion that such structures may be developmentally flexible and able to form either shoots or roots (53). But this has not yet been proven.

Whether or not the group of cells that forms the meristem originated from a single cell has been the subject of much discussion. Stewart (43), based on observations of chimeras, suggested that the formation of an apical meristem always involves a group of 3 or more cells which may or may not have been derived from a single cell. However, the absence or extremely low frequency of chimeras in most adventitious shoots, along with the presence of large numbers of solid, nonchimeric mutants following irradiation of the mother plant, has suggested their derivation ultimately from one cell (10).

Whether of single cell origin or not, the groups of cells eventually organize into meristems that develop further into typical shoots and roots. Their development usually parallels that of adventitious shoots and roots in vivo. In similar fashion to those arising in vivo, both roots and shoots arising in culture have vascular tissue extending into the explant or callus mass (19).

Embryogenesis. Somatic embryos arising in culture also appear as groups of small, densely cytoplasmic cells; however, these are often very distinctive in appearance, occurring as highly, discrete groups of cells, surrounded by darker-staining cell walls (31,34,51). These groups of cells may grow into one embryo or a cluster of embryos with additional, secondary, or accessory embryos often appearing on larger, more mature embryos. The process is highly repetitive under inductive conditions (1,20).

At the base of the developing somatic embryos, or surrounding them if they are buried in the tissue, are larger, more vacuolate cells. In the basal cells of some somatic embryos, there is a gradual transition to progressively larger cells. An analogous structure in zygotic embryos is the suspensor. However, there has been a general reluctance to consider these large cells at the base of somatic embryos as the equivalent of suspensor cells. Haccius (19) considers the proliferation of somatic proembryos or embryos in culture to be analogous to cleavage polyembryony observed in gymnosperm zygotic embryos where embryos arise from a proembryonal cell complex. In her interpretation, the suspensor found in zygotic embryos is a rudimentary, proembryonal cell complex. In her opinion, the two are clearly analogous, but she prefers reserving the term suspensor for those cases, either zygotic or nonzygotic (i.e., somatic), where it is "morphologically constant."

Somatic embryos are generally considered to originate from single cells. This point has been open to question from the earliest days of research. Somatic embryos have been observed to arise from single isolated cells or protoplasts (25), but often a mass of cells forms prior to embryo development. However, histological studies of somatic embryos, such as those developing in suspension cultures of carrot (19,31) and celery (16), on leaves of *Dactylis glomerata* (13), and on excised embryo, leaf, and inflorescence callus of *Pennisetum* (58) and other millets (see later chapters in this volume), support the idea that single cells can give rise to somatic embryos.

Their identification as *embryos* arising from somatic or gametic cells rather than shoots and roots together in one meristemoid has been based on the following:

(A) *Morphological observations*, in particular (a) a sequence of development that may closely resemble that found in zygotic embryos and that progresses from globular stage to heart, torpedo, and cotyledonary stages; (b) the initial development of cotyledons rather than leaves at the shoot end; and (c) the formation of a bipolar structure with a shoot at one end and a root at the other.

(B) *Anatomical evidence*, and in particular (a) the interconnection of the shoot and root apices by a solid procambium ultimately generating vascular tissue, and (b) the end of the radicle being "closed," i.e., it is not connected by vascular tissue to the remaining cells of the callus tissue (19).

(C) *Chemical evidence*, such as the demonstration of (a) embryo-specific storage proteins in *Brassica napus* somatic embryos (14), and (b) embryo-specific proteins in embryogenic cultures of oil palm, *Elaeis guineensis* (24), detected by immunofluorescence techniques.

The overall information is convincing enough to counter a recent suggestion that all somatic embryos are in reality shoot-root structures masquerading as pseudo-embryos (47). However, there are reports where there is insufficient evidence--morphological, anatomical, or biochemical--to conclude that the structures shown are in reality true somatic embryos.

INITIATION OF MORPHOGENESIS

Some of the most intriguing questions facing investigators today concern the conditions that reprogram hitherto differentiated or otherwise committed plant cells into forming adventitious structures in culture, and that determine the path taken, embryogenic or organogenic, shoot or root. It is not within the purview of this paper to review the factors that have proven successful in this endeavor. However, a basic approach in promoting morphogenesis in culture is to manipulate hormonal and environmental conditions. For organogenesis (17), a shift in the total and relative amounts of cytokinins and auxins may promote unorganized callus growth, shoot formation, or root formation. For embryogenesis (4), growth in a medium with sufficient reduced nitrogen and high auxins appears essential for the induction of embryonic growth (embryogenesis). Kept on a high-auxin medium, the embryogenic cells and proembryos produce additional cells and proembryos. A reduction or omission in the auxin supply apparently allows for the maturation of the somatic embryos. There are many variations on these basic schemes and cases where strikingly different manipulations have led to morphogenesis in culture. Not only have a multitude of species responded to a variety of treatments, but there is little indication as to why some species generate somatic embryos so readily (such as carrot, celery, and other members of the Umbelliferae), while others generally produce adventitious organs [such as tobacco, although there are a few reports of somatic embryos and microspores typically producing excellent embryos in culture (4)].

Sites of Morphogenesis

Reports on development in culture have generally stressed the treatments leading to organogenesis or embryogenesis. Receiving less emphasis have been the sites of initiation and differentiation. But from these studies, it appears that not all cells within the initial explant or within the subsequent callus tissue or cell suspension go on to form organs or embryos. Also, not all explants respond equally to conditions promoting morphogenesis. These factors produce considerable variation in the pattern of development in culture.

Sites within an explant: An example. Somatic embryos and plants will form directly on carrot petiole segments when they are placed on semisolid medium containing the Murashige and Skoog basal medium (33) supplemented with 10 μM naphthaleneacetic acid (MS + NAA). Even if the petioles are removed from a small number of young plantlets that were raised at the same time, under the same conditions, and from the same seed lot, not all petioles in the same treatment will form somatic embryos (Fig. 1A). Within any one segment, only some of the cells go on to produce plants and these are generally at one end of the segment (a classic case of polarity in development). A second morphogenetic response is root formation, which occurs at the other cut end of the petiole, or it may arise from both ends in those pieces lacking embryogenesis. Thus, in any petiole segment, or population of petiole segments, under given conditions, only certain cells appear competent to develop as somatic embryos.

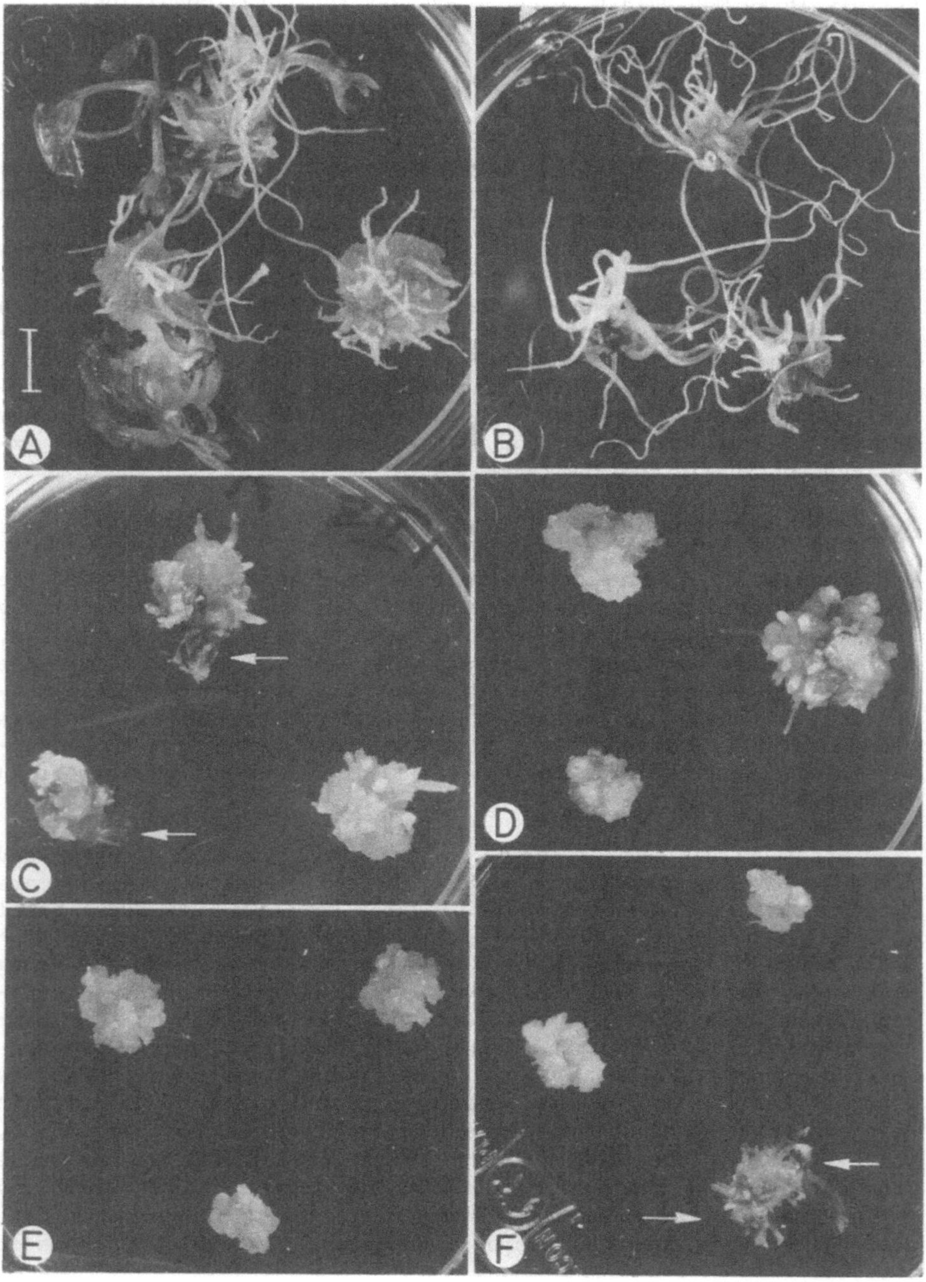
A
B
C
D
E
F

Fig. 1. Induction of development in carrot (Daucus carota) petiole explants. [Scale in A (applies to A-F) = 3 mm] (A) Petiole segments after 6 weeks on NAA-containing medium. Two of the segments have formed somatic embryos and plants at one end and roots on the other; the third segment has formed only roots. (B) Petioles that were grown for one day on NAA-containing medium, then transferred to hormone-free medium for 6 weeks. There has been extensive initiation and growth of roots throughout the explants. (C) Petiole segments that were grown for one day on 2,4-D-containing medium, then grown for 6 weeks on hormone-free medium. Two explants have initiated somatic embryos from one pole (arrows) and roots on the other; the third explant has initiated only roots. (D) Explants that were pulsed for 2 days with 2,4-D. One explant has initiated roots from one pole and callus tissue from the other; 2 explants have formed only callus. (E) Explants following treatment with 2,4-D for 3 days; only callus tissue formed. (F) Explants treated with 2,4-D for 4 days. One explant in this replicate initiated somatic embryos from both poles (arrows).

Competence and determination. More and more, researchers in plant development are using terminology originally developed by animal embryologists and developmental biologists (44). Competence is the capacity to express an inherent potential. One can speak of certain cells within a tissue as being competent to grow as embryos. Determination is the process by which the developmental potential within any cell becomes limited to a particular pathway, and induction is either the change in competence of particular cells ("directive induction") or the triggering of a particular developmental response from competent tissue ("permissive induction"). These terms are often difficult to apply to plant development. However, their introduction to this area reflects recent attempts to dissect and understand underlying mechanisms. Later papers in this volume will return to this topic. The reader is also directed to some excellent recent reviews (22,30,46).

The variability in response seen in the petioles shown in Fig. 1A is certainly not suprising. As complex organs, petioles have a mixture of cell types. Somatic embryos have been seen to arise from only certain cells or certain tissues within an explant (13,58). The petioles were taken from a number of plants. Carrot is a natural outbreeder, and in the absence of directed hybridization (such as with this variety, Half Nantes), each plant has a different genotype. The role of the genotype in affecting regeneration from cultured plant cells has been noted in an ever increasing number of species including Medicago sativa (26) and Trifolium pratense (27). Finally, petioles were taken from different positions along the axis.

Position effects. The type and frequency of morphogenesis in an explant, or cells derived from it, have also been shown to depend upon the original position of the explant within the plant. This has been seen in studies of somatic embryogenesis such as one comparing the responses of juvenile and adult stem tissues from Hedera helix (8) and in the de novo formation of meristems in various stem segments from Nicotiana tabacum (23,54). The position on the plant from which the explant is taken has proven to be critical in many monocotyledonous cultures, especially of the Gramineae (the grains and grasses). Embryonic, meristematic (as in young leaves) and inflorescence tissue have shown a greater propensity for morphogenesis than cells removed from other sites (59). There are a number of excellent recent reviews on position effects in development (9,30).

Even within what appears to be fairly uniform tissue, such as tobacco epidermal peels (54), tobacco pith explants, or tobacco callus cultures (49,50), only a small percentage of cells develop into organized structures. It may be that only a number of cells within a tissue are competent at any one time to form an organized structure; or particular physiological conditions are required for morphogenesis and, due to gradients within the tissue, only specific regions, or positions, within the callus have those conditions.

In a quantitative study of bud initiation in cultured tobacco tissues, Meins et al. (32) showed that (a) the incidence of buds was very low when the tissue was transferred from cytokinin-auxin medium to cytokinin medium to induce bud formation; (b) bud-free regions did not form buds when incubated a second time on the inductive medium; and (c) bud-free regions did eventually form buds after they were first grown on the complete medium with auxin and cytokinin, and then transferred to the inductive medium. A statistical analysis of the distribution of buds among explants and subcloning experiments showed that the low frequency of buds resulted from neither negative interactions among bud-forming centers nor a paucity of cells with the potential for organogenesis. Their results were interpreted as being consistent with either hypothesis (that the frequency of buds is determined by the number of competent cells or by position effects), or with both. The definitive answer, if there is one, is still forthcoming. However, the development of buds from nonbud-forming segments after pretreatment on the complete medium shows that cells can change their competence in response to changes in hormonal conditions.

Induction

In the carrot petioles described earlier, the competence of some of the cells to grow into somatic embryos and complete plants is not present at the time of excision but develops during the course of incubation on the hormone-containing medium. Explants transferred directly to semisolid medium with no growth regulators do not generate callus or organized structures. In one experiment, petioles were placed on the inductive medium [MS + NAA; MS + 2,4-dichlorophenoxyacetic acid (2,4-D)] removed at one-day intervals to a hormone-free medium (MS), and observed after 6 and then 12 weeks. Those remaining on NAA-containing medium produced plants as seen earlier (Fig. 1A); those on 2,4-D generated what appeared to be unorganized callus tissue with no visible somatic embryos.

After 6 weeks' growth, those pulsed with NAA for one day and then transferred to hormone-free medium formed roots at one end of most segments. Those on the NAA-medium for 2, 3, and 4 days and then removed developed numerous adventitious roots from both ends of each explant (Fig. 1B). There were no shoots or embryos. Short treatments with NAA did not induce somatic embryogenesis after 6 weeks' growth. (However, when examined after 12 weeks, somatic embryos and plants were visible; the responses for 1-, 2-, 3-, and 4-day pulses were 0, 22, 56, and 67%, respectively. It is difficult to interpret these results at this time since the plants appeared so late and after early and extensive root formation).

For those pulsed with 2,4-D for one day, 33% of the segments developed somatic embryos/shoots at one end and roots at the other; the remaining explants produced only roots, generally only at one end (Fig. 1C). For those on 2,4-D for 2 days, 22% developed roots on one end; callus tissue appeared at the other end, and at the other segments (Fig. 1D). There was extensive callus formation after a 3-day pulse, but no root, shoot, or embryo formation (Fig. 1E). After a 4-day treatment, there was somatic embryo formation, on both ends, on 44% of the segments (Fig. 1F).

There appear to be at least 2 populations of cells in the explants and they have different time courses in response to the inductive treatments. Polarity during regeneration in organ segments is of course a well-known phenomenon. These results show a propensity for root formation at one end of each segment and for somatic embryogenesis at the other. Rhizogenesis appears to be readily induced with short hormone treatments; prolonged exposure with 2,4-D inhibits root formation, a response often observed (4,17). With NAA as the auxin, if somatic embryogenesis was induced, it occurred only at one end, with root formation at the other. Prolonged exposure to NAA (or a long incubation time afterwards) appears necessary for induction. With 2,4-D as the auxin, a short induction time (one day) can promote some somatic embryogenesis. However, a longer exposure

(4 days) is needed to induce somatic embryogenesis at both ends. The absence of any embryo formation in the 2- and 3-day treatments suggests that a major change in the competence of the cells was in progress but incomplete. [Interestingly, in studies of the induction of somatic embryogenesis in alfalfa, a 4-day treatment is typically used (60)]. The induction of competence for rhizogenesis and embryogenesis depends on the region in the explant, the type of hormone, and the length of time of exposure to the hormone.

In a study of shoot organogenesis in Convolvulus arvenis leaf explants, Christianson and Warnick (12) demonstrated that the induction process includes at least 5 separate transient sensitivities to the inhibitors tri-iodobenzoic acid, sorbitol, ribose, ammonium ion, and acetylsalicylic acid. Thus, the induction process for shoot formation in this system has begun to be dissected into a series of discrete steps.

These observations reinforce the concept that morphogenesis in culture is directly influenced by sequential changes in variables that are both extrinsic and intrinsic (42,48). The hormonal milieu is only one of the external factors effecting and affecting patterns of development. It is becoming increasingly clear that more complex sequences of changes will probably be needed to promote development in morphogenetically recalcitrant cultures and to better control development in those instances where it can be fostered.

MORPHOGENESIS AND DEVELOPMENT: SOMATIC EMBRYOGENESIS

Studies on somatic embryogenesis in culture have increased dramatically over the past few years, in part because of the increasing application of plant cell culture to practical problems of agriculture and agribusiness. Somatic embryos present certain advantages over plants regenerated via organogenesis (4). Since they are bipolar structures, each bearing shoot and root apices, both meristems needed for plant growth are present. (Media changes which are necessary for shoot and root formation in organogenic systems are costly in terms of labor and materials.) In addition, somatic embryos can be generated in large numbers, often freely floating, thereby making them amenable to bioreactor technology and mechanical handling, including mechanized delivery. As natural organs of perennation, embryos could be made dormant, included in artificial seeds, and otherwise efficiently handled, stored, and shipped.

The growing interest in and importance of somatic embryogenesis are reflected in other papers in this volume. For this reason, the remaining discussion will be devoted to this area.

External Patterns of Development: Morphology

Although early work with somatic embryogenesis often stressed the great similarity between somatic and zygotic embryos, closer examination revealed some striking differences as well as a great variability in the patterns of somatic embryo maturation (7,21,28).

Precocious germination. Within most populations of somatic embryos, the embryos, in contrast to their zygotic counterparts, do not stop developing when embryonic maturation is completed, but continue into plantlet formation (Fig. 2A). Rapidly growing and elongating plants can present a practical problem. For one, the elongated roots tangle easily and make separation of the somatic embryos and subsequent planting into other culture media or soil difficult. In a number of cases, the embryos can germinate precociously, resulting in stunted or abnormal embryos. These responses are also observed in cultures of excised zygotic embryos (37,62).

Abnormal development. When populations of somatic embryos are examined, there is a range of abnormal forms. Many of these are developmental or epigenetic changes rather than genetic changes, resulting from the diversion of embryonic development away from the normal path (Fig. 2A and 2B). Normal plants can be grown from them. However, these structural abnormalities may create problems, as follows.

(a) Multiple embryos. In most somatic embryo populations, twin or multiple embryos may be seen. Each part is a complete embryo with root and shoot apices but usually linked or fasciated for part of the axis. Single embryos with 2 or more shoots can be seen. Large embryos with a smaller one emanating from the hypocotyl region, the cotyledons, or the radicle have also been observed. Embryogenesis in culture is repetitive; embryos form other embryos. If this occurs early in development, twins or multiple twins arise; a bit later, the 2 shooted embryos or "double headers" form (Fig. 2C). The late initiation of new embryos along the flanks of developing embryos results in accessory or secondary embryos (Fig. 2D). Multiple embryos, or those with many shoots, can create a practical problem if the ultimate goal is to obtain single plants, which is critical in many crops, e.g., multicrown carrot plants produce poor taproots.

(b) Malformed shoots. Many embryos show abnormalities in shoot formation, including the absence of cotyledons, too few or too many cotyledons, or fused cotyledons (Fig. 2D). The shoot meristem may also fail to develop at the base of the cotyledon. Such embryos will often not grow into plants and there are papers reporting the development of somatic embryos and their failure to grow, or convert, into plants.

(c) Premature shoot emergence. Somatic embryos have also been seen where cotyledons develop incompletely, or not at all, and the shoot apex grows out to form the first mature leaves. This often occurs at reasonably high levels of cytokinins (2) and such structures look more like shoots forming on callus than zygotic embryos.

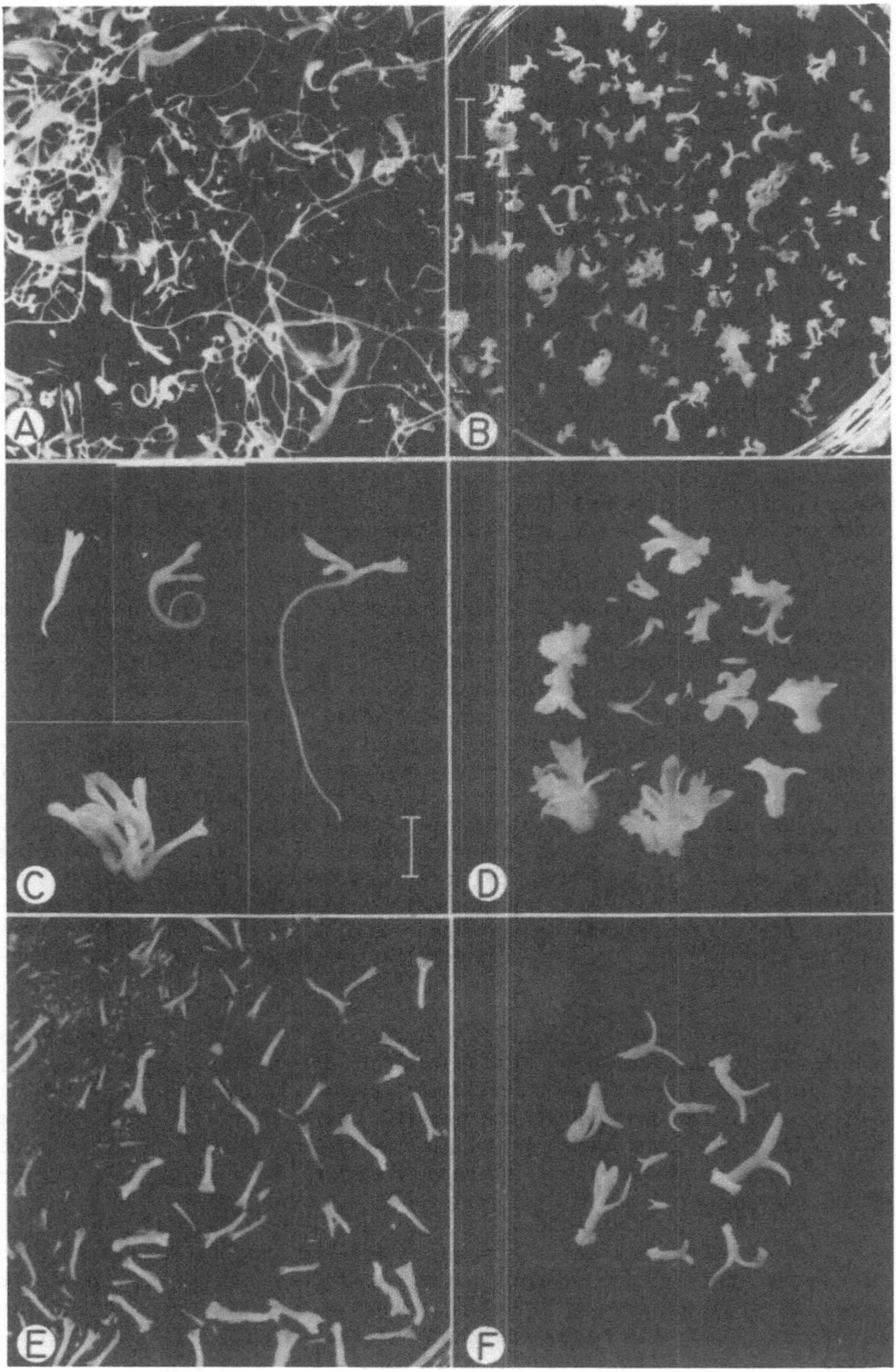
A
B
C
D
E
F

←

Fig. 2. Populations of somatic embryos showing external morphology. [Scale in B (for A and B) = 5 mm; scale in C (for C-F) = 3 mm] (A) Carrot somatic embryos showing elongated radicles and cotyledons. (B) Caraway somatic embryos. Radicles have not grown out and hypocotyls and cotyledons have not elongated. (C) A sample of carrot somatic embryos showing, clockwise from upper left: single embryo, twin, two-headed embryos, and multiple embryo. (D) A sample of caraway somatic embryos showing some abnormal forms surrounding a single dicotyledonous embryo. Included are embryos with more than 2 cotyledons (pluricotyledony), multiple embryos, and those with numerous accessory embryos along the main axis. (E) Carrot somatic embryos grown with 0.1 μM ABA in the medium [compare with (A) and (C)]. (F) Caraway somatic embryos grown with 0.1 μM ABA in the medium [compare with (B) and (D)].

The fact that normal embryonic development may be so diverted as to result in shoot formation may suggest that organogenesis in cultures is a form of somatic embryogenesis, i.e., when cells are induced, they become competent to develop either as embryos or meristems and depend upon other factors to determine the ultimate pathway. It is an intriguing idea currently being discussed. The alternative concept is that embryogenic and organogenic competences are distinct states (see Ref. 22). However, given the plasticity seen in populations of somatic embryos and their responsiveness to external variables, it is clear that one may have embryogenic suspensions but fail to provide the conditions that will allow embryo maturation.

<u>Differences among species.</u> By comparing somatic embryogenesis in members of the carrot or parsley family, the Umbelliferae, one can see that even closely related members of a family can differ in the pattern of embryo maturation. For example, the somatic embryos of carrot (<u>Daucus carota</u>) shown in Fig. 2A and caraway (<u>Carum carvi</u>), in Fig. 2B, were grown under similar conditions and using similar protocols. Somatic embryos of carrot show a rapid germination (or conversion) in culture resulting in long radicles, extended green cotyledons (in the light), and a low frequency of accessory embryos. In contrast, somatic embryos of caraway have not elongated, radicles have not grown out, and cotyledons are shorter, although also green. They show a high propensity for accessory embryo formation. The pattern of maturation in carrot is similar to that seen in celery and anise, while that of caraway is more similar to water parsnip (3), although there may be some subtle differences. The reasons for the different patterns are not understood. One implication, however, is that treatments to both promote and control

development may be expected to differ from species to species, even if only in terms of concentrations of regulatory substances, if not in the substances themselves.

Factors modifying developmental patterns. Early studies with excised zygotic embryos grown in culture demonstrated that embryo maturation could be significantly modified by changing various cultural and environmental factors. Similar manipulations have been applied to populations of somatic embryos, including investigations of reduced nitrogen, other inorganic nutrients, carbohydrates, various growth regulators, light, darkness, and culture method (see, for example, Ref. 4). A number of factors stand out as strongly influencing the pattern of somatic embryo development.

(A) Elevated osmolarity. Raising the osmotic concentration of the culture medium by elevating the sucrose levels or by adding hexitols such as inositol or sorbitol was shown to substantially inhibit precocious germination of carrot and water parsnip somatic embryos (7). The resulting cotyledons were short and fleshy, more similar to those found on zygotic embryos, and, with those grown with high sucrose, exhibited substantial anthocyanin formation. These somatic embryos did show an increased tendency to accessory embryo formation. In recent years, elevation in the osmolarity, usually effected by increasing sucrose content, has been used to advantage in both obtaining somatic embryos (29,59) and in altering synthetic pathways, as in the fatty acid composition in somatic embryos of *Theobroma cacao* (36).

(B) Abscisic acid (ABA). Working with caraway cultures, Ammirato (1,2) reported that abscisic acid, when added at concentrations that did not totally inhibit growth and development, could selectively inhibit certain aspects of somatic embryo development. Similar responses were reported for carrot (3). Among its effects, ABA was shown to (a) prevent precocious germination, (b) inhibit accessory embryo formation along the embryonic axis, and (c) increase the frequency of embryos with the normal 2 cotyledons. ABA selectively promoted normal somatic embryo maturation. The use of a pulse of ABA at the time proembryos began to progress through the various embryonic stages resulted in a high percentage (over 90%) of normal dicotyledonous embryos. Maintenance of the fully formed embryos on ABA could effectively keep the embryos quiescent. Removal from ABA allowed their growth into plants. Some carrot somatic embryos grown with ABA in the medium are shown in Fig. 2E; caraway somatic embryos grown under similar treatments are shown in Fig. 2F. The external morphology in both is more similar to normal zygotic embryos than those grown without ABA. Caraway somatic embryos closely resemble their zygotic counterparts in both size and relative proportion of cotyledons to radicle; cotyledonary development is poorer on the carrot somatic embryos.

Internal Patterns: Anatomy

Early investigations of somatic embryogeny reported the great similarity between the internal structure of somatic and zygotic embryos, but also pointed out certain differences (20,21,45,57). Somatic embryos typically showed (a) the absence of a well-defined epidermal layer, having been sloughed off in liquid medium; (b) premature vacuolation in most of the cells and particularly in the cortical region; and (c) premature cell differentiation, especially of vascular tissue. This has been seen in more recent research (34,51).

Figure 3A shows a longitudinal section of a zygotic embryo of caraway 2 days after germination has begun. Although vacuolation and cell enlargement have begun, most cells are still small and densely cytoplasmic (Fig. 3B). Rows of cells, the product of orderly development, are evident. There are no mature, differentiated cells.

A somatic embryo of caraway, taken from a population grown in hormone-free medium similar to that seen earlier (Fig. 2B) is shown in Fig. 3C. Although of similar size to the zygotic embryo (see, for example, Fig. 3A), it shows the anatomical changes typical of somatic embryos. There is extensive cell enlargement (Fig. 3D), especially in the cortical region, and also distortion of the cell lineages following such enlargement. The epidermal layer is not intact, some cells having sloughed off. Others have divided; one in this section has formed a small proembryo. There are also mature vascular tissue tracheary elements (Fig. 3D, arrow). Also visible is the extensive tissue at the radicular end of the embryo, from which some additional somatic embryos are developing.

Caraway somatic embryos grown with 0.1 µM ABA in the medium show not only normal external morphology (Fig. 2F), but also internal anatomy (Fig. 3E) more similar to zygotic embryos. Cells are smaller and more highly cytoplasmic, and distinct linear rows of cells are present (Fig. 3F). The epidermal layer is intact, and although some divisions can be seen, accessory embryos are lacking. There are no mature tracheary elements, although the procambium is distinct. There is substantially less growth of the suspensor-like cells at the base of the embryos and no additional embryos are forming. ABA not only promotes a more normal maturation of embryonic organs, but also a pattern of cell maturation that more closely resembles zygotic embryos.

This modification of the pattern of embryonic cell differentiation by ABA is also manifest in changes in the biochemistry of embryonic cells. When immature, excised zygotic embryos of Brassica napus are excised, the level of embryo-specific storage proteins drops unless exogenous ABA is present (15). These embryo-specific storage proteins have also been detected in Brassica napus somatic

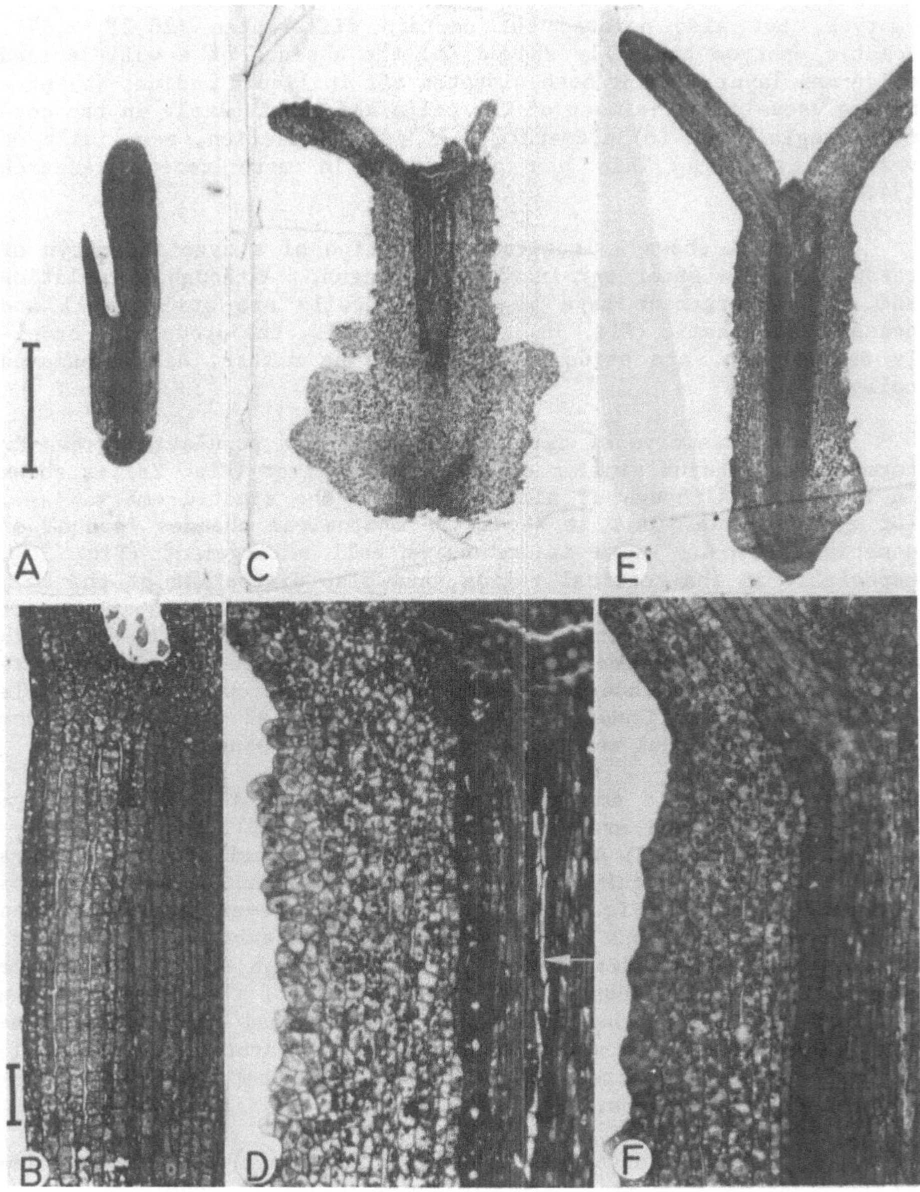
A
C
E
B
D
F

Fig. 3. Thin sections of somatic embryos of caraway (Carum carvi). Embryos were fixed in glutaraldehyde, imbedded in Spurr's and sectioned at 1 μm. [Scale in A (for A,C,E) = 0.5 mm; scale in B (for B,D,F) = 50 μm]

(A and B) Longitudinal section of an excised zygotic embryo 2 days following imbibition. The cells are still small and have just begun to vacuolate.

(C and D) Somatic embryo grown on hormone-free medium. (C) Note the accessory embryo along the axis and the extensive proliferations at the base in the suspensor-like cells. (D) There is extensive enlargement, especially in the cortical region, and mature tracheary elements (arrow).

(E and F) Somatic embryo grown on 0.1 μM ABA. (E) Note the absence of accessory proliferations along the axis and the greatly reduced proliferation of the suspensor. (F) The epidermis is intact; cells are appreciably smaller than controls [see, for example, (D)] and no vascular elements have matured. (The clear areas are tears in the plastic.)

embryos (14), but at levels 10 times lower than mature zygotic levels. Although somatic embryos and zygotic embryos are remarkably similar, there are substantial differences, both morphological and anatomical, that may arise under certain conditions. The condition that generates the most normal pattern of somatic embryo development is likely to produce the most similar biochemical pattern.

CONCLUSIONS

Continued investigations of plant development in tissue culture have reaffirmed the basic patterns of organogenesis and somatic embryogenesis (and the factors promoting their formation), as well as increased our appreciation of the complexities that are inherent in plant morphogenesis. The concepts of competence, determination, and induction are being used in efforts to dissect the multistep processes leading to development. Position effects reveal how the processes unfold. Development is plastic and the processes promoting development can be diverted or modulated, producing a variety of patterns. At present, we do not fully understand how flexible these systems are or how much we can regulate their development in culture. However, it is important that this plasticity in development and variation in pattern be appreciated if we are to both understand and apply these phenomena.

REFERENCES

1. Ammirato, P.V. (1974) The effects of abscisic acid on the development of somatic embryos from cells of caraway (Carum carvi L.). Bot. Gaz. 135:328-337.
2. Ammirato, P.V. (1977) Hormonal control of somatic embryo development from cultured cells of caraway: Interactions of abscisic acid, zeatin and gibberellic acid. Plant Physiol. 59:579-586.
3. Ammirato, P.V. (1983) The regulation of somatic embryo development in plant cell cultures: Suspension culture techniques and hormone requirements. Bio/Technology 1:68-74.
4. Ammirato, P.V. (1983) Embryogenesis. In The Handbook of Plant Cell Culture. Volume 1. Techniques for Propagation and Breeding, D.A. Evans, W.R. Sharp, P.V. Ammirato, and Y. Yamada, eds. Macmillan Publishing Co., New York, pp. 82-123.
5. Ammirato, P.V. (1985) Control and expression of morphogenesis in vitro. In Plant Tissue Culture and its Agricultural Applications, L.A. Withers and P.G. Alderson, eds. Butterworths, London (in press).
6. Ammirato, P.V., D.A. Evans, C.E. Flick, R.J. Whitaker, and W.R. Sharp (1984) Biotechnology and agricultural improvement. Trends in Biotechnology 2(3):1-6.
7. Ammirato, P.V., and F.C. Steward (1971) Some effects of the environment on the development of embryos from cultured free cells. Bot. Gaz. 132:149-158.
8. Banks, M.S. (1979) Plant regeneration from callus of two growth phases of English ivy, Hedera helix. Z. Pflanzenphysiol. 92: 349-353.
9. Barlow, P., and D.J. Carr (1983) Positional Controls in Plant Development, Cambridge University Press, Cambridge.
10. Broertjes, C., and A. Keen (1980) Adventitious shoots: Do they develop from one cell? Euphytica 29:73-87.
11. Bunning, E. (1952) Morphogenesis in plants. Survey of Biological Progress 2:105-140.
12. Christianson, M.L., and D.A. Warnick (1984) Phenocritical times in the process of in vitro shoot organogenesis. Devel. Biol. 101:382-390.
13. Conger, B.V., G.E. Hanning, D.J. Gray, and J.K. McDaniel (1983) Direct embryogenesis from mesophyll cells of orchardgrass. Science 221:850-851.
14. Crouch, M.L. (1982) Non-zygotic embryos of Brassica napus contain embryo-specific storage proteins. Planta 156:520-524.
15. Crouch, M.L., and I.M. Sussex (1981) Development and storage-protein synthesis in Brassica napus L. embryos in vivo and in vitro. Planta 153:64-74.
16. Dunstan, D.I., K.C. Short, M.A. Merrick, and H.A. Collin (1982) Origin and early growth of celery embryoids. New Phytol. 91: 121-128.
17. Flick, C.E., D.A. Evans, and W.R. Sharp (1983) Organogenesis. In The Handbook of Plant Cell Culture. Volume 1. Techniques

for Propagation and Breeding, D.A. Evans, W.R. Sharp, P.V. Ammirato, and Y. Yamada, eds. Macmillan Publishing Co., New York, pp. 13-81.

18. Fosket, D.E. (1980) Hormonal control of morphogenesis in cultured tissues. In Plant Growth Substances, F. Skoog, ed. Springer-Verlag, Berlin, pp. 362-369.
19. Haccius, B. (1978) Question of unicellular origin of non-zygotic embryos in callus cultures. Phytomorphology 28:74-81.
20. Halperin, W. (1966) Alternative morphogenetic events in cell suspensions. Am. J. Bot. 53:443-453.
21. Halperin, W., and D.F. Wetherell (1964) Adventive embryony in tissue cultures of the wild carrot, Daucus carota. Am. J. Bot. 51:274-283.
22. Henshaw, G.G., J.F. O'Hara, and K.J. Webb (1982) Morphogenetic studies in plant tissue cultures. In Differentiation in Vitro, M.M. Yeoman and D.E.S. Truman, eds. Cambridge University Press, Cambridge, pp. 231-251.
23. Hillson, T.D., and C.E. LaMotte (1977) In vitro formation and development of floral buds on tobacco stem explants. Plant Physiol. 60:881-884.
24. Hulme, J., and W.A. Hughes (1985) Biochemical markers of somatic embryogenesis in oil palm tissue culture. In Plant Tissue Culture and its Agricultural Applications, L. Withers and P. Alderson, eds. Butterworths, London (in press).
25. Kameya, Y., and H. Uchimiya (1972) Embryoids derived from isolated protoplasts of carrot. Planta 103:356-360.
26. Kao, K.N., and M.R. Michayluk (1980) Plant regeneration from mesophyll protoplasts of alfalfa. Z. Pflanzenphysiol. 96:135-141.
27. Keyes, C.J., G.B. Collins, and N.L. Taylor (1980) Genetic variation in tissue cultures of red clover. Theor. Appl. Genet. 58:265-271.
28. Konar, R.N., E. Thomas, and H.E. Street (1972) The diversity of morphogenesis in suspension cultures of Atropa belladonna L. Ann. Bot. 36:249-258.
29. Litz, R.E., and R.A. Conover (1982) In vitro somatic embryogenesis and plant regeneration from Carica papaya L. ovular callus. Plant Sci. Lett. 26:153-158.
30. McDaniel, C.N. (1984) Competence, determination and induction in plant development. In Pattern Formation, G.M. Malacinski and S.V. Bryant, eds. Macmillan Publishing Co., New York, pp. 393-412.
31. McWilliam, A.A., S.M. Smith, and H.E. Street (1974) The origin and development of embryoids in suspension cultures of carrot (Daucus carota). Ann. Bot. 38:243-250.
32. Meins, Jr., F., R. Foster, and J. Lutz (1982) Quantitative studies of bud initiation in cultured tobacco tissues. Planta 155:473-477.
33. Murashige, T., and F. Skoog (1962) A revised medium for rapid growth and bioassays with tobacco tissue cultures. Physiol. Plant. 15:473-497.

34. Nessler, C.L. (1982) Somatic embryogenesis in the opium poppy, Papaver somniferum. Physiol. Plant. 55:453-458.
35. Nobécourt, P. (1939) Sur les radicelles naissant des cultures de tissus de tubercule de Carotte. Compt. Rend. Soc. Biol. 130:1271-1272.
36. Pence, V.C., P.M. Hasegawa, and J. Janick (1981) Sucrose-mediated regulation of fatty acid composition in asexual embryos of Theobroma cacao. Physiol. Plant. 53:378-384.
37. Raghavan, V. (1980) Embryo culture. In Perspectives in Plant Cell and Tissue Culture, I.K. Vasil, ed. International Rev. Cytology Suppl. 11B:209-240.
38. Reinert, J. (1958) Morphogenese und ihre Kontrolle an Gewebekulturen aus Karotten. Naturwissenschaften 45:344-345.
39. Reinert, J. (1959) Uber die Kontrolle der Morphogenese und die Induktion von Advientiveembryonnen an Gewebekulturen aus Karotten. Planta 58:318-333.
40. Skoog, F., and C.O. Miller (1957) Chemical regulation of growth and organ formation in plant tissue cultured in vitro. Symp. Soc. Exp. Biol. 11:118-131.
41. Steward, F.C., M.O. Mapes, and K. Mears (1958) Growth and organized development of cultured cells. II. Organization in cultures grown from freely suspended cells. Am. J. Bot. 45: 705-708.
42. Steward, F.C., A.E. Kent, and M.O. Mapes (1967) Growth and organization in cultured cells: Sequential and synergistic effects of growth regulating substances. Ann. N.Y. Acad. Sci. 144:326-334.
43. Stewart, R.N. (1978) Ontogeny of the primary body in chimeral forms of higher plants. In The Clonal Basis of Development, S. Subtelny and I.M. Sussex, eds. Academic Press, London, pp. 131-160.
44. Street, H.E. (1979) Embryogenesis and chemically induced organogenesis. In Plant Tissue and Cell Culture: Principles and Applications, W.R. Sharp, P.O. Larsen, E.F. Paddock, and V. Raghavan, eds. Ohio State University Press, Columbus, pp. 127-153.
45. Sussex, I.M. (1972) Somatic embryos in long-term carrot tissue cultures: Histology, cytology and development. Phytomorphol. 22:50-59.
46. Sussex, I.M. (1983) Determination of plant organs and cells. In Genetic Engineering in Plants. An Agricultural Perspective, T. Kosuge, C.P. Meredith, and A. Hollaender, eds. Plenum Press, New York, pp. 443-451.
47. Swamy, B.G.L., and K.V. Krishnamurthy (1981) On embryos and embryoids. Proc. Indian Acad. Sci. 90:401-411.
48. Thorpe, T.A. (1978) Physiological and biochemical aspects of organogenesis in vitro. In Frontiers of Plant Tissue Culture 1978, T.A. Thorpe, ed. The International Association of Plant Tissue Culture 1978, Calgary, pp. 49-58.
49. Thorpe, T.A., and T. Murashige (1968) Starch accumulation in shoot-forming tobacco callus cultures. Science 160:421-422.

50. Thorpe, T.A., and T. Murashige (1970) Some histochemical changes underlying shoot initiation in tobacco callus cultures. Can. J. Bot. 48:277-285.
51. Tisserat, B., and D.A. DeMason (1980) A histological study of development of adventive embryos in organ cultures of Phoenix dactylifera L. Ann. Bot. 46:465-472.
52. Tisserat, B., B.B. Esan, and T. Murashige (1979) Somatic embryogenesis in angiosperms. Hortic. Rev. 1:1-78.
53. Torrey, J.G. (1966) The initiation of organized development in plants. Adv. Morphogenesis 5:39-91.
54. Tran Thanh Van, K. (1973) Direct flower neoformation from superficial tissue of small explants of Nicotiana tabacum L. Planta 115:87-92.
55. Tran Thanh Van, K. (1980) Thin cell layers: Control of morphogenesis by inherent factors and exogenously applied factors. Intern. Rev. Cytol. Suppl. 11A:175-194.
56. Tran Thanh Van, K. (1981) Control of morphogenesis in in vitro cultures. Ann. Rev. Plant Physiol. 32:291-311.
57. Vasil, I.K., and A.C. Hildebrandt (1966) Variations of morphogenetic behaviour in plant tissue cultures. I. Cichorium endiva. Am. J. Bot. 53:860-869.
58. Vasil, V., and I.K. Vasil (1982) The ontogeny of somatic embryos of Pennisetum americanum (L.) K. Schum. I. in cultured immature embryos. Bot. Gaz. 143:454-465.
59. Vasil, V., and I.K. Vasil (1984) Induction and maintenance of embryogenic callus cultures of Gramineae. In Cell Culture and Somatic Cell Genetics of Plants. Vol. 1. Laboratory Procedures and their Applications, I.K. Vasil, ed. Academic Press, New York, pp. 36-42.
60. Walker, K.A., M.L. Wendeln, and E.G. Jaworski (1979) Organogenesis in callus tissue of Medicago sativa. The temporal separation of induction processes from differentiation processes. Plant Sci. Lett. 16:23-30.
61. White, P.R. (1939) Controlled differentiation in a plant tissue culture. Bull. Torrey Bot. Club 66:505-513.
62. Yeung, E.C., T.A. Thorpe, and C.J. Jensen (1981) In vitro fertilization and embryo culture. In Plant Tissue Culture: Methods and Applications in Agriculture, T.A. Thorpe, ed. Academic Press, New York, pp. 253-271.

SOMATIC EMBRYOGENESIS AND ITS CONSEQUENCES IN THE GRAMINEAE

Indra K. Vasil

Department of Botany
University of Florida
Gainesville, Florida 32611

There are 2 principal pathways of plant regeneration in tissue cultures: organogenesis (shoot morphogenesis) and somatic embryogenesis. Organogenesis is accomplished by the development of axillary buds in cultured shoot segments or by the de novo organization of shoot meristems in callus tissues. The shoots are later rooted to give rise to plants. Somatic embryos are formed both in callus and cell suspension cultures and directly give rise to plants. The development of axillary shoot buds is perhaps the most common phenomenon, and is widely exploited in the large-scale clonal propagation of horticultural plants.

Most accounts of plant regeneration from tissue cultures of the Gramineae before 1980 described shoot morphogenesis (18). Shoots developed either by the de novo organization of shoot meristems in callus cultures (43,55,59,60), or by the derepression of presumptive shoot primordia in cultures that consisted largely of proliferating meristems (15,27). In a majority of the reports, the cultures retained their morphogenetic competence only for a brief period of time and plant regeneration was sporadic and transient, the number of plants formed was rather limited, and regenerable cultures could be obtained only from a few selected genotypes.

Shoot meristems in vivo as well as in vitro are believed to be multicellular in origin (3,11,13,38,44,46,62,71). Plants of multicellular origin can often be chimeras (2,3,6,41,45,47,58,88) and, hence, unsuitable for mutation research, genetic analyses, maintenance of genetic stocks, breeding, etc. The formation of shoot meristems in callus cultures enhances further the chances of recovering abnormal plants because of the high incidence of cytological anomalies in callus tissues.

Since 1980, thanks to the use of immature embryos, inflorescences and leaves as initial explants, and 2,4-dichlorophenoxyacetic acid (2,4-D) as the principal plant growth regulator, embryogenic tissue cultures have been established from all of the major species of cereals and grasses (66-69). Plant regeneration in these cultures takes place by the organization of somatic embryos. Indeed the phenomenon of somatic embryogenesis has been found to be so widespread in tissue cultures of the Gramineae that it has been suggested that this may be the most common and even the preferred method of in vitro morphogenesis in this important group of crop plants (78).

The work presented and discussed in this paper was started in 1978. We chose to work initially with Pennisetum americanum (pearl millet) because reports of generally poor plant regeneration in most of the other major species of cereals had appeared already previously, but more importantly because in my graduate work 30 years ago I had found this species to be unique among the Gramineae in the response of its pollen to germination in vitro and to storage (64,65). This choice turned out to be a fortunate one as it enabled us to establish the first truly embryogenic callus, cell suspension, and protoplast cultures within the family Gramineae. One of the major reasons for this success was also our long and sustained interest in developmental studies of higher plants and in plant embryology, and hence the ability to distinguish not only embryogenic tissues but also the earliest stages of somatic embryogenesis (73,75-77). Our findings would have been interesting but nothing more than of academic importance had they been limited to or were a unique feature of this particular species only, and could not be extended to other important crop species within the family. Accordingly, we initiated a deliberate, extensive, and ambitious effort aimed at extending the success with pearl millet to other major cereal and grass crops.

This paper summarizes the work carried out during the last 5 years in my laboratory on the regeneration of plants by the formation of somatic embryos in tissue cultures of 12 species and one interspecific hybrid of the Gramineae (Tab. 1), and discusses the consequences of somatic embryogenesis in the developing biotechnology for the genetic improvement and modification of cereal and grass crops. The purpose of this report then is to present an overall summary and discussion of our work. It is recommended that for details of particular experiments or species reference be made to the original research publications listed here. No attempt has been made to review the literature on tissue cultures of the Gramineae. Such accounts can be found elsewhere (12,14,66-70).

REGENERATION FROM CALLUS CULTURES

Embryogenic callus tissue cultures were established from all the species listed in Tab. 1, irrespective of the genotype used. Thus, embryogenic cultures were induced from each of the about 70

Tab. 1. Regeneration of plants by the formation of somatic embryos in tissue culture.

Plant	Regeneration Sources	References
Panicum maximum Jacq. (Guinea Grass)	a,b,c,d,e	Lu and Vasil (31-34), Lu et al. (36), Hanna et al. (19)
Panicum miliaceum L. (Proso or Common Millet)	b	Rangan and Vasil (56)
Panicum miliare Lamk. (Little Millet)	b	Rangan and Vasil (56)
Paspalum notatum Flugge (Bahia Grass)	b	Waters et al. (unpub. data)
Pennisetum americanum (L.) K. Schum. (Pearl Millet)	a,b,d,e	V. Vasil and Vasil (75-79), Rangan and Vasil (57), Botti and Vasil (4,5), Swedlund and Vasil (63)
Pennisetum purpureum Schum. (Napier Grass)	b,c,e	Haydu and Vasil (21), Wang and Vasil (87), V. Vasil et al. (83), Chandler and Vasil (7,8)
Pennisetum americanum x Pennisetum purpureum (Hybrid Napier Grass)	b,c	V. Vasil and Vasil (76), Chandler et al. (9)
Saccharum officinarum L. (Sugarcane)	c,d	Ho and Vasil (24,25)
Secale cereale L. (Rye)	a	Lu et al. (30)
Sorghum arundinaceum (Desv.) Stapf. (Sudan Grass)	b	Boyes and Vasil (6)
Sorghum bicolor L. (Sorghum)	a,b	Boyes and Vasil (unpub. data)
Triticum aestivum L. (Wheat)	a,b	Ozias-Akins and Vasil (49-52)
Zea mays L. (Maize)	a	Lu et al. (35,37), V. Vasil et al. (84-86)

a Regeneration from cultured immature embryos.
b Regeneration from cultured immature inflorescence segments.
c Regeneration from cultured immature leaf segments.
d Regeneration from cell suspension cultures.
e Regeneration from protoplast cultures.

genotypes randomly selected among the various species. Immature embryos (from open- and self-pollinated commercial hybrid cultivars and inbred lines), and segments of young inflorescences and the basal portions of leaves proved to be the ideal explants. A single nutrient formulation, that of Murashige and Skoog (42), was used successfully in all cases. In most instances 2,4-D gave the best response and was the only plant growth regulator added to the basal medium.

Embryogenic callus was formed only by specific tissues in each of the 3 different explants used. In the case of immature embryos the callus was formed by divisions in the peripheral cells of the scutellum at its coleorhizal end (34,49,79). In inflorescences, the meristematic cells of the floral primordia as well as the cells surrounding the peripheral vascular bundles in the inflorescence axis proliferated to form the callus (5,87). In cultured leaf segments embryogenic callus originated from cells of the lower epidermis and mesophyll tissue between the vascular bundles (21,24,31).

The developmental stage of the explant was found to be the most critical factor in obtaining the optimal response and in the establishment of vigorously growing embryogenic callus tissues. The immature embryos were cultured 10-14 days after pollination (33,49,51, 76). At this time all the characteristic organs of the embryo had already been formed. The endosperm had begun to become cellular and the storage of reserve food materials in the scutellar cells had barely started. It was necessary to place the embryos in a face down position, where the scutellum was away from the medium and exposed. The first mitotic divisions always started around the scutellar provascular strand in the region of the scutellar node, but later these spread to the surface layers of the scutellum. The bulk of the embryogenic callus was formed by divisions in 2-4 peripheral cell layers of the scutellum.

Immature inflorescences in which primordia of individual floral organs had just begun to be formed proved to be the best (5,6,56,80, 87). Fewer of the younger and older inflorescences formed embryogenic callus; only a limited amount of embryogenic callus was formed from these explants which gave rise to more of the nonmorphogenic friable type of callus, or formed many roots.

Leaves of the Gramineae have a basal meristem. The activity of this meristem ceases early during the life of the leaf. Segments of tissue excised from the basal-most portions of the second to the fifth youngest leaves formed the most embryogenic callus. Older and younger leaves were not suitable. Even in those leaves from which embryogenic callus was initiated, the degree of response decreased as more mature portions of the leaf--away from the base--were cultured. There was a clear gradient in the response from the base to the apex (21,24,31). It was found also that leaf segments in which vascular tissue had fully differentiated were not suitable for obtaining embryogenic cultures.

Regardless of the explant used, the first cell divisions always started near developing vascular tissues. Whether the presumed presence of higher levels of plant growth regulators and nutrients in the vascular tissues is related to such activity is not known.

Embryogenic calli can be easily identified by their characteristic morphology and cytology. They are whitish to pale yellow in color, and are compact and nodular in appearance (76,80). They are often surrounded by a friable and transluscent nonembryogenic callus. In some instances pockets of embryogenic cells are randomly distributed within a soft and friable callus (4,78). Soft embryogenic calli can also be obtained (85,86), particularly when low concentrations of sucrose are used in the medium. The embryogenic cells are small, isodiametric, thin-walled, richly cytoplasmic, strongly basophilic, and contain many plastids with starch grains and a large nucleus with prominent nucleoli. They divide frequently.

As stated earlier, only specific tissues of the original explant form embryogenic calli. This competence is attained by the target cells only during the first few days of excision and culture. Once an embryogenic callus has been initiated, it perpetuates itself in culture by continued divisions of its cells, and sometimes by the formation of distinct meristematic layers. The cells can easily lose their embryogenic nature if the levels of 2,4-D fall below a certain threshold. The conversion of the large and vacuolated nonembryogenic cells into embryogenic cells is not known.

In the presence of high levels of 2,4-D in the medium the embryogenic nature of the callus can be maintained by subculture for long periods of time, several years in some species. Rapid transfer of calli to fresh media every 2 weeks is thus helpful; in Napier grass such a strategy was shown to have the capacity to produce 25,000 green plants from a single leaf explant in 6-7 months (Fig. 1; Ref. 7 and 9). Lowering of the 2,4-D levels not only causes cell enlargement, vacuolation, and loss of starch grains, but also induces the formation of somatic embryos. In many instances an abrupt lowering of the 2,4-D level to as low as 0.1 mg/l may be needed to initiate embryogenesis. Complete removal of 2,4-D is not possible in tissue cultures of most of the Gramineae, as they require at least a minute amount for continued cell divisions and growth. In many cases somatic embryos are formed without any lowering of 2,4-D levels.

Somatic embryos formed in callus cultures germinate easily to give rise to plantlets. In some instances they must be transferred to media containing gibberellic acid to promote germination. In others, abscisic acid is required to encourage normal maturation and germination. Almost every plantlet transferred to soil survives and can be grown to maturity (7,80).

There has been considerable emphasis recently on the relationship between the genotype of the explant and its response in vitro, particularly its ability to form regenerable cultures. In our experience, however, the physiological and developmental stages of the explant are the most critical factor in eliciting a favorable response. These conclusions are based on experiments with about 70 genotypes from 13 species used in our experiments, and include data

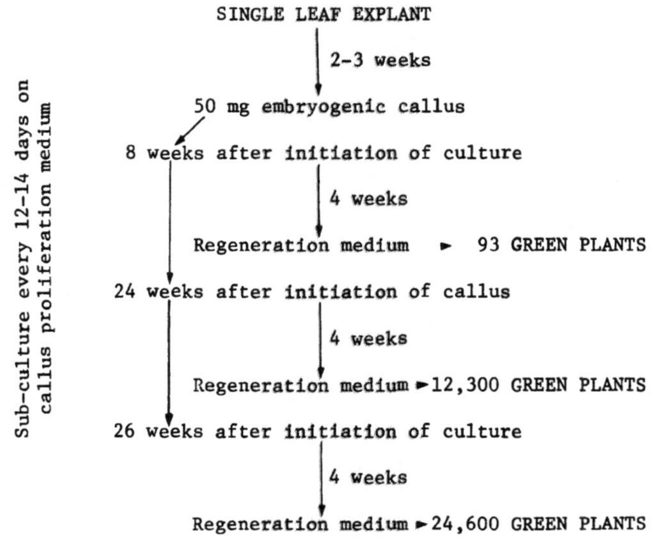

Fig. 1. Rate of regeneration of green plants (albinos excluded) of Napier grass (Pennisetum purpureum Schum.), based on rates of embryogenic callus proliferation and plant regeneration attained in vitro by somatic embryogenesis.

not only from immature embryos from open-pollinated hybrid cultivars, but also from self-pollinated hybrid as well as inbred genotypes, and from inflorescence and leaf segments in many instances.

A reevaluation of the early reports of plant regeneration in the Gramineae that described only shoot morphogenesis indicates that in many of these instances embryogenically competent tissue cultures were also formed. However, these were inadvertently or deliberately discarded during subculture because of their slow growth in favor of the more common friable and nonembryogenic calli (see also Ref. 24, 25,56,68,69). Often the process of regeneration was either not understood or interpreted incorrectly to be shoot morphogenesis. Some of the problems were apparently caused by the formation of green leafy structures prior to the appearance of shoots. These have now been shown to be the enlarged and greening scutella of precociously germinating somatic embryos (24,25,49,78,79,87). The often reported association of green spots on the surface of regenerating calli with shoots is also the result of greening somatic embryos at very early stages during precocious germination.

REGENERATION FROM CELL SUSPENSION CULTURES

Plant regeneration from suspension cultures of the Gramineae is rare (81). Embryogenic calli obtained from a variety of explants have been used to establish stable and long-term embryogenic suspension culture lines in Panicum maximum (32), Pennisetum americanum (77), P. purpureum (83), and Saccharum officinarum (25). These cell lines are highly dispersed, are fast growing with a mean cell generation time of 27-30 hr, and contain aggregates of about 50 cells. Each of the aggregates is actually made up of individual and discrete groups of 2-6 cells (Karlsson and Vasil, unpub. data). Each such group appears to arise from a single cell. No compact callus pieces or organized meristems and meristemoids are present. Growth takes place entirely by rapid cell divisions in the individual cell groups, and their continued dissociation or fragmentation into smaller groups.

The embryogenic suspensions contain 10-20 million cells/ml and are maintained by subculture every 3-5 days. Lowering of the 2,4-D levels by continued and rapid metabolism during prolonged subculture intervals or by transfer to low 2,4-D media, causes the organization of somatic embryos. They develop only up to about the globular or early scutellar stage in suspension culture. Thereafter, the cultures must be plated on agar media to obtain further development and mature somatic embryos. In comparison to callus cultures, the number of somatic embryos formed in suspension cultures is enormously large. However, only a small number are able to develop to maturity. A majority of them undergo secondary callusing or germinate precociously to give rise to weak plantlets. This problem must be overcome before the enormous potential of the embryogenic suspensions in large-scale clonal propagation can be fully realized.

Some recent reports have described morphogenic cell suspension cultures of the Gramineae which contain a substantial number of enlarged, thick-walled, poorly cytoplasmic, and vacuolated cells, along with many large and compact callus pieces or proliferating root meristems. The growth of such suspensions is not sustained by the large and vacuolated cells which have ceased dividing, but by the proliferating callus pieces accompanied by continued sloughing off of senescent cells into the medium. Such cultures are not true suspension cultures or lines as they can not be maintained by divisions in the freely dispersed cells within the population. The calli in such cultures often form only roots, but sometimes also somatic embryos which may develop to maturity and are later released into the medium. No further growth or cell divisions take place when the freely dispersed part of such cultures is plated, after removal of the callus pieces. Such cultures also do not yield viable populations of protoplasts as the walls of both the thick-walled vacuolated cells as well as the large callus pieces are difficult to remove, enzymatically.

REGENERATION FROM PROTOPLASTS

Regeneration of plants from protoplasts of the Gramineae has proven to be very difficult (70,72,82). One of the principal reasons for this is the fact that based on the remarkable success in the regeneration of plants from mesophyll protoplasts of the Solanaceae, all of the early as well as the major efforts directed toward the culture of gramineous protoplasts used mesophyll protoplasts only. In spite of tens of thousands of variations in nutrient media and other parameters of protoplast isolation and culture, there is as yet no convincing record of sustained cell divisions in mesophyll protoplasts of any species of the Gramineae (14,17,53,54). These unfortunate and frustrating experiences have led to premature and unfounded conclusions about the presence of a mitotic block and even the suggestion that cereal protoplasts are irreversibly arrested and constitutionally incapable of mitotic divisions (16,20).

There are many reports of sustained cell divisions in protoplasts isolated from nonmorphogenic cell suspension cultures of the Gramineae (14,70,73,82). The protoplast-derived calli, as expected, also failed to undergo any morphogenesis.

Embryogenic calli, somatic embryos, and green plantlets were first regenerated from protoplasts isolated from embryogenic cell suspension cultures of Pennisetum americanum (75). Similar success was later obtained in Panicum maximum (36) and Pennisetum purpureum (83).

Heyser (23) also cultured protoplasts isolated from a suspension culture of Panicum miliaceum but obtained only albino plantlets. In an earlier report Lorz et al. (29) confirmed the results of V. Vasil and Vasil (75) on the regeneration of somatic embryos and plantlets from protoplasts of Pennisetum americanum and stated that "protoplasts isolated from cell cultures, originally derived from inflorescences and embryos, can be cultured and regenerated. Cell divisions and plant regeneration however are infrequent... Special emphasis was made to observe real protoplast derived cell divisions, and to exclude non-digested, meristematic cell clumps."

Although these few instances of success are encouraging, major problems still remain. For example, in none of the above reports were the protoplast-derived plantlets successfully transferred to soil or grown to maturity. The reasons for this are not understood although physiological inhibition or dormancy may be responsible. Furthermore, there are very few species in which embryogenic suspensions have been established successfully. As such suspensions are the only current source of totipotent protoplasts of the Gramineae, serious and sustained efforts need to be made to isolate embryogenic suspensions from other representative species of the family.

UNIFORMITY OF EMBRYOGENIC TISSUE CULTURES AND THEIR REGENERANTS

Generation of variability is common in plant tissue cultures (1). Very often at least some of this variability can also be seen in the regenerated plants. In almost all cases where this problem has been studied or described, no attempt has been made to relate the absence or the presence or variability to the nature of the tissue cultures and the method of plant regeneration. In our preliminary observations on plants regenerated from embryogenic tissue cultures of gramineous species we noticed a remarkable degree of morphological uniformity and the absence of polyploids, aneuploids, and albinos (74). In more comprehensive cytogenetic analyses of regenerants of Panicum maximum (19) and Pennisetum americanum (63) it was found that the regenerants were morphologically similar to the parent plants and showed no evidence of any structural changes in their chromosomes. In a study of the explants, callus tissue, and regenerated plants of P. americanum, it was demonstrated that the initial inflorescence explants themselves contained some polyploid and aneuploid cells (63). During callus formation and maintenance in culture, an increase in the frequency of polyploidy and aneuploidy was noticed, but no new classes of chromosome changes appeared. It is thus clear that at least in this case the few cytologically atypical cells present in the initial explants divided faster than the normal diploid cells in vitro. More than 75% of the cells in 6-month-old cultures remained diploid. However, of the 101 regenerants analyzed, only one plant was tetraploid and all the rest were diploids; 4 plants, including the lone tetraploid, were albinos. This shows that there was a strong selection in favor of the normal cells during the formation of somatic embryos and plants.

Cytophotometric and flowcytometric analyses of 3- to 4-year-old embryogenic cell suspension cultures of Panicum maximum and Pennisetum americanum also showed that more than 75% of the cells in such cultures were still diploid (Karlsson and Vasil, unpub. data).

In many other reports, relative uniformity, as well as extensive variability in plants regenerated from tissue cultures of the Gramineae, have been reported (10,22,26,28,39,40,48).

The generation and/or the presence of variability in vitro can be a serious problem during clonal propagation of plants when absolute fidelity of the genotype is implied as well as required. Nevertheless, this problem has largely been ignored in the current enthusiasm for the professed uses of such variability in plant improvement programs.

"It is virtually an axiom of cytogenetics that meristematic or 'germ line' tissues should be genetically stable and that the daughter cells of a mitosis should contain identical chromosome complements" (1). The degree of cellular and tissue organization has an important influence on the regularity of mitosis. Thus the polarized cellular environment of meristematic cells ensures normal mitoses. Therefore, very rare, if any, cytological changes take place in organ or meristem cultures, and in cultured meristematic cells or cell groups such as the embryogenic tissue cultures of the

Gramineae. On the other hand, the large and vacuolated cells characteristically found in friable tissue cultures provide a less polarized and less organized cellular environment, and are far more susceptible to abnormalties during mitosis. It is not surprising, therefore, that in a majority of the reports describing variability, friable and less organized tissue cultures were used. Although these large and vacuolated cells have more tolerance for chromosomal aberrations, an excess of such anomalies can result in the complete loss of regenerative capacity. In comparison, the embryogenic cells have a low threshold for chromosomal changes resulting in the selective development of somatic embryos from cytologically normal cells only.

During the development of the angiosperm embryo 2 primary meristems--the shoot and the root meristem--are organized. These not only maintain their morphogenetic competence, but also their genomic integrity throughout the life of the plant. They are thus in a continued "embryonic phase" until senescence and death of the plant. The embryogenic calli are also derived from active and totipotent meristematic cells that can be considered to be direct descendants of the shoot meristem. The embryogenic tissue cultures may thus be considered to represent a continuation of the embryonic phase of plant development. The high degree of uniformity seen even in long-term embryogenic cultures may, therefore, be related more to their embryogenic/meristematic nature rather than any other factors. It should also be pointed out that in general structure and cytoplasmic organization, the embryogenic cells are very similar to meristematic cells and cells of the shoot meristem.

Although no evidence was found of readily detectable changes in the morphological characteristics of plants regenerated from somatic embryos or in the number and structure of their chromosomes, the occurrence of point mutations and other subtle and minor genetic changes can not be entirely ruled out. More sophisticated analyses and technology will be needed to discover them. It should also be understood that such minor genetic changes are more likely to yield variants of a useful nature, rather than alterations in the number, or major changes in the structure, of chromosomes which may involve many genes or gene families.

CONSEQUENCES OF SOMATIC EMBRYOGENESIS

The extensive information reported in this paper on embryogenic tissue cultures and regenerated plants in 13 species and over 70 genotypes highlights the advantages of somatic embryogenesis in the Gramineae. Embryogenic tissue cultures have been shown to be a stable, long-term, and highly efficient source for rapid and truly clonal propagation of plants. Plants are regenerated from somatic embryos which develop either directly, or indirectly, from single cells. The presence of an integrated root-shoot axis and a closed vascular system in the regenerants allows easy and successful transplantation to soil. The embryogenic callus as well as suspension cultures are cytologically quite stable and give rise to nonchimeral plants that do not show any evidence of major cytological or genetic changes. Lastly, embryogenic suspensions currently provide the only source of morphogenically competent protoplasts in the Gramineae. It is to be hoped that these attributes of embryogenic cultures will assist in the application of biotechnology towards the genetic modification and improvement of this very important group of crop plants.

ACKNOWLEDGEMENTS

The research reported and discussed in this paper was supported by the University of Florida, by a joint project between the Gas Research Institute (Chicago, Illinois) and the Institute of Food and Agricultural Sciences (University of Florida), and by the Monsanto Co. (St. Louis, Missouri). However, this work could not have been done without the dedicated efforts of a large and brilliant group of graduate students and post-doctoral associates. To them all, I am thankful and grateful.

REFERENCES

1. Bayliss, M.W. (1980) Chromosomal variation in plant tissues in culture. In Perspectives in Plant Cell and Tissue Culture, I.K. Vasil, ed. Int. Rev. Cytol. Suppl. 11A:113-144.
2. Bennici, A. (1979) Cytological chimeras in plants regenerated from *Lilium longiflorum* tissues grown *in vitro*. Z. Pflanzenzüchtg. 82:349-353.
3. Bennici, A., and F. D'Amato (1978) *In vitro* regeneration of *Durum* wheat plants. I. Chromosome numbers of regenerated plantlets. Z. Pflanzenzüchtg. 81:305-311.
4. Botti, C., and I.K. Vasil (1983) Plant regeneration by somatic embryogenesis from parts of cultured mature embryos of *Pennisetum americanum* (L.) K. Schum. Z. Pflanzenphysiol. 111:319-325.
5. Botti, C., and I.K. Vasil (1984) The ontogeny of somatic embryos of *Pennisetum americanum* (L.) K. Schum. II. In immature inflorescences. Can. J. Bot. 62:1629-1635.
6. Boyes, C.J., and I.K. Vasil (1984) Plant regeneration by somatic embryogenesis from cultured young inflorescences of *Sorghum arundinaceum* (Desv.) Stapf. var. *sudanense* (Sudan grass). Plant Sci. Lett. 35:153-157.
7. Chandler, S.F., and I.K. Vasil (1984a) Optimization of plant regeneration from long term embryogenic callus cultures of *Pennisetum purpureum* Schum. (Napier grass). J. Plant Physiol. 117:147-156.

8. Chandler, S.F., and I.K. Vasil (1984b) Selection and characterization of NaCl tolerant callus from embryogenic cultures of Pennisetum purpureum Schum. (Napier grass). Plant Sci. Lett. (in press).
9. Chandler, S.F., K. Rajasekaran, and I.K. Vasil (1984) Large scale propagation of Napier grass and Giant Napier grass by tissue culture. In Proc. 1984 International Gas Research Conference, Gas Research Institute, Chicago, pp. 359-364.
10. Chen, C.H., P.F. Lo, and J.G. Ross (1981) Cytological uniformity in callus culture-derived big bluestem plants (Andropogon gerardii Vitman). Proc. S.D. Acad. Sci. 60:39-43.
11. Coe, Jr., E.H., and M.G. Neuffer (1978) Embryo cells and their destinies in the corn plant. In The Clonal Basis of Development, S. Subtelny and I.M. Sussex, eds. Academic Press, New York, pp. 113-129.
12. Conger, B.V. (1981) Agronomic crops. In Cloning Agricultural Plants via in vitro Techniques, B.V. Conger, ed. CRC Press, Boca Raton, Florida, pp. 165-215.
13. Crooks, D.M. (1933) Histological and regenerative studies on the flax seedling. Bot. Gaz. 95:209-239.
14. Dale, P.J. (1983) Protoplast culture and plant regeneration of cereals and other recalcitrant crops. Experientia Supp. 46:31-41.
15. Dunstan, D.I., K.C. Short, H. Dhaliwal, and E. Thomas (1979) Further studies on plantlet production from cultured tissues of Sorghum bicolor. Protoplasma 101:355-361.
16. Flores, H.E., R. Kaur-Sawhney, and A.W. Galston (1981) Protoplasts as vehicles for plant propagation and improvement. Adv. Cell Cult. 1:241-279.
17. Galston, A.W. (1978) The use of protoplasts in plant propagation and improvement. In Propagation of Higher Plants Through Tissue Culture, K.W. Hughes, R. Henke, and M. Constantin, eds. U.S. Dept. of Energy, Knoxville, Tennessee, pp. 200-212.
18. Green, C.E. (1978) In vitro plant regeneration in cereals and grasses. In Frontiers of Plant Tissue Culture 1978, T.A. Thorpe, ed. Univ. of Calgary, pp. 411-418.
19. Hanna, W.W., C. Lu, and I.K. Vasil (1984) Uniformity of plants regenerated from somatic embryos of Panicum maximum Jacq. (Guinea grass). Theoret. Appl. Genet. 67:155-159.
20. Harms, C.T. (1982) Maize and cereal protoplasts - facts and perspectives. In Maize for Biological Research, W.F. Sheridan, ed. Plant Molecular Biology Association, Charlottesville, Virginia, pp. 373-384.
21. Haydu, Z., and I.K. Vasil (1981) Somatic embryogenesis and plant regeneration from leaf tissues and anthers of Pennisetum purpureum. Theoret. Appl. Genet. 59:269-273.
22. Heinz, D.J., M. Krishnamurthi, L.G. Nickell, and A. Maretzki (1977) Cell, tissue and organ cultures in sugarcane improvement. In Plant Cell, Tissue and Organ Culture, J. Reinert and Y.P.S. Bajaj, eds. Springer-Verlag, Heidelberg, pp. 3-17.

23. Heyser, J.W. (1984) Callus and shoot regeneration from protoplasts of Proso millet (Panicum miliaceum L.). Z. Pflanzenphysiol. 113:293-299.
24. Ho, W., and I.K. Vasil (1983a) Somatic embryogenesis in sugarcane (Saccharum officinarum L.). I. The morphology and physiology of callus formation and the ontogeny of somatic embryos. Protoplasma 118:169-180.
25. Ho, W., and I.K. Vasil (1983b) Somatic embryogenesis in sugarcane (Saccharum officinarum L.). II. The growth of and plant regeneration from embryogenic cell suspension cultures. Ann. Bot. 51:719-726.
26. Karp, A., and S.E. Maddock (1984) Chromosome variation in wheat plants regenerated from cultured immature embryos. Theoret. Appl. Genet. 67:249-255.
27. King, P.J., I. Potrykus, and E. Thomas (1978) In vitro genetics of cereals: Problems and perspectives. Physiol. Veg. 16:381-399.
28. Larkin, P.J., S.A. Ryan, R.I.S. Brettell, and W.R. Scowcroft (1984) Heritable somaclonal variation in wheat. Theoret. Appl. Genet. 67:443-455.
29. Lorz, H., R.I.S. Brettell, and I. Potrykus (1981) Protoplast culture of Pennisetum americanum. XIII Int. Bot. Congr. (Poster Section 12), Sydney, Australia.
30. Lu, C., S.F. Chandler, and I.K. Vasil (1984) Somatic embryogenesis and plant regeneration in cultured immature embryos of rye (Secale cereale L.). J. Plant Physiol. 115:237-244.
31. Lu, C., and I.K. Vasil (1981a) Somatic embryogenesis and plant regeneration from leaf tissues of Panicum maximum Jacq. Theoret. Appl. Genet. 59:275-280.
32. Lu, C., and I.K. Vasil (1981b) Somatic embryogenesis and plant regeneration from freely suspended cells and cell groups of Panicum maximum in vitro. Ann. Bot. 47:543-548.
33. Lu, C., and I.K. Vasil (1982) Somatic embryogenesis and plant regeneration in tissue cultures of Panicum maximum Jacq. Amer. J. Bot. 69:77-81.
34. Lu, C., and I.K. Vasil (1984) Histology of somatic embryogenesis in Panicum maximum Jacq. (Guinea grass). (Submitted for publication.)
35. Lu, C., I.K. Vasil, and P. Ozias-Akins (1982) Somatic embryogenesis in Zea mays L. Theoret. Appl. Genet. 62:109-112.
36. Lu, C., V. Vasil, and I.K. Vasil (1981) Isolation and culture of protoplasts of Panicum maximum Jacq. (Guinea grass): Somatic embryogenesis and plantlet formation. Z. Pflanzenphysiol. 104:311-318.
37. Lu, C., V. Vasil, and I.K. Vasil (1983) Improved efficiency of somatic embryogenesis and plant regeneration in tissue cultures of maize (Zea mays L.). Theoret. Appl. Genet. 62:285-290.
38. Lupi, M.C., A. Bennici, S. Baroncelli, D. Gennari, and F. D'Amato (1981) In vitro regeneration of Durum wheat plants. II. Diplontic selection of aneusomatic plants. Z. Pflanzenzüchtg. 87:167-171.

39. McCoy, T.J., and R.L. Phillips (1982) Chromosome stability in maize (Zea mays) tissue cultures and sectoring in some regenerated plants. Can. J. Genet. Cytol. 24:559-565.
40. McCoy, T.J., R.L. Phillips, and H.W. Rines (1982) Cytogenetic analysis of plants regenerated from oat (Avena sativa) tissue cultures: High frequency of partial chromosome loss. Can. J. Genet. Cytol. 24:37-50.
41. Mix, G., H.W. Wilson, and B. Foroughi-Wehr (1978) The cytological status of plants of Hordeum vulgare L. regenerated from microspore callus. Z. Pflanzenzüchtg. 80:89-99.
42. Murashige, T., and F. Skoog (1962) A revised medium for rapid growth and bioassays with tobacco tissue cultures. Physiol. Plant. 15:473-497.
43. Nakano, H., and E. Maeda (1979) Shoot differentiation in callus of Oryza sativa L. Z. Pflanzenphysiol. 93:449-458.
44. Norris, R., R.H. Smith, and K.C. Vaughn (1983) Plant chimeras used to establish de novo origin of shoots. Science 220:75-76.
45. Novak, F.J., and B. Vyskot (1975) Karyology of callus cultures derived from Nicotiana tabacum L. haploids and ploidy of regenerants. Z. Pflanzenzüchtg. 75:62-70.
46. Ogihara, Y. (1981) Tissue culture in Haworthia. IV. Genetic characterization of plants regenerated from callus. Theoret. Appl. Genet. 60:353-363.
47. Ogura, H. (1976) Cytological chimeras in original regenerates from tobacco tissue cultures and their offsprings. Jap. J. Genet. 51:161-174.
48. Oono, K. (1978) Test tube breeding of rice by tissue culture. Trop. Agric. Res. Ser. 11:109-123.
49. Ozias-Akins, P., and I.K. Vasil (1982) Plant regeneration from cultured immature embryos and inflorescences of Triticum aestivum L. (wheat): Evidence for somatic embryogenesis. Protoplasma 110:95-105.
50. Ozias-Akins, P., and I.K. Vasil (1983a) Callus induction and growth from the mature embryo of Triticum aestivum (wheat). Protoplasma 115:104-113.
51. Ozias-Akins, P., and I.K. Vasil (1983b) Proliferation of and plant regeneration from the epiblast of Triticum aestivum (wheat; Gramineae) embryos. Amer. J. Bot. 70:1092-1097.
52. Ozias-Akins, P., and I.K. Vasil (1983c) Improved efficiency and normalization of somatic embryogenesis in Triticum aestivum (wheat). Protoplasma 117:40-44.
53. Potrykus, I. (1980) The old problem of protoplast culture: Cereals. In Advances in Protoplast Research, L. Ferenczy, G.L. Farkas, and G. Lazar, eds. Akademiai Kiado, Budapest, pp. 243-254.
54. Potrykus, I., C.T. Harms, and H. Lorz (1976) Problems in culturing cereal protoplasts. In Cell Genetics in Higher Plants, D. Dudits, G.L. Farkas, and P. Maliga, eds. Akademiai Kiado, Budapest, pp. 129-140.
55. Rangan, T.S. (1974) Morphogenic investigations on tissue cultures of Panicum miliaceum. Z. Pflanzenphysiol. 72:456-459.

56. Rangan, T.S., and I.K. Vasil (1983a) Somatic embryogenesis and plant regeneration in tissue cultures of Panicum miliaceum L. and Panicum miliare Lamk. Z. Pflanzenphysiol. 109:49-53.
57. Rangan, T.S., and I.K. Vasil (1983b) Sodium chloride tolerant embryogenic cell lines of Pennisetum americanum (L.) K. Schum. Ann. Bot. 52:59-64.
58. Sacristan, M.D., and G. Melchers (1969) The karyological analysis of plants regenerated from tumorous and other callus cultures of tobacco. Molec. Gen. Genet. 105:317-333.
59. Shimada, T., and Y. Yamada (1979) Wheat plants regenerated from embryo cell cultures. Jap. J. Genet. 54:379-385.
60. Springer, W.D., C.E. Green, and K.A. Kuhn (1979) A histological examination of tissue culture initiation from immature embryos of maize. Protoplasma 101:269-281.
61. Sree Ramulu, K., M. Devreux, G. Ancora, and U. Laneri (1976) Chimerism in Lycopersicum peruvianum plants regenerated from in vitro cultures of anthers and stem internodes. Z. Pflanzenzüchtg. 76:299-319.
62. Steffensen, D.M. (1968) A reconstruction of cell development in the shoot apex of maize. Amer. J. Bot. 55:354-369.
63. Swedlund, B., and I.K. Vasil (1984) Cytogenetic characterization of embryogenic callus and regenerated plants of Pennisetum americanum (L.) K. Schum. (Submitted for publication.)
64. Vasil, I.K. (1960) Pollen germination in some Gramineae: Pennisetum typhoideum. Nature 187:1134-1135.
65. Vasil, I.K. (1962) Studies on pollen storage of some crop plants. J. Indian Bot. Soc. 41:178-196.
66. Vasil, I.K. (1982a) Plant cell culture and somatic cell genetics of cereals and grasses. In Plant Improvement and Somatic Cell Genetics, I.K. Vasil, W.R. Scowcroft, and K.J. Frey, eds. Academic Press, New York, pp. 179-203.
67. Vasil, I.K. (1982b) Somatic embryogenesis and plant regeneration in cereals and grasses. In Plant Tissue Culture 1982, A. Fujiwara, ed. Maruzen, Tokyo, pp. 101-104.
68. Vasil, I.K. (1983a) Regeneration of plants from single cells of cereals and grasses. In Genetic Engineering in Eukaryotes, P.F. Lurquin and A. Kleinhofs, eds. Plenum, New York, pp. 233-252.
69. Vasil, I.K. (1983b) Toward the development of a single cell system for grasses. In Cell and Tissue Culture Techniques for Cereal Crop Improvement, Science Press, Beijing, pp. 131-144.
70. Vasil, I.K. (1983c) Isolation and culture of protoplasts of grasses. In Plant Protoplasts, K.L. Giles, ed. Int. Rev. Cytol. Suppl., 14:79-88.
71. Vasil, I.K., and A.C. Hildebrandt (1966) Variations of morphogenetic behaviour in plant tissue cultures. I. Cichorium endivia. Amer. J. Bot. 53:860-869.
72. Vasil, I.K., and V. Vasil (1980a) Isolation and culture of protoplasts. In Perspectives in Plant Cell and Tissue Culture, I.K. Vasil, ed. Int. Rev. Cytol. Supp., 11B:1-19.

73. Vasil, I.K., and V. Vasil (1980b) Embryogenesis and plantlet formation from protoplasts of pearl millet (Pennisetum americanum). In Advances in Protoplast Research, L. Ferenczy, G.L. Farkas, and G. Lazar, eds. Akademiai-Kiado, Budapest, pp. 255-259.
74. Vasil, I.K., V. Vasil, C. Lu, P. Ozias-Akins, Z. Haydu, and D. Wang (1982) Somatic embryogenesis in cereals and grasses. In Variability in Plants Regenerated from Tissue Culture, E. Earle and Y. Demarly, eds. Praeger Press, New York, pp. 3-21.
75. Vasil, V., and I.K. Vasil (1980) Isolation and culture of cereal protoplasts. II. Embryogenesis and plantlet formation from protoplasts of Pennisetum americanum. Theoret. Appl. Genet. 56:97-99.
76. Vasil, V., and I.K. Vasil (1981a) Somatic embryogenesis and plant regeneration from tissue cultures Pennisetum americanum and P. americanum x P. purpureum hybrid. Amer. J. Bot. 68:864-872.
77. Vasil, V., and I.K. Vasil (1981b) Somatic embryogenesis and plant regeneration from suspension cultures of pearl millet (Pennisetum americanum). Ann. Bot. 47:669-678.
78. Vasil, V., and I.K. Vasil (1982a) Characterization of an embryogenic cell suspension culture derived from inflorescences of Pennisetum americanum (pearl millet; Gramineae). Amer. J. Bot. 69:1441-1449.
79. Vasil, V., and I.K. Vasil (1982b) The ontogeny of somatic embryos of Pennisetum americanum (L.) K. Schum.: In cultured immature embryos. Bot. Gaz. 143:454-465.
80. Vasil, V., and I.K. Vasil (1984a) Induction and maintenance of embryogenic callus cultures of the Gramineae. In Cell Culture and Somatic Cell Genetics of Plants, Vol. I, Laboratory Procedures and Their Applications, I.K. Vasil, ed. Academic Press, New York, pp. 36-42.
81. Vasil, V., and I.K. Vasil (1984b) Isolation and maintenance of embryogenic cell suspension cultures of Gramineae. In Cell Culture and Somatic Cell Genetics of Plants, Vol. I, Laboratory Procedures and Their Applications, I.K. Vasil, ed. Academic Press, New York, pp. 152-158.
82. V. Vasil, and I.K. Vasil (1984c) Isolation and culture of embryogenic protoplasts of cereals and grasses. In Cell Culture and Somatic Cell Genetics of Plants, Vol. I, Laboratory Procedures and Their Applications, I.K. Vasil, ed. Academic Press, New York, pp. 398-404.
83. Vasil, V., D. Wang, and I.K. Vasil (1983a) Plant regeneration from protoplasts of Pennisetum purpureum Schum. (Napier grass). Z. Pflanzenphysiol. 111:319-325.
84. Vasil, V., C. Lu, and I.K. Vasil (1983b) Proliferation and plant regeneration from the nodal region of Zea mays L. (maize; Gramineae) embryos. Amer. J. Bot. 70:951-954.
85. Vasil, V., I.K. Vasil, and C. Lu (1984a) Somatic embryogenesis in long-term callus cultures of Zea mays L. (Gramineae). Amer. J. Bot. 71:158-161.

86. Vasil, V., C. Lu, and I.K. Vasil (1984b) Histology of somatic embryogenesis in cultured immature embryos of maize (*Zea mays* L.). (Submitted for publication.)

87. Wang, D., and I.K. Vasil (1982) Somatic embryogenesis and plant regeneration from inflorescence segments of *Pennisetum purpureum* Schum. (Napier or Elephant grass). *Plant Sci. Lett.* *25*:147-154.

88. Yamabe, M., and T. Yamada (1973) Studies on differentiation in culture cells. II. Chromosomes of *Haworthia* callus and of the plants grown from callus. *La Kromosome* (Tokyo) 94:2923-2931.

SOMATIC EMBRYO ONTOGENY IN TISSUE CULTURES OF ORCHARDGRASS

D.J. Gray* and B.V. Conger

Department of Plant and Soil Science
University of Tennessee
P.O. Box 1071
Knoxville, Tennessee 37901-1071

INTRODUCTION

A plant embryo can be defined as a new individual arising from a single cell and having no vascular connection with maternal tissue (12). Embryos formed during sexual reproduction and attached to adjacent tissue by suspensors, obviously develop from single cells (zygotes) and, thus, fit this definition. However, somatic (nonzygotic or adventive) embryos produced in plant tissue cultures often appear as buds, broadly attached to underlying tissue, and this has caused some confusion as to whether or not they originated from single cells.

Although somatic embryogenesis has been described in tissue cultures of numerous plant species (20), the development of an embryo from a single somatic cell has been documented only in carrot (*Daucus carota* L.) with time lapse photomicrography (1). In the Gramineae, somatic embryo ontogeny has been described for relatively few species (8,15,21). In these studies, somatic embryogenesis was shown to be similar to zygotic embryogenesis. However, in sorghum (*Sorghum bicolor* L.), somatic embryos developed with a broad basal attachment to underlying zygotic embryo scutellar tissue prompting the authors to postulate that these embryos may have originated from more than one cell (8). Similar studies of pearl millet [*Pennisetum americanum* (L.) K. Schum.] demonstrated that single, isolated, densely cytoplasmic cells occurred during early culture stages in

* Present address: Agricultural Research and Education Center, University of Florida, P.O. Box 388, Leesburg, Florida 32749-0388.

the same scutellar regions that gave rise to broadly attached embryos at later stages (21). This provided strong evidence for a single cell origin.

In our laboratory, somatic embryos of orchardgrass (Dactylis glomerata L.) develop with either a broad attachment to underlying tissue (11) or with a fine, suspensor-like connection (6) that is more characteristic of zygotic embryogenesis (9). The former occurs chiefly in callus cultures whereas the latter is common during direct embryogenesis in leaf cultures. The direct response provides an ideal system to follow the process of somatic embryogenesis because the cultured leaf is easily manipulated and single developing embryos are highly conspicuous. Microscopic examination of the attachment of suspensor to leaf could shed more light on the exact cellular origin of the somatic embryos. The purpose of this presentation is to summarize the process of orchardgrass somatic embryo ontogeny based on our previously published papers and ongoing studies. Early stages of embryogenesis occurring in cultured leaves are emphasized. The relevance of this information to current concepts of embryogeny is discussed.

MATERIALS AND METHODS

Media and cultural methods employed to obtain somatic embryogenesis from various types of orchardgrass tissue cultures were previously described (6,11,14,17). Small pieces of leaves bearing embryos or single embryos were prepared for scanning electron microscopy (SEM) and light microscopy as previously described (6,11).

RESULTS

Cultured segments from the basal leaf meristems of greenhouse grown orchardgrass plants produced embryos either directly from mesophyll regions or from highly embryogenic callus (6,14). Histological sampling, over a period of 3 weeks, of embryos produced from leaves plated on medium containing the auxin, dicamba (3,6-dichloro-o-anisic acid), allowed for reconstruction of the cell and tissue level events leading to embryo formation.

Small, densely cytoplasmic cells, indicative of early stages of embryogenesis, were found in mesophyll and vascular regions within a week after plating leaf sections on medium with dicamba. These were identical in appearance to "embryogenic cells" reported in a number of previous studies (7,16,21) and were never found in leaf sections not introduced into culture. Divisions ensued in these cells, leading to the production of either uni- or multiseriate suspensors (Fig. 1 and 2) or masses of densely cytoplasmic cells. The most basally oriented cells of suspensors were often surrounded by necro-

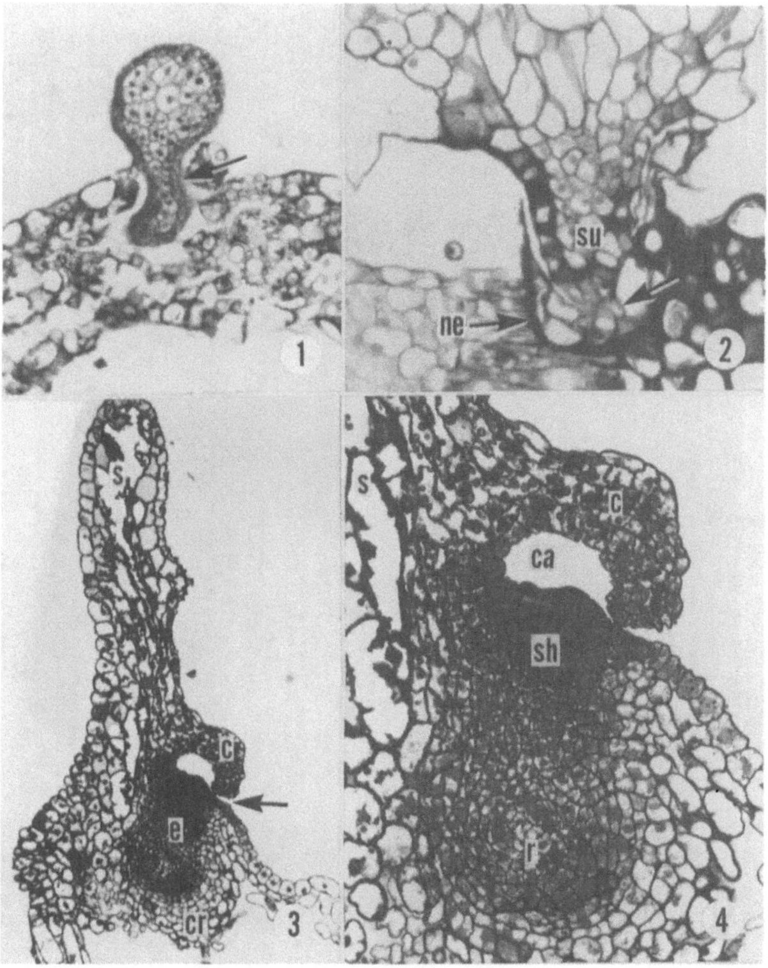

Fig. 1-4. Light microscopy of somatic embryo ontogeny. Fig. 1. Globular stage embryo attached to leaf mesophyll region by a multiseriate suspensor (arrow) (X 150). From Conger et al. (Ref. 6; copyright 1983 by the A.A.A.S.). Fig. 2. Embryo suspensor (su) detailing basal attachment to a small mass of cells (arrow). Necrotic material (ne) surrounds the cell mass (X 375). Fig. 3. Well-developed embryo. Note scutellum (s), embryo axis (e) coleoptile (c), coleorhiza (cr), and notch (arrow) (X 90). From Gray et al. (11). Fig. 4. Higher magnification of Fig. 3 detailing shoot (sh) and root (r) apices of embryo axis. Note attachment of coleoptile (c) to scutellum (s) and cavity (ca) created by c enclosing sh (X 230).

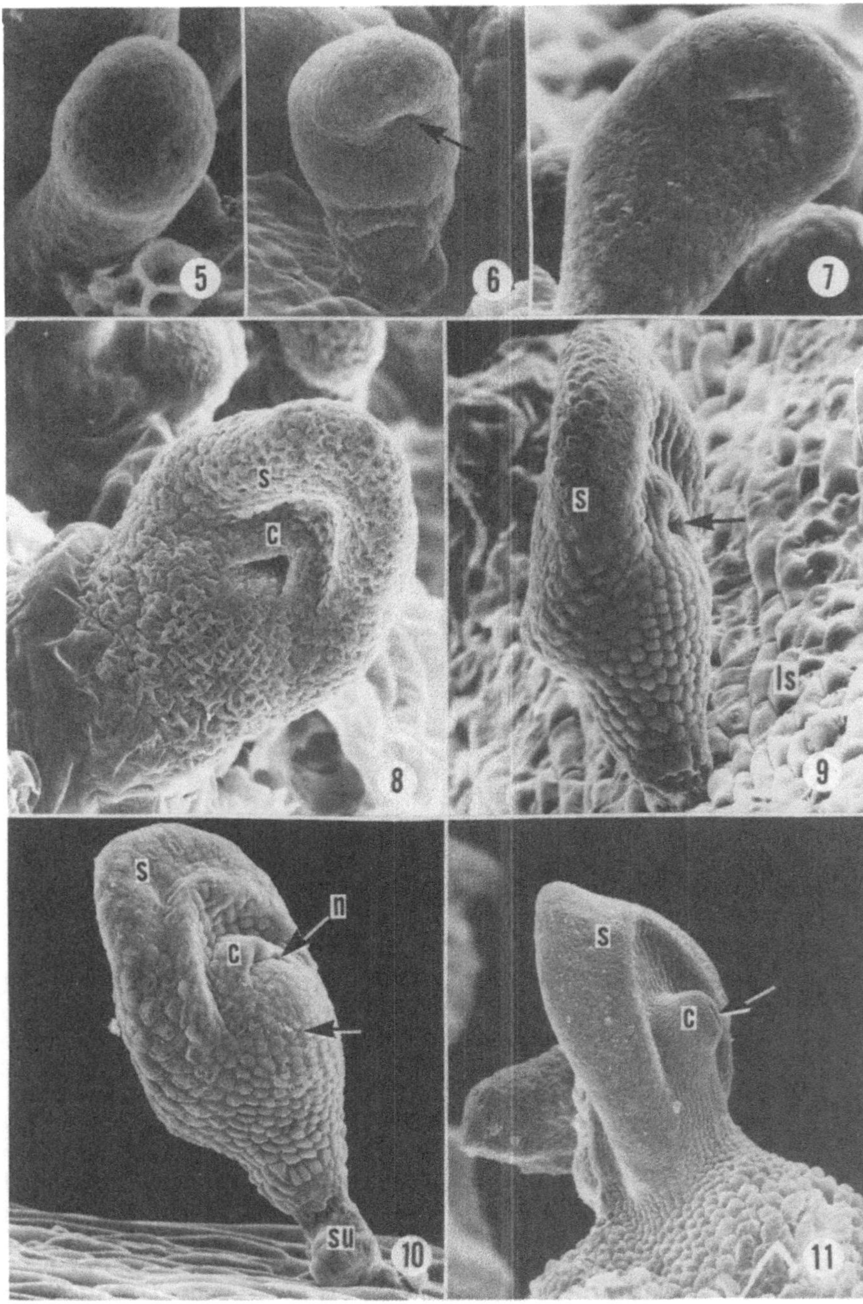
5
6
7
s
c
8
s
ls
9
s
n
c
su
10
s
c
11

Fig. 5-11. Scanning electron microscopy of somatic embryo ontogeny. Fig. 5. Globular stage embryo growing from cultured leaf surface (X 200). Fig. 6. Clavate embryo with rudimentary notch (arrow) development (X 100). Fig. 7. Further development of notch (X 130). Fig. 8. Embryo with well-formed notch. Note early coleoptile (c) and scutellum development (s) (X 110). Fig. 9. Outward growth of coleoptile begins to close notch (arrow). Scutellum (s), leaf surface (ls) (X 100). Fig. 10. Fully-formed embryo attached to leaf by a narrow suspensor (su). Note swelled appearance of embryo body (arrow) due to internal development of embryo axis. Scutellum (s), coleoptile (c), remnants of notch (n) (X 100). From Conger et al. (Ref. 6; copyright 1983 by the A.A.A.S.). Fig. 11. Fully-developed embryo with broad attachment to underlying callus tissue. Scutellum (s), coleoptile (c), notch (arrow) (X 90). From Gray et al. (11).

tic tissue (Fig. 2), suggesting that the entire embryo may have originated from one or few cells. Continued growth of suspensors and cell masses caused the leaf epidermis to rupture in localized areas, and globular-shaped embryos typically developed on the epidermal surface (Fig. 1 and 5). Cells of globular embryos were relatively densely cytoplasmic and undifferentiated from each other (Fig. 1).

During further development, the embryos elongated and became clavate (Fig. 6) due to cell division and enlargement, and a degree of cellular differentiation occurred. At this stage, large vacuolate cells containing starch, destined to compose the scutellum, were evident, as well as a region of relatively surface-oriented, densely cytoplasmic cells indicative of nonscutellar tissues. Further elongation of the embryos and organized divisions of the cytoplasmic cells produced a notch (Fig. 6-8), marking the position of the developing coleoptile and shoot apex. The notch was formed partially by differential, schizogenous divisions of some cells.

As the scutellum enlarged by divisions of cells on its surface, the coleoptile developed outwardly by similar means and eventually closed the notch (Fig. 3,9,10, and 11), creating a cavity enclosing the developing shoot apex (Fig. 4). At this point, many small, deep-seated meristematic cells in the position of the embryo axis were actively dividing, causing that region of the embryo to swell greatly [Fig. 3 and 10 (arrow)]. Scutellar cells typically contained an abundance of starch and lipids. Coleoptile cells appeared to possess only starch, whereas those of the embryo axis contained only lipids.

Further development and germination were stimulated by transfer of leaves bearing embryos to medium lacking dicamba. A number of simultaneous events typically occurred at that time. Cells in lower portions of the embryo, probably analogous to the coleorhiza and epiblast of zygotic embryos, became swollen and vacuolate, and many produced root hairs. The scutellum became depleted of starch and degenerated as the coleoptile became highly elongated and pigmented with anthocyanins. The embryo axis differentiated due to continued rapid divisions of the deep-seated meristematic cells to yield a shoot apex with leaf primordia enclosed by the coleoptile and one or more root apices embedded in the coleorhiza. Embryos typically remained attached to the leaf by vacuolate suspensor cells throughout their development.

Our ongoing investigations of embryo ontogeny in orchardgrass callus cultures have determined that the process is essentially the same as from leaf explants except that a narrow suspensor (Fig. 10) is usually not present. Instead, the embryos are often broadly attached to small masses of underlying callus tissue (Fig. 11) as is the case with somatic embryos of other grass callus cultures.

Orchardgrass suspension cultures produced embryos capable of germination directly in liquid medium. Although free embryos in every stage were present, they usually had a suspensor-like appendage suggesting that they were at one time attached to and arose from other cell masses (11).

Studies of suspension (11) and callus culture growth dynamics demonstrated that somatic embryos in various stages of development became disorganized into cell masses capable of producing more embryos in the presence of dicamba. In addition, individual somatic embryos on solid or in liquid medium containing auxin produced highly embryogenic cell masses (6). Thus, it is clear that a cyclic growth pattern, from embryogenesis to disorganization and back, routinely occurs. This suggests that the cultures may exhibit infinite embryogenic potential because a source of new cells capable of embryogenesis is continually produced.

DISCUSSION

The developmental morphology of orchardgrass somatic embryos was similar to descriptions of zygotic gramineous embryogenesis (3,4). The interrelationships of coleoptile, embryo axis, scutellum, and suspensor were typical of their zygotic counterparts. Although not well supported in the literature, Brown (2) suggested that the grass coleoptile and scutellum arose from proembryo tissue and not from the shoot apex. Our studies demonstrated that these organs developed at a stage when the embryo axis was not yet formed, thus corroborating Brown's observations. The germination events of orchardgrass somatic embryos, including the development of root hairs and elongated, pigmented coleoptiles, were also similar to that of zygotic grass embryos (2,19).

Although only typical ontogeny has been described here, much developmental variation occurs in the orchardgrass system. Time lapse photomicrography (10) revealed that some embryos initially developed normally only to cease further growth. Other embryos spontaneously disorganized to produce an embryogenic callus, and one out of 12 examined developed into, otherwise normal, twin embryos attached to the leaf by a common suspensor. Studies of germinating suspension culture-derived embryos demonstrated that some produced only a shoot, whereas others developed only a root resulting in no further growth. Other embryos from suspension did not produce a coleoptile, indicative of precocious germination (18), or developed twin shoot apices (11). Both of these latter germination patterns resulted in plantlets capable of survival.

Investigations of orchardgrass somatic embryogenesis have not yet directly demonstrated a single cell origin of embryos. The proof, which is currently lacking in the Gramineae, will come only by observing the initial division of a living cell and its eventual development into an embryo as was done for carrot (1). However, previous studies of isolated meristematic cells in scutellar tissues of pearl millet (21) and our observations of uniseriate suspensors surrounded completely by necrotic leaf tissue provide strong evidence for a single cell origin.

Speculation upon the single cell origin of somatic embryos has also continued because many species that develop suspensors during zygotic embryogenesis produce embryos broadly attached to underlying tissue during somatic embryogenesis from callus cultures. However, the underlying tissue can be regarded as a proembryonal complex which was itself derived from a single cell, and this type of development is termed cleavage polyembryony (12). Various types of cleavage polyembryony occur commonly in nature during zygotic embryogenesis in gymnosperms (5) as well as in some angiosperms (13), and are considered to be indicative of a more primitive form of development. In fact, Haccius (12) considered the suspensor of evolutionarily more-advanced species to be nothing more than a rudimentary proembryonal complex. Thus, it appears that, for a given species, a more primitive type of embryogenesis often occurs in tissue culture. In some species (e.g., orchardgrass) both primitive and advanced types are present (compare Fig. 10 and 11). This extreme flexibility of embryogenetic patterns within a given species brings into question the significance of using such indicators in considerations of plant evolution.

In conclusion, although some developmental variation occurs during orchardgrass somatic embryogenesis, the overall process is very similar to zygotic embryogenesis as previously described for the Gramineae. Ongoing research seeks to increase uniformity in the orchardgrass tissue culture system. Mass in vitro cloning may be possible when somatic embryo uniformity approaches that of their zygotic counterparts.

REFERENCES

1. Backs-Husemann, D., and J. Reinert (1970) Embryobildung durch isolierte Einzelzellen aus Gewebekulturen von Daucus carota. Protoplasma 70:49-60.
2. Brown, W.V. (1959) The epiblast and coleoptile of the grass embryo. Bull. Torrey Bot. Club 86:13-16.
3. Brown, W.V. (1960) The morphology of the grass embryo. Phytomorphology 10:215-223.
4. Brown, W.V. (1965) The grass embryo - A rebuttal. Phytomorphology 15:274-284.
5. Buchholz, J.T. (1933) Determinant cleavage polyembryony with special reference to Dacrydium. Bot. Gaz. 94:579-588.
6. Conger, B.V., G.E. Hanning, D.J. Gray, and J.K. McDaniel (1983) Direct embryogenesis from mesophyll cells of orchardgrass. Science 221:850-851.
7. Dos Santos, A.V.P., E.G. Cutter, and M.R. Davey (1983) Origin and development of somatic embryos in Medicago sativa L. (alfalfa). Protoplasma 117:107-115.
8. Dunstan, D.I., K.C. Short, and E. Thomas (1978) The anatomy of secondary morphogenesis in cultured scutellum tissues of Sorghum bicolor. Protoplasma 97:251-260.
9. Gray, D.J., and B.V. Conger (1984) Nonzygotic embryogenesis in tissue cultures of forage grasses. In Proc. 40th Southern Pasture and Forage Crop Imp. Conf. (in press).
10. Gray, D.J., and B.V. Conger (1984) Time lapse light photomicrography and scanning electron microscopy of somatic embryo ontogeny from cultured leaves of Dactylis glomerata (Gramineae). Trans. Amer. Microsc. Soc. (in press).
11. Gray, D.J., B.V. Conger, and G.E. Hanning (1984) Somatic embryogenesis in suspension and suspension-derived callus cultures of Dactylis glomerata. Protoplasma 122:196-202.
12. Haccius, B. (1978) Question of unicellular origin of non-zygotic embryos in callus cultures. Phytomorphology 28:74-81.
13. Haccius, B., and N.N. Bhandari (1975) Delayed histogen differentiation as a common primitive character in all types of nonzygotic embryos. Phytomorphology 25:91-94.
14. Hanning, G.E., and B.V. Conger (1982) Embryoid and plantlet formation from leaf segments of Dactylis glomerata L. Theoret. Appl. Genet. 63:155-159.
15. Ho, W.J., and I.K. Vasil (1983) Somatic embryogenesis in sugarcane (Saccharum officinarum L.) I. The morphology and physiology of callus formation and the ontogeny of somatic embryos. Protoplasma 118:169-180.

16. Konar, R.N., E. Thomas, and H.E. Street (1972) Origin and structure of embryoids arising from epidermal cells of the stem of Ranunculus sceleratus L. J. Cell Sci. 11:77-93.
17. McDaniel, J.K., B.V. Conger, and E.T. Graham (1982) A histological study of tissue proliferation, embryogenesis and organogenesis from tissue cultures of Dactylis glomerata L. Protoplasma 110:121-128.
18. Norstog, K., and R.M. Klein (1972) Development of cultured barley embryos. II. Precocious germination and dormancy. Can. J. Bot. 50:1887-1894.
19. Norstog, K. (1955) Responses of the oat coleorhiza to various treatments in culture. Ohio J. Sci. 55:340-342.
20. Tisserat, B., E.B. Esan, and T. Murashige (1978) Somatic embryogenesis in angiosperms. Hort. Rev. 1:1-78.
21. Vasil, V., and I.K. Vasil (1982) The ontogeny of somatic embryos of Pennisetum americanum (L.) K. Schum. I. In cultured immature embryos. Bot. Gaz. 143:454-465.

FACTORS AFFECTING DEVELOPMENTAL PROCESSES IN ALFALFA CELL CULTURES

David A. Stuart, Janet Nelsen,
Steven G. Strickland, and James W. Nichol

Plant Genetics, Inc.
1930 Fifth Street
Davis, California 95616

ABSTRACT

The development of somatic embryos in alfalfa is strongly influenced by the presence of a reduced nitrogen source during embryogeny. Amino acids such as proline, alanine, arginine, and glutamine at high concentrations (30 mM and above) improve embryo yields. The effect of proline on embryo yield is the result of a synergistic interaction of proline and ammonium. The quality of embryogenesis is also enhanced by amino acids as measured by improved morphology and conversion of embryos to plantlets. An additional measure of somatic embryo quality may be the expression of the seed-specific 7S and 11S storage proteins common to legume seeds. Somatic embryos of alfalfa show little expression of these storage proteins following induction on high 2,4-D medium. If low 2,4-D induction is used, embryo morphology and conversion to plantlets are improved and expression of storage protein is high for both the 7S and 11S proteins. These results are discussed with respect to the developmental program of zygotic embryos.

STIMULATION OF SOMATIC EMBRYOGENESIS BY REDUCED NITROGEN

Introduction

In their studies of carrot somatic embryogenesis, Halperin and Wetherell (1) were the first to recognize the importance of ammonium in determining the developmental fate of differentiating cell cultures. Since that report a number of other investigations in diverse species have also concluded that there is an ammonium requirement for effective somatic embryogenesis (for recent reports, see Ref. 2 and 3). Perhaps one of the few exceptions to these observations is the work of Tazawa and Reinert (4) who found that high levels of nitrate in the absence of ammonium would stimulate embryo formation in carrot cultures. They found that intracellular levels of ammonium increased as did embryogenesis in cultures regenerated on 60 mM nitrate as a sole nitrogen source. Ammonium ion, either added exogenously or reduced endogenously from nitrate, seems to be required for development of somatic embryos in vitro.

Work in carrot with other forms of reduced nitrogen has focused on the application of single amino acids in the absence of other sources of nitrogen in the culture medium (5,6). These studies reported that alanine and glutamine were far superior to ammonium as a nitrogen source when nitrate was also present. Glutamine was the best source of nitrogen for embryogenesis when no other nitrogen source was present.

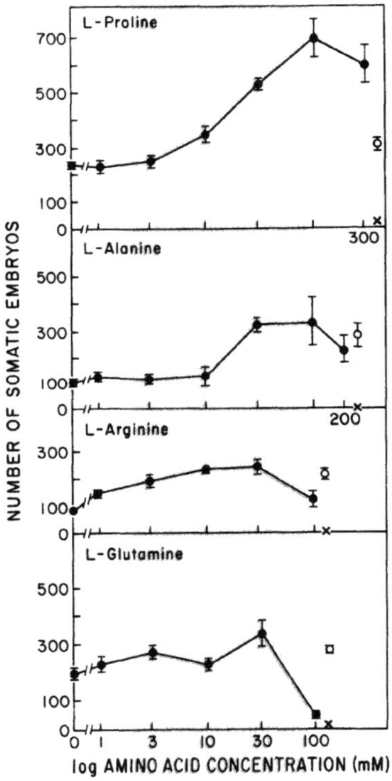

Fig. 1. Yield of somatic embryos incubated for 21 days on agar-solidified SH medium containing various amino acids or NH_4^+ additions. This figure represents separate experiments graphed together. Included in each experiment are the 0 NH_4^+ (x), 2.6 mM NH_4^+ (.), and 25 mM NH_4^+ (0) control values with no amino acid addition. Bars are the standard error of the mean.

Tab. 1. The effect of adding NH_4^+ and L-proline to the regeneration medium of alfalfa. The regeneration medium is SH-modified medium which contains no NH_4^+ (7). $(NH_4)_2SO_4$ was used as the source of NH_4^+.

Proline (mM)	Ammonium (mM)	Embryo Yield + Standard Error
0	0	6 ± 3
0	2.6 (SH level)	146 ± 4
0	25.0	197 ± 10
100	2.6	652 ± 35
100	0	44 ± 24

Amino Acid and Ammonium Effects on Somatic Embryogenesis

The ammonium ion requirement for alfalfa somatic embryogenesis has been previously characterized and shown to be optimal in the range of 10-25 mM NH_4^+ (2). In the initial experiments reported here, single amino acids were added over a 1-50 mM concentration range to the regeneration medium of alfalfa cells in order to screen the effectiveness of other reduced nitrogen sources (7,8). The regeneration medium used was based on the formulation of Schenk and Hildebrandt (SH) (9) which contains 2.6 mM NH_4^+ and 24.6 mM NO_3^-. Thus, these amino acids were not tested as a sole source of reduced nitrogen. The best responses which emerged from the initial screen are shown in Fig. 1. In each case the amino acids shown here exceed the 25 mM NH_4^+ control in yield of embryos. L-proline emerges as the most stimulatory amino acid tested. The optimal level for L-proline in enhancing embryo yield lies near 100 mM. Two earlier reports of amino acid stimulation of somatic embryogenesis in carrot (5,6) did not explore amino acid concentrations in excess of 30 mM and may have missed important amino acid activity as a result. Other amino acids have been screened over the range of 1-50 mM. With the exception of serine and lysine, which are less stimulatory for embryo yield than 25 mM NH_4^+, all other amino acids tested are either neutral or toxic.

The Effect of Ammonium and Proline on Embryo Yield

When proline-enhanced embryogenesis is examined more closely in modified SH medium it is clear that the response is due to a synergistic interaction of proline and ammonium ion. Table 1 shows the effect of adding 100 mM L-proline to medium with and without 2.6 mM NH_4^+. In the absence of NH_4^+, 100 mM L-proline is less active in stimulating embryogenesis than the treatment containing only 25 mM NH_4^+. Addition of 2.6 mM NH_4^+ improves the response of cultures treated with 100 mM L-proline by 15-fold. A further characterization of L-proline versus ammonium concentration is summarized in Tab. 2. The highest yield of somatic embryos in this experiment occurred at 100 mM L-proline and 25 mM NH_4^+ although other proline and NH_4^+ concentrations could be used to obtain very high regeneration. Further increases in NH_4^+ and L-proline above 25 mM and 300 mM, respectively, are deleterious to alfalfa somatic embryo yields (data not shown). Other amino acids which have a positive interaction with ammonium are L-arginine and L-alanine; the L-glutamine response is inhibited by high NH_4^+ (8). It is interesting to note that full restoration of the embryo yields in medium containing 100 mM L-proline without NH_4^+ is achieved if 25 mM L-glutamine is used in place of 2.6 mM NH_4^+ (7).

The important feature of these results is the discovery of L-proline as an important amino acid for stimulating embryogenesis in alfalfa. Proline is only effective if high levels (30 mM or higher) are included in the regeneration medium and if used in combination with the appropriate NH_4^+ level or L-glutamine. This discovery would not have been made if low concentrations of amino acids had been tested or if the amino acids had been screened as a sole reduced nitrogen source as had been done in carrot (5,6). Thus, we recommend that screening for amino acid effects on embryogenesis be done over a broad concentration range and in the presence of several ammonium concentrations.

The Effect of Amino Acids on Embryo Quality

From the initial screen it was apparent that some of the effective amino acids had a great effect on embryo size. For example,

Tab. 2. The effect of varying ammonium and proline concentrations in the regeneration medium on the somatic embryo yield of alfalfa cultures. Regeneration medium is SH-medium modified as in Ref. 7.

EMBRYO YIELD ± STANDARD ERROR OF MEAN				
	L-proline (mM)			
NH_4^+ (mM)	0	30	100	300
0	3 ± 1	327 ± 91	470 ± 53	133 ± 36
1.0		502 ± 84	747 ± 138	542 ± 120
2.6	150 ± 29	753 ± 106	459 ± 95	823 ± 120
10.0		888 ± 129	811 ± 146	572 ± 92
25.0	396 ± 91	744 ± 87	1042 ± 95	844 ± 122

L-glutamine-treated cultures appeared to have many more embryos than 25 mM NH_4^+-treated cultures. Counting the embryos under the dissecting scope shows that actual embryo yields are not significantly different between these 2 treatments, but that embryo development is changed. This difference in embryo quality, as distinguished from embryo yield, is the result of greater embryo size and the development of lobed cotyledon-like lateral appendages. Table 3 shows the length and width of a sample of embryos produced in response to regeneration on various amino acids. The amino acids L-glutamine and L-arginine give the largest embryos but not the higher embryo yields of L-proline. Proline-treated embryos are not as large as those produced with the other amino acids, but are larger than embryos generated on ammonium.

In these types of experiments we have noted that as embryo size increases the embryos tend to assume a more "mature" or "normal" appearance as is depicted in Fig. 2. With ammonium-treatment most of the embryos are globular or ellipsoid in appearance. Regeneration on L-glutamine gives a higher proportion of embryos which are ellipsoid- to bottle-shaped. Embryos which reach this stage are less likely to recallus and are more likely to continue development and produce leaves if left on the primary medium than are NH_4^+-treated cultures.

A second quality measurement which can be recognized is the ability to form plantlets (conversion) upon transfer to a secondary medium in which we halve the SH salts and sucrose. Inclusion of hormones, although sometimes used, is not necessary for high-frequency conversion. The conversion assay shows that amino acids, especially glutamine, improve the rate of plantlet formation compared to ammonium. Comparison of the results of embryo sizing (Tab. 3) and conversion (Tab. 4) shows that there is an approximate correlation between embryo size from the primary regeneration medium and the subsequent recovery of a plantlet on conversion medium.

Tab. 3. Effect of reduced nitrogen source on somatic embryo size. Randomly chosen embryos from these treatments were measured after 21 days of regeneration.

Treatment	Length (mm)	Width (mm)
SH + 25 mM NH_4^+ total	0.80 ± 0.08	0.38 ± 0.02
SH + 100 mM L-proline	1.14 ± 0.08	0.87 ± 0.05
SH + 100 mM L-alanine	1.19 ± 0.10	0.83 ± 0.05
SH + 30 mM L-arginine	1.52 ± 0.14	0.66 ± 0.05
SH + 30 mM L-glutamine	1.34 ± 0.12	0.77 ± 0.05

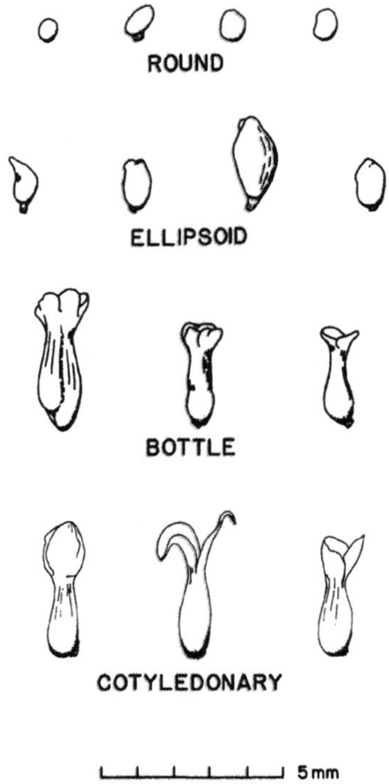

Fig. 2. Drawings of embryo shapes and sizes described in text.

PROSPECTS FOR A BIOCHEMICAL MARKER OF EMBRYO QUALITY

Background

With experience one is able to predict, using morphological features, which treatments will give better conversion of embryos to plantlets. Unfortunately, this assay is, at best, highly subjective. The conversion assay suffers from the fact that it is time consuming to manually transfer individual embryos to secondary medium and then wait 30 days for the conversion results.

We decided to develop a biochemical marker for the later stages of embryo development which might be diagnostic for embryo quality. One class of candidate molecules which immediately comes to mind is seed-specific secondary products which are formed in embryogenic cultures, such as in celery (10) or cocoa (11,12). Another class of products is seed storage proteins which accumulate to high levels in seed, are developmentally regulated, and are expressed in somatic embryos (13,14). Storage protein deposition in alfalfa embryos might be a good marker for embryo quality since storage proteins are deposited in the middle to late stages of zygotic seed maturation

Tab. 4. The effect of nitrogen source on the conversion of embryos to plants. Embryos were regenerated for 3 weeks on SH-media containing the additions listed. Then, 100 of the largest embryos from each treatment were transferred to ½ strength SH-plus GA- and NAA-medium (7). Final measurements were made after 30 days on secondary medium.

Initial Treatment	Plants with Trifoliates (%)
NH_4^+ 25 mM	33.3% ± 4.2
L-proline 100 mM	54.0% ± 6.4
L-alanine 50 mM	63.5% ± 4.4
L-arginine 30 mM	59.0% ± 6.2
L-glutamine 30 mM	67.0% ± 3.4

(15,16) and are the food reserve during germination. Also, few secondary products have been identified which are seed specific in alfalfa.

Isolation of Alfalfa Storage Protein from Cotyledons and Seed

While there are numerous reports on storage protein isolation and purification from large-seeded legumes, the only previous work on alfalfa storage protein was done by Danielsson in 1949 (17). He extracted seed with 0.2 M NaCl at pH 7, precipitated the soluble material with 70% $(NH_4)_2SO_4$, redissolved this pellet in 0.2 M NaCl at pH 7, dialyzed against water, and repeated the last 2 steps. The precipitable material was analyzed in the ultracentrifuge for sedimentation characteristics. He was able to resolve alfalfa seed protein into proteins with sedimentation values of 7S and 11S. Using these procedures we found that most of the alfalfa seed proteins were either not extracted initially from the seed meal or cotyledons, or were insoluble when we attempted to dissolve them in 0.2M NaCl at pH 7.

It turned out that most of what we know to be 11S protein was not extracted by Danielsson's methods, while the 7S protein was soluble. A key to solubilization of 11S alfalfa storage protein is the inclusion of dithiothreitol (DTT) in the extraction buffer. Seed meal or one day-imbibed cotyledons of Saranac AR alfalfa were extracted in 10 volumes of 1 M NaCl, 25 mM phosphate buffer (pH 7) and 10 mM DTT. The extract was loaded onto a 5 to 20% sucrose gradient containing extraction buffer and centrifuged at 38k rpm for 12 hr in a Sorvall AH-650 rotor (18). Fractions were collected by bottom puncture, analyzed for total protein (19), and by sodium dodecyl

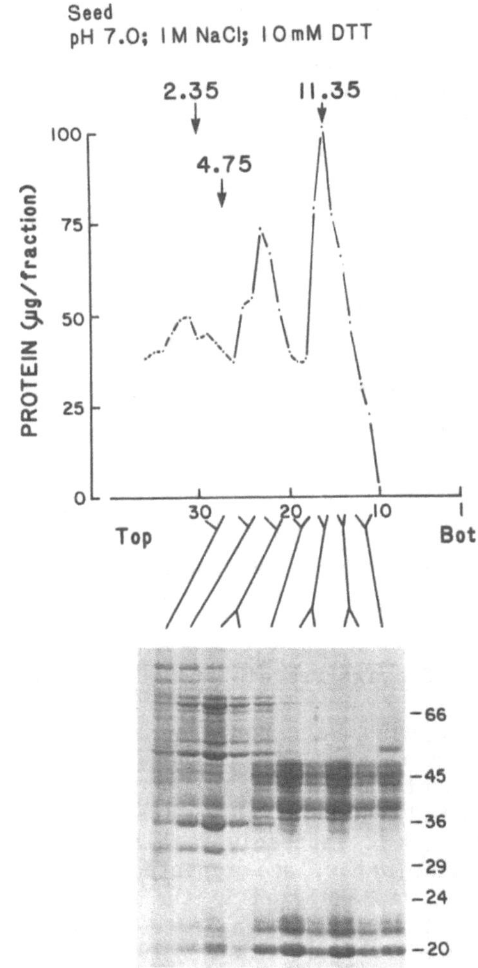

Fig. 3. Sedimentation of SDS-PAGE profiles of extracted alfalfa seed proteins. Seed meal was extracted under the conditions noted, sedimented in a sucrose gradient, fractionated, and analyzed for protein. Protein-laden fractions were pooled as noted and separated on a 10% SDS-PAGE gel system and stained for protein. Three pooled fractions were run at a 1:5 dilution as noted, otherwise all fractions were loaded at the same rate onto SDS-PAGE. Molecular weight standards in the sucrose gradient were soybean trypsin inhibitor (2.35S), bovine serum albumin (4.75S), and bovine liver catalase (11.35S) sedimented in separate tubes during the same centrifuge run. SDS-PAGE molecular weight standards are also noted in kd.

sulfate electrophoresis (SDS-PAGE; Ref. 20). The results of this extraction are shown in Fig. 3. The protein sedimenting with a value of 7S is composed of a series of bands around 68 kd, and single bands at 52, 36, 32, and 20 kd. The protein sedimenting at 11S is composed of a triplet at 47 k, a doublet at 39 k, and single bands at 22 and 20 kd.

Table 5 summarizes additional features of the 7S and 11S proteins. The 11S protein can be purified to virtual homogeneity by exhaustively extracting seed meal or imbibed cotyledons with pH 7 1 M NaCl and solubilizing the protein by adding DTT or lowering the pH to 3. The 7S protein can be purified by low salt extraction, sizing, and ion exchange chromatography.

Expression of Storage Protein in Somatic Embryos of Alfalfa

By knowing these solubility characteristics and by assuming that storage protein deposition in somatic embryos occurs in a manner similar to that in alfalfa zygotic embryos and seed, 200 somatic embryos regenerated on SH + 30 mM L-proline and 10 mM NH_4^+ were sequentially extracted at 0.2 M NaCl, pH 7 and then in 1 M NaCl (both with pH 7, 10 mM DTT and 200 μM PMSF) followed by centrifugation and

Tab. 5. Characteristics of alfalfa seed storage proteins.

7S Protein

Soluble at low salt, pH 7.0

Aggregate composed of
-- 68K (triplet), 52K, 36K, 32K

Purification
-- Gel filtration (A-0.5 M) or sedimentation
-- Ion exchange (DEAE, pH 5.5)

11S Protein

Soluble
-- pH 7, NaCl, 1 to 10 mM DTT
-- pH 3, low salt
-- pH 9, low salt

Aggregate composed of
-- 47K (triplet), 39K, 22K, 20K

Purification
-- Extensive pellet washing followed by specific solubilization
-- Gel filtration (A-1.5 M, pH 3 or pH 9) or sedimentation

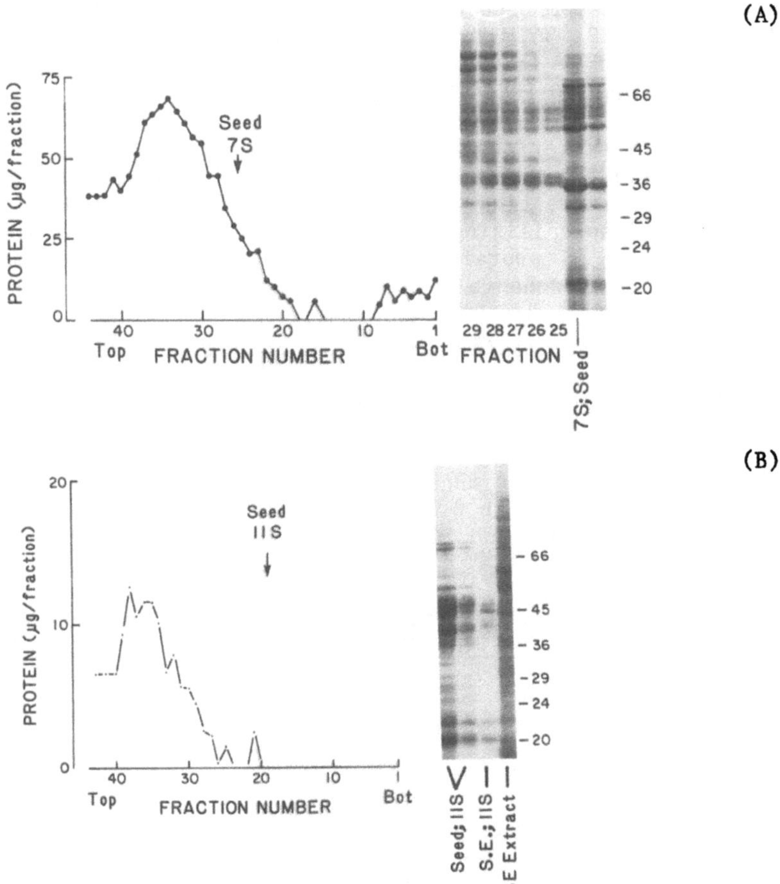

Fig. 4. Sedimentation and SDS-PAGE analysis of proteins extracted from somatic embryos induced with 50 µM 2,4-D and regenerated on SH + 10 mM NH_4^+ + 50 mM L-proline which were sequentially extracted with low (A) and high salt (B). Fig. 4A. Extracts and gradients were in 25 mM PO_4^{-2} buffer, 0.2 M NaCl, 10 mM DTT, and 200 µM PMSF, at pH 7.0. Authentic 7S seed protein was extracted and run in a separate tube as a standard. In somatic embryos, fractions near the 7S standard sedimentation were analyzed by SDS-PAGE. Insoluble material of above extract was reextracted with the same buffer system with 1 M NaCl. Fig. 4B. As above except extracts and buffers had 1 M NaCl. Fractions around the 11S standard were pooled from the somatic embryo extract gradient.

analysis by SDS-PAGE and total protein (Fig. 4A and 4B). No convincing 7S expression was found and only a low level of 11S protein was evident. Attempts to pulse-label embryos with H^3-leucine for one day followed by extraction, SDS-PAGE, and fluorography produced no better results in detecting the storage proteins in these embryos. With these procedures the detection of protein is limited to 10 ng of protein per embryo. Expression of 11S may be higher than 10 ng per embryo but 7S expression is certainly lower.

Low 2,4-D Exposure Improves Embryo Morphological Quality

Walker et al. had shown earlier that the optimum embryo yields are achieved after a 3-day exposure to 50 µM 2,4-D. At the time we were doing embryo protein extraction, an experiment investigating a variety of auxins and their dosages on the induction of embryogenesis revealed that distinctly improved embryo morphology was achieved if 10 µM 2,4-D or 50 µM p-chlorophenoxyacetic acid were used for induction, albeit with poorer embryo yields (Fig. 5). The 21-day old embryos induced from low auxin cultures fell into a new morphological class called "cotyledonary" embryos (Fig. 2) with fleshy cotyledon-like lobes which were at least one-third the overall length of the entire embryo. These embryos also converted to plantlets at a higher frequency than the high auxin-induced embryos (Fig. 5).

Comparison of Embryos Formed on Low- and High-2,4-D-induced Cultures

Embryos were induced in cultures exposed to either 10 or 50 µM 2,4-D and regenerated on SH with 30 mM L-proline and 10 mM NH_4^+.

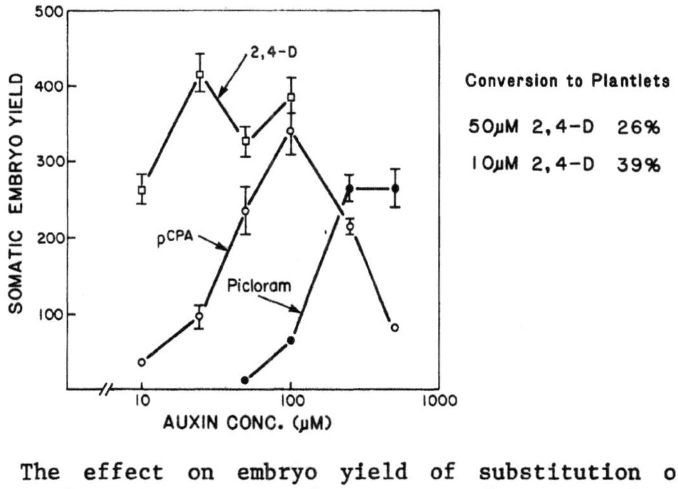

Fig. 5. The effect on embryo yield of substitution of various auxin structures and concentrations to the induction medium of alfalfa cultures.

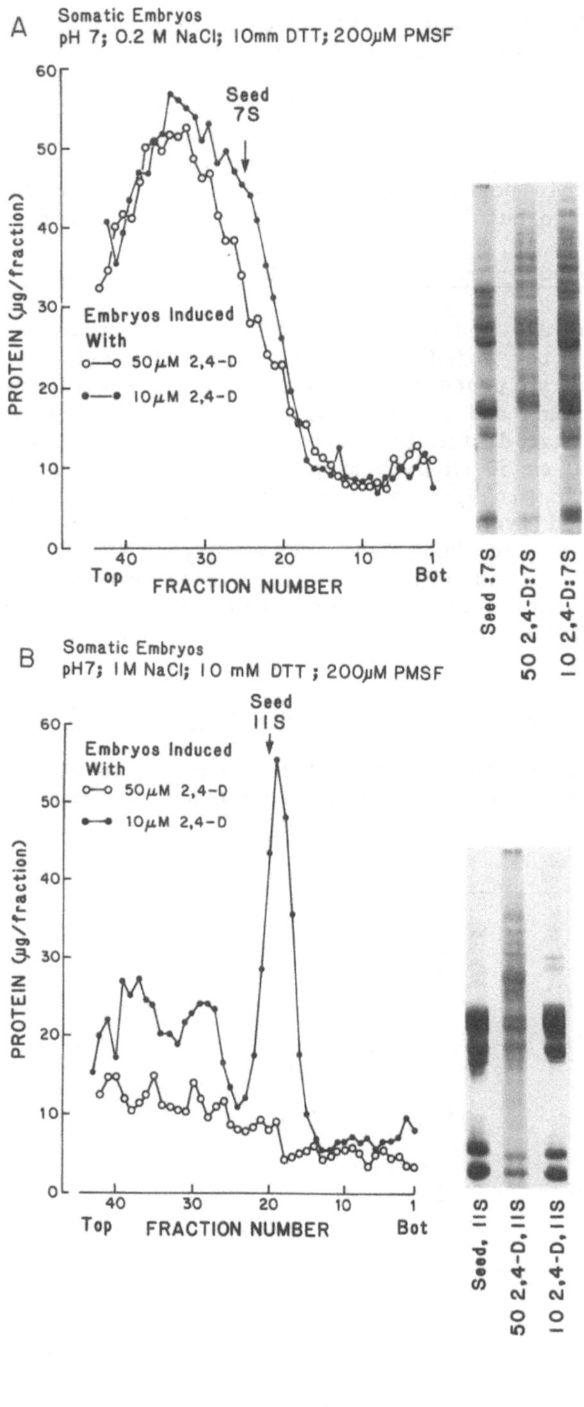
A
Somatic Embryos
pH 7; 0.2 M NaCl; 10mm DTT; 200μM PMSF
Seed 7S
Embryos Induced With
50μM 2,4-D
10μM 2,4-D
PROTEIN (μg/fraction)
Top
FRACTION NUMBER
Bot
Seed :7S
50 2,4-D:7S
10 2,4-D:7S
B
Somatic Embryos
pH7; 1M NaCl; 10 mM DTT ; 200μM PMSF
Seed 11S
Embryos Induced With
50μM 2,4-D
10μM 2,4-D
PROTEIN (μg/fraction)
Top
FRACTION NUMBER
Bot
Seed, 11S
50 2,4-D, 11S
10 2,4-D, 11S

Fig. 6. Sedimentation and SDS-PAGE analysis of proteins extracted from somatic embryos under conditions used in Fig. 4. Embryos were induced by either 10 μM (●) or 50 μM 2,4-D (○). Proteins extracted from seed for 7S (A) or 11S (B) were sedimented in separate tubes during the same run. Fig. 6A. Sedimentation of proteins extracted at low salt. Fig. 6B. Proteins extracted at high salt.

The 24-day old embryos were extracted and analyzed as earlier (Fig. 6A and 6B) for 7S and 11S protein deposition. The gradient profile of the low auxin-induced material shows a distinct shoulder at 7S for the low auxin-treated material which is absent in the high auxin-induced embryos. The major peptides from this shoulder migrate at the same molecular weight as authentic 7S seed protein. If the area of the added protein accumulation is assumed to represent 7S protein exclusively, then low 2,4-D induced embryos accumulate an average of 2.75 μg of 7S protein per embryo. The profile of the high salt extract of somatic embryos shows that 10 μM 2,4-D-induced embryos have a peak of protein at 11S in the gradient (Fig. 6B) whereas the extract from 50 μM 2,4-D-induced embryos does not (Fig. 4B and 6B). On SDS-PAGE analysis the subunits of protein from the 11S peak have mobility identical to that of the subunits of the 11S protein from authentic seed extracts. This represents an average accumulation of 2.1 μg of protein per embryo in the low 2,4-D-treated material.

Perspectives for Analysis of Protein Expression in Somatic Embryos

From the above results we conclude that expression of both the 7S and 11S storage protein of alfalfa zygotic embryos (seed) also occurs in somatic embryos. This conclusion is based on the coincidence of: a) low salt (7S) and high salt plus reductant solubility (11S); b) assembly of subunit polypeptides into aggregates with identical sedimentation patterns; and c) comigration of the subunit polypeptides on SDS-PAGE for seed and somatic embryo proteins. Antibody prepared to the seed 11S protein of alfalfa also precipitates with somatic embryo 11S protein in Ochterlony double diffusion tests (results not shown). Crouch (14) has also concluded that *Brassica napus* somatic embryos accumulate the 12S storage protein characteristic of seed.

These results also suggest that storage protein deposition may represent a marker which is diagnostic of subsequent embryo quality or vigor. The differences between low- and high-2,4-D-treated embryos in morphology, ability to convert to plantlets, and expression of storage protein all correlate. The possibility exists that development of a serological technique for detection of 11S or 7S protein in alfalfa may aid in the quantitation of quality of individual somatic embryos. Establishment of such an assay may also help in the identification and optimization of procedures which would improve somatic embryo vigor. Since storage protein is assumed to represent nutrient reserves for the seedling development, optimization of its accumulation may improve embryo performance. The use of such a probe may also aid our understanding of precocious germination in somatic embryos.

Finally, while it is generally recognized that a strong auxin such as 2,4-D is necessary to induce callus cultures to form embryos at high yields, the subsequent development of embryos may be hindered by the induction treatment. This suggests to us that the process(es) necessary for embryo induction is distinct from the process of somatic embryo maturation. If we use the developmental timetable proposed for zygotic embryogeny as a model for somatic embryogeny, the strong auxin and ammonium/reduced nitrogen requirement may represent exogenous signals necessary for the initial events of embryogenesis (Phase I, see Ref. 14). Factors such as 2,4-D carryover may prevent embryo storage protein deposition which is characteristic of Phase II of zygotic embryogeny. Understanding of the process(es) which controls somatic embryo development will depend on the availability of assays for identification of factors which affect the maturation process.

REFERENCES

1. Halperin, W., and D.F. Wetherell (1965) Ammonium requirement for somatic embryogenesis _in vitro_. _Nature_ 205:519-520.
2. Walker, K.A., and S.J. Sato (1981) Morphogenesis in callus tissue of _Medicago sativa_: The role of ammonium ion in somatic embryogenesis. _Plant Cell Tissue Organ. Culture_ 1:109-121.
3. Gleddie, S., W. Keller, and G. Setterfield (1983) Somatic embryogenesis and plant regeneration from leaf explants and cell suspensions of _Solanum melongena_ (eggplant). _Can. J. Bot._ 61:656-666.
4. Tazawa, M., and J. Reinert (1969) Extracellular and intracellular chemical environments in relation to embryogenesis _in vitro_. _Protoplasma_ 68:157-173.
5. Wetherell, D.F., and D.K. Dougall (1976) Sources of nitrogen supporting growth and embryogenesis in cultivated wild carrot tissue. _Physiol. Plant._ 37:97-103.
6. Kamada, H., and H. Harada (1979) Studies on organogenesis in carrot tissue culture. II. Effects of amino acids and inorganic nitrogenous compounds on somatic embryogenesis. _Z. Pflanzenphysiol._ 91:453-463.
7. Stuart, D.A., and S.G. Strickland (1984) Somatic embryogenesis from cell cultures of _Medicago sativa_ L. I. The role of amino

acid additions to the regeneration medium. Plant Sci. Lett. 34:165-174.

8. Stuart, D.A., and S.G. Strickland (1984) Somatic embryogenesis from cell cultures of Medicago sativa L. II. The interaction of amino acids with ammonium. Plant Sci. Lett. 34:175-181.
9. Schenk, R.U., and A.C. Hildebrandt (1972) Medium and techniques for induction and growth of monocotyledonous and dicotyledonous plant cell cultures. Can. J. Bot. 29:199-204.
10. Al-abta, S., I. J. Galpin, and H.A. Collin (1979) Flavour compounds in tissue cultures of celery. Plant Sci. Lett. 16:129-134.
11. Jala, M.A.F., and H.A. Collin (1979) Secondary metabolism in tissue cultures of Theobroma cacao. New Phytol. 83:343-349.
12. Janick, J., D.C. Wright, and P.M. Hasegawa (1982) In vitro production of cocoa seed lipids. J. Am. Soc. Hort. Sci. 107:919.
13. Crouch, M.L., and I.M. Sussex (1981) Development and storage protein synthesis in Brassica napus L. embryos in vivo and in vitro. Planta 153:64-74.
14. Crouch, M.L. (1982) Non-zygotic embryos of Brassica napus L. contain embryo specific storage proteins. Planta 156:520-524.
15. Sun, S.M., M.A. Mutschler, F.A. Bliss, and T.C. Hall (1978) Protein synthesis and accumulation in bean cotyledons during growth. Plant Physiol. 61:918-923.
16. Sussex, I.M., and R.M.K. Dale (1979) Hormonal control of storage protein synthesis in Phaseolus vulgaris. In The Plant Seed, I. Rubinstein, R.L. Phillips, C.E. Green, and B.G. Gegenbach, eds. Academic Press, New York, pp. 129-141.
17. Danielsson, C.E. (1949) Seed globulins of the Gramineae and Leguminosae. Biochem. J. 44:387-400.
18. Martin, G.G., and B.N. Ames (1961) A method for determining the sedimentation behavior of enzymes: Application to protein mixtures. J. Biol. Chem. 236:1372-1379.
19. Bradford, M.M. (1976) A rapid and sensitive method for the quantitation of microgram quantities of protein utilizing the principle of protein-dye binding. Annals Biochem. 72:248-254.
20. Laemmli, U.K. (1970) Cleavage of structural proteins during the assembly of the head of bacteriophage T4. Nature 227:680-685.
21. Walker, K.A., M.L. Wendeln, and E.G. Jaworski (1979) Organogenesis in callus tissue of Medicago sativa. The temporal separation of induction processes from differentiation processes. Plant Sci. Lett. 16:23-30.

FACTORS INFLUENCING TOMATO PROTOPLAST DEVELOPMENT

Elias A. Shahin and Mayar Yashar

Genetics and Tissue Culture Group
ARCO Plant Cell Research Institute
6560 Trinity Court
Dublin, California 94568

INTRODUCTION

The availability of a protoplast regeneration system together with the new emerging techniques for genetic manipulation offer plant breeding specific advantages for developing new crop cultivars. The techniques for producing somatic hybrids, cybrids, and plant mutants require the ability to isolate large numbers of viable protoplasts capable of division and plant regeneration. Protoplast culture techniques are relatively new and there are still numerous obstacles to overcome before they are fully applicable to major horticultural and agronomic crops. Plants have been obtained from protoplasts in a number of crop species (1), but in many instances their yield and reliability have not been satisfactory for genetic manipulation.

The tomato is an important horticultural crop which is being improved through the use of protoplasts. Studies have been undertaken in our laboratory to develop procedures for protoplast isolation, and to identify factors which influence shoot formation. In this paper, we will report our studies on those factors. The procedures and media utilized for protoplast isolation and plant regeneration have been described previously (2,3).

PHYSIOLOGICAL CONDITIONS OF THE PROTOPLAST DONORS

A number of studies have shown that the physiological condition of the source tissue affects protoplast development in potato (4) and tobacco (5). Our studies with tomato protoplasts indicate that plants grown in vitro appear to be the most suitable for protoplast isolation, culture, and regeneration (2). In the tomato cultivar,

"H-2152," greater protoplast yield and high viability was obtained with plants grown in vitro than with growth chamber of field-grown material (Fig. 1). In some instances, we were able to produce protoplasts from greenhouse and field-grown plants, but with varied enzyme combinations. The fact that tomato leaves undergo hardening with increased light intensity necessitates the varying of enzyme concentrations and combinations continuously. With our system, however, protoplast release can be achieved consistently, and with all the tested cultivars, using a single enzyme treatment.

The optimum age, after placing the excised shoots on TM-1 medium, for protoplast release and cell division was 10-12, 14, and 21 days for cotyledons, stems, and leaves, respectively (Fig. 2). Investigations on the effect of growth temperature revealed that higher protoplast yield and cell division was obtained if in vitro plants were grown in a 25°C constant temperature rather than 25°C/20°C, or 25°C/15°C day/night cycle (unpub. data). Another study on the role of light intensity and photoperiod, under 25°C constant temperature, demonstrated that the best in vitro donors were grown under 4,800 lux and 16 hr/8 hr photoperiod (unpub. data).

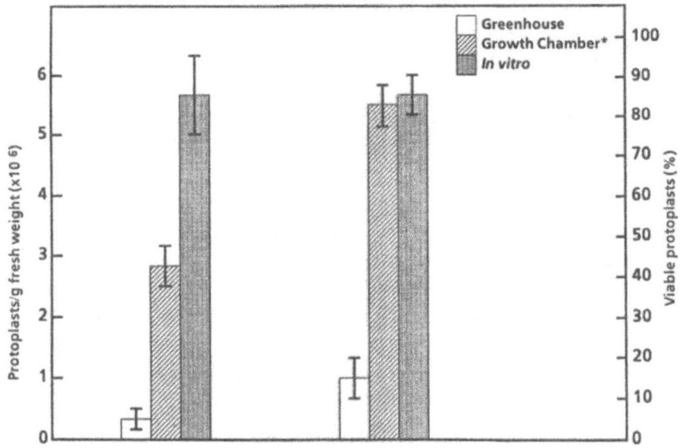

Fig. 1. Effect of source material on yield and viability of protoplast isolated from tomato leaves. Plants were grown in the growth chamber under 25°C constant temperature with a 16 hr/18 hr photoperiod, and light intensity of 7,000 lux. In vitro-grown plants were maintained as previously described (2). Viability of the protoplast was determined by FDA stain.

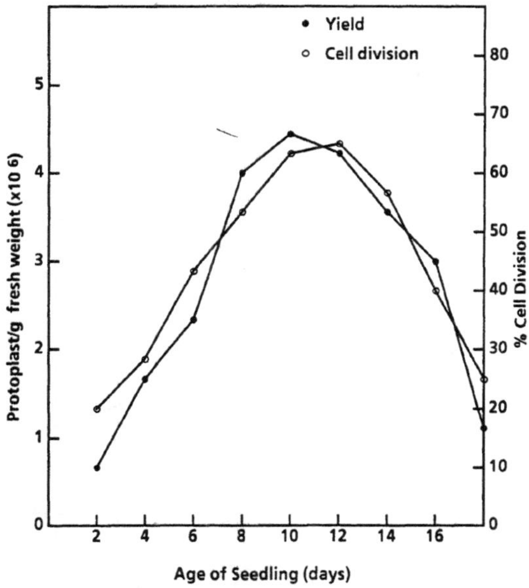

Fig. 2. Effect of age of seedling on yield and cell division of protoplasts isolated from tomato cotyledons. Protoplasts were isolated as previously described (2).

PREENZYME COLD TEMPERATURE TREATMENT (PET)

In tomato, we evaluated the effect of preplasmolysis on the release and viability of protoplasts obtained from tomato donor tissue grown under different temperatures. Maximal yields (4.6×10^6 protoplasts/g) and cell division (56%) were obtained when excised cotyledons maintained in preplasmolysis solution had been pretreated for 12 hr at 10°C in the dark (Fig. 3). Figure 4 shows that preenzyme cold treatment of the donor tissue enhanced the yield and division of protoplasts.

We were unable to isolate viable protoplasts from freshly excised leaves of 30-day-old plants growing in the greenhouse, but - totipotency of the excised tomato leaf cells can easily be induced before culture by pretreatment. Pretreatment also enabled the isolation of viable protoplasts from the leaves of greenhouse-grown plants (Tab. 1). The protoplasts underwent cell division and later regenerated plants at a rate comparable to those protoplasts obtained from in vitro plants. This pretreatment has allowed us to consistently isolate viable protoplasts, using the same enzyme treatment, from donor tissue of plants grown in the greenhouse. These results suggest that it may not be necessary to grow plants in vitro to achieve consistent results if the donor tissue is pretreated; however, in vitro grown plants have many other experimental advantages.

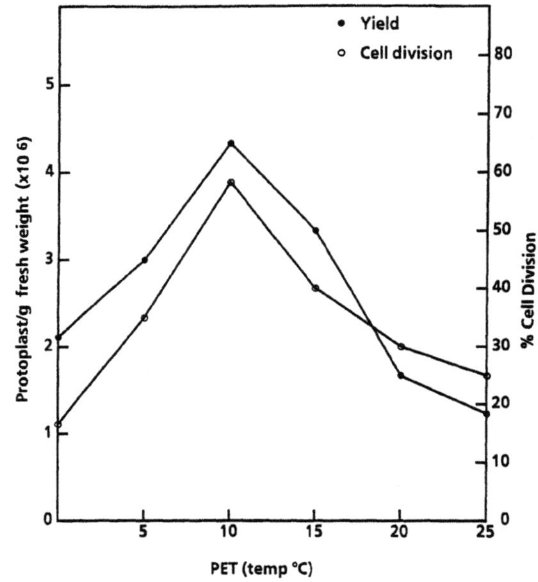

Fig. 3. Effect of temperature during pre-enzyme treatment on yield and cell division of protoplasts isolated from tomato cotyledon. The culture procedures and medium employed were as described by Shahin (2,3).

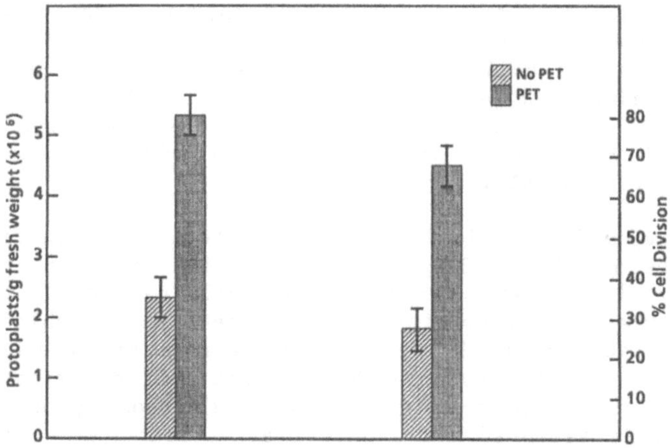

Fig. 4. Effect of pre-enzyme cold treatment (PET) on yield and division of protoplasts isolated from tomato cotyledon. Protoplasts were isolated and cultured as previously described (2).

Tab. 1. Effect of pretreatment of tomato leaves on cell division and plant formation of tomato protoplasts.

	Protoplasts from fresh leaves			Protoplasts from pretreated leaves*		
Plant Age	Yield Protoplast/g	% Cell Division	Shoot Formation	Yield Protoplast/g	% Cell Division	Shoot Formation
90 days	0	0	0	0	0	0
60 days	0	0	0	0	0	0
30 days	0.45×10^6	0	0	2.0×10^6	30	+++
30 days (in vitro)	1.15×10^6	15	+++	2.5×10^6	35	+++

* Leaflets were removed and placed in TM-1 medium (3), without sucrose, supplemented with 0.25 mg/l carbenicillin and then placed in the dark at 25°C. Two days later, the leaves were sterilized for 15 min in 1% Ca-hypochlorite solution and then washed 4 times in sterile distilled water. Thereafter, sterilized leaves were gently brushed with a nylon brush, cut into small pieces, and placed in PET for 12 hr at 10°C; the culture procedures and media employed were as described by Shahin (2,3).

+++ More than 75% of the calli formed multiple shoots.

Although the beneficial effect of preplasmolysis, prior to enzyme treatment, of the donor tissue has been adequately demonstrated (2,6), information regarding the biochemical events during this treatment is still lacking. It is well known that plant tissue undergoes physiological stress during the enzymatic digestion process. Preplasmolysis, prior to enzyme incubation, shrinks the protoplasts and disrupts the plasmadesmata connecting adjacent cells. This in turn prevents leakage of the cellular constituents which can cause protoplast lysis and senescence.

The requirement for the chilling step in the PET is well established, but not understood. Cold temperatures in the range of 0-4°C may shut down the physiological stress response which would otherwise have harmful consequences with the osmotically induced plasmolysis.

COMPOSITION OF THE CULTURE MEDIA

Plating Medium (TM-2)

The composition of the plating medium is a key factor influencing protoplast division. In culturing tomato protoplasts, we have observed that the presence of NH_4NO_3 is damaging to the protoplasts. Consequently, ammonium nitrate was replaced by reduced nitrogen in the form of glutamine (2). The presence of all the vitamins was also very critical.

The auxin NAA and cytokinin zeatin riboside were used with the culture medium (TM-2) for all of the tomato cultivars tested in our laboratory (2,3). However, we have also used other hormones in the basal medium with success (2).

Since agar may be toxic to some protoplasts, there has been interest in low-temperature gelling agarose (7,8,9). However, in our studies we have observed significant damage to the protoplasts when plated in agarose. Instead, liquid medium has been used for culturing protoplasts and maintaining their division and proliferation. At the same time, liquid medium allowed for the feeding, dilution, and rinsing of the cell colonies when necessary (2).

In tomato protoplast culture we have observed that the use of 0.25 M sucrose as the carbon source is well suited for all the varieties tested. Protoplasts from some cultivars do aggregate in the presence of sucrose. When 0.25 M sucrose was replaced by 0.25 M glucose, the protoplast aggregation and clumping ceased (unpub. results).

Shooting Medium (TM-4)

The hormonal requirements for shoot regeneration in all 14 cultivars tested appeared to be similar: TM-4 medium supplemented with 1.0 mg/l zeatin riboside and 0.2 mg/l gibberellic acid. Nonetheless, we were able to induce profuse shoots by employing several combinations of auxins (IAA, GA_3) and cytokinin (kinetin, 2iP, BAP, and zeatin riboside) in the basal TM-4 medium (2). Therefore, it appears that the hormonally-induced growth requirements for shoot induction are not highly specific.

It should be stressed that the use of zeatin riboside is by far superior to that of zeatin (Tab. 2). This phenomenon can be attributed to the ease of translocation of zeatin riboside into the tissues.

Tab. 2. Percentage of mini-calli developing shoots after 3 weeks of culture on TM-4 medium containing different form of zeatin. (Abbrev.: pCPA, parachlorophenoxyacetic acid.)

	Plating Medium (TM-2) Growth hormones (mg/l)			Shooting Medium (TM-4) Growth hormones (mg/l)			
	p-CPA	NAA	Zeatin	GA_3	Zeatin	Zeatin Riboside	Shoot formation (%)
Stem	1.0	-	0.5	0.2	-	1.0	76
protoplasts	1.0	-	0.5	0.2	1.0	-	25
	1.0	-	0.5	0.2	-	1.0	75
Leaf	1.0	-	0.5	0.2	1.0	-	30
protoplasts	-	1.0	0.5	0.2	-	1.0	65
	-	1.0	0.5	0.2	1.0	-	27

VF-36 tomato protoplasts were used in this experiment. Protoplasts were obtained from both stem and leaf materials. The culture procedures were as described by Shahin (2,3).

Multiple shoots will appear within 2-3 weeks after placing the mini-calli on the shooting medium. These shoots may require subculturing on the same medium (TM-4) to enhance further development (2). It is worth noting that a tomato callus, once induced to form shoots, will maintain its totipotency as long as it is being subcultured once every 3 weeks. In our laboratory, we were able to maintain shoot regeneration from protoplast calli of cultivars UC-82 and H-2152 for almost 8 months. During that period, a new crop of multiple shoots (3-8 per callus) was harvested every 3 weeks. In a few cases, depending upon the cultivar, some calli failed to produce normal shoots, but did produce leaf-like structures; nevertheless, shoots were easily obtained by placing those leaf-like structures on the surface of the agar of TM-4 medium (unpub. results).

In the shoot induction medium, we have observed that either sucrose or glucose can be used as the carbon source. Sucrose has been considered to be a good source with levels of 3% (w/v) (2,3), but glucose at a level of 2% was far superior. Shoots regenerated on TM-4 medium, supplemented with 2% glucose, were more vigorous and normal than those produced on medium supplemented with sucrose, and usually appeared in a shorter time (unpub. results).

GENOTYPE

The role of the donor genotype in protoplasts was previously shown to be critical in a number of plant species. There is now considerable evidence for genetic control of tissue and protoplast culture response (10,11,12,13). In tomato, however, we have obtained plants from all the cultivars tested although there were differences in degree of regeneration among the cultivars (2). It is worth mentioning that these differences reflect the fact that the culture procedures and media are not optimal for all tomato cultivars, but it is possible to optimize them with minor adjustments.

CONCLUSIONS

The above-mentioned results indicate that the physiological condition of the plant material for protoplast isolation was critical for achieving cell division and further plant regeneration. Although we have identified some influences of culture media on protoplast development, much attention should be given to pretreatment of plants prior to protoplast isolation. The endogenous environment is easier to manipulate than the culture media. Physiological conditions of the donor plants, and stress pretreatment (chilling and plasmolysis) could alter the hormonal components of the cells and/or other mechanisms that are involved in determining totipotency. On the other hand, less attention should be given to the approach of selecting for responsive genotypes. Each line or cultivar presents its own particular challenges, therefore researchers will probably find it more efficient to work with one genotype while finding ways of maintaining competence in this protoplast donor. Emphasis should be placed on learning how to grow donor plants and to prepare them to become "healthy donors."

REFERENCES

1. Davey, M.R. (1983) Recent developments in the culture and regeneration of plant protoplasts. Experienta Supp. 46:19-30.
2. Shahin, E.A. (1984) Totipotency of tomato protoplasts. Theor. Appl. Gent. (in press).
3. Shahin, E.A. (1984) Isolation and culture of protoplasts: Tomato. In Cell Culture and Somatic Cell Genetics of Plants, I.K. Vasil, ed. Academic Press, New York, pp. 371-380.
4. Shepard, J.F., and R.E. Totten (1975) Isolation and regeneration of tobacco cell protoplasts under low osmotic conditions. Plant Physiol. 55:689-694.
5. Cassels, A.C., and F.M. Cocker (1982) Seasonal and physiological aspects of the isolation of tobacco protoplasts. Plant Physiol. 56:69-79.
6. Donn, G. (1978) Cell division and callus regeneration from leaf protoplasts of Vicia narbonensis. Z. Pflanzenphysiol. 86:65-75.
7. Shepard, J.F. (1980) Mutant selection and plant regeneration from potato mesophyll protoplasts. In Emergent Techniques for the Genetic Improvement of Crops, I. Rubenstein, B. Gengenbach, and C.E. Green, eds. University of Minnesota Press, pp. 185-219.
8. Shahin, E.A., and J.F. Shepard (1980) Callus formation and organogenesis from mesophyll protoplasts of cassava. Plant Sci. Lett. 17:459-465.
9. Shahin, E.A. (1984) Isolation, culture and regeneration of potato leaf protoplasts from plants preconditioned in vitro. In Cell Culture and Somatic Cell Genetics of Plants, Vol. 1, I. Vasil, ed. Academic Press, New York, pp. 381-390.
10. Ohki, S., C. Bigot, and J. Mousseau (1978) Analysis of shoot-forming capacity in vitro in two lines of tomato (Lycopersicon esculentum Mil.) and their hybrids. Plant and Cell Physiol. 19:27-42.
11. Keyes, G.J., G.B. Collins, and N.L. Taylor (1980) Genetic variation in tissue cultures of red clover. Theor. Appl. Genet. 58:265-271.
12. Dale, P.J., and M.G.K. Jones (1982) Studies in callus and plant regeneration from tissues and protoplasts of the forage grass Lollium multiforum. In Proc. 5th Int. Cong. Plant Tissue and Cell Culture, Tokyo, A. Fuyiwara, ed., pp. 579-580.
13. Bingham, T.E., L.V. Hurley, D.M. Kaatz, and J.W. Saunders (1975) Breeding alfalfa which regenerates from callus tissue in cultures. Crop Sci. pp. 719-721.

AN EMBRYOGENIC CULTURE OF SOYBEAN: TOWARDS A GENERAL THEORY OF SOMATIC EMBRYOGENESIS

M.L. Christianson

Zoecon Corporation
975 California Avenue
Palo Alto, California 94304

PREFACE

It is a simple matter of definition to say that an embryogenic cell culture is composed of embryogenic cells, and quite another matter to understand why certain cells are embryogenic or have embryogenic potential. We know "what" they are, of course: they are cells which will make somatic embryoids in response to an appropriate stimulus, typically the removal of the synthetic auxin, 2,4-dichlorophenoxyacetic acid (2,4-D) from the culture medium. In many cases, this knowledge may be enough. Practical applications do not require that we know "why" they make embryos, but only that we can recognize such embryogenic cells and can obtain or produce cultures containing large numbers of such cells or cell masses. Certainly, somatic embryogenesis has been described in a large number of species, including species of major and minor agronomic importance (1), even down to the last recalcitrant ones, cotton (9) and soybeans (7). But this is not somatic embryogenesis with ease, and not in every cultivar, and not from any tissue source of any physiological state or age. This is, I believe, a direct consequence of our lack of understanding about the "whys" of embryogenic competence even though we understand the "whats" and the "hows." Previous papers in this volume describe the importance of preconditioning the plant or explant. In many cases this is crucial to success in achieving somatic embryogenesis. This paper will deal in part with another kind of precondition, the mind-set of the investigator. I believe this is no less important for our eventual success in achieving somatic embryogenesis from any species, cultivar or tissue, and in anyone's laboratory. What is needed is a general theory of somatic embryogenesis. I can't supply that, but I will make a contribution towards that end. I can outline what I feel some of the salient features of such a theory must be. That accomplished, I will describe the development and behavior of a morphogenetically competent culture of soybean, and experiments designed to flesh out a general theory of somatic embryogenesis.

INTRODUCTION TO PHYSIOLOGICAL GENETICS

The subject of somatic embryogenesis has been reviewed and catalogued a large number of times in the last few years by an almost equally large number of tissue culture specialists and plant physiologists (1,13). These reviews treat the very real factors governing somatic embryogenesis: hormones, carbon sources, isolation, pH, and so forth. Geneticists, on the other hand, are concerned with the intangible factors, those apocryphal genes for somatic embryogenesis whose action is supposedly required for somatic embryogenesis to occur. Genes are, in large part, still mysteries, despite quantum leaps forward in that branch of qualitative biochemistry known as "molecular genetics," the isolation and sequencing of certain transcriptional units and their flanking regions. One hopes that the genes involved in somatic embryogenesis are a large subset of the genes used in the process of zygotic embryogenesis. Why else would selection over the past few million years retain the ability to make embryoids in vitro if this ability were not due to the pleiotropic effects of genes for some essential whole plant function (41)? One also hopes that the real factors manipulated by plant physiologists function to trigger this subset of the genes of zygotic embryogenesis.

There are formal genetic approaches to describe this triggering of gene action. Goldschmidt (11) discusses it in more general terms, while Rendel (25) approaches it in a quantitative way; this area falls under the general heading of "control theory" and texts do exist for those interested in further details (27). A simple example can illustrate the point.

Imagine a gene whose activity is controlled by the binding of an "activator" to a particular "receptor site." The state of gene expression, on or off, is governed, then, by the concentration of the activator in the immediate vicinity of the receptor site (Fig. 1). Since this gene is to be regulated, binding of the activator must be rapidly reversible; the binding constant for the activator cannot be such that the gene is irreversibly activated once it sees a single activator molecule. The concentration of the activator in the "sensitive" vicinity of the receptor site will vary from cell to cell, and, over time, within any one cell due to metabolic fluxes and the stochastic considerations of compartmentation and of mixing in a cell. It will, however, be distributed around some mean value.

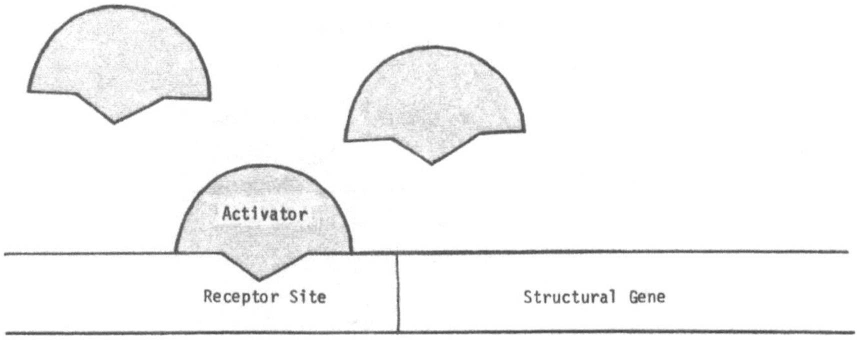

Fig. 1. A schematic gene under positive control. In order for the structural gene to be transcribed, an activator molecule must be bound to the receptor site. The activity of this gene is a function of the binding constant and the local concentration of activator molecules. As the binding constant is invariant for a system this simple, the activity of the gene is a direct function of the local concentration of the activator molecules. In genes under negative control, the inactivity of the gene is a direct function of the local concentration of repressor molecule. Well-characterized real genes have been shown to be controlled in a complex way, with both positive and negative effectors.

This is also true for complex regulatory systems although the distribution will be of some artificial variable constructed from several components. This "activator" makes the gene active, and Rendel (25) calls it "Make" for that reason. There will be values of "Make" too low to keep the gene active, of course. The least upper bound of these values is the threshold value of "Make," and it divides the distribution of cells into 2 subpopulations: cells with an "active" gene and cells with an "inactive" gene (Fig. 2a).

The mean value of "Make" in a cell or a group of cells can and does change. This can occur under the influence of external factors, or be a consequence of the action of some gene within the cell. Whatever its cause, the end result is the same: a change in the proportion of cells with this particular gene in an active state (Fig. 2b). This is the causal molecular basis for cellular differentiation. In the final analysis, it is the small metabolites that determine whether genes are active or not. The epigenetic factors control the on/off state of the genes; "hypogenetic" factors, the gene products, actually participate in metabolism (Fig. 3).

A developmental system that involves several genes, each one activated by sets of factors, just extends this one-dimensional or-

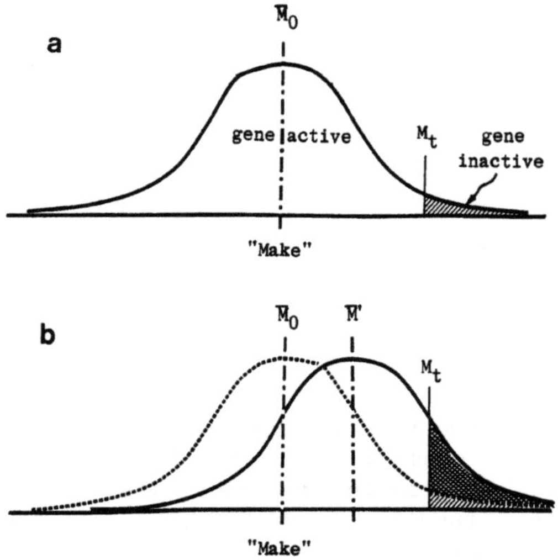

Fig. 2. Distribution of gene activating molecules. a) The concentration of a gene activator in the local vicinity of the receptor site will vary from cell to cell as well as over time within an individual cell. This distribution of "Make" can be represented as a Gausian curve, with a mean value of M_o. There will be a threshold value of "Make," M_t, that divides cells into 2 groups, those with the gene active, and those with the gene inactive. b) Change in cell physiology can shift the mean value of "Make" from M_o to M'. The distribution around the new mean will remain Gausian and the result is an increase (or decrease) in the number of cells with an inactive (or active) gene. This diagram shows a small change; a large change would have a correspondingly larger effect.

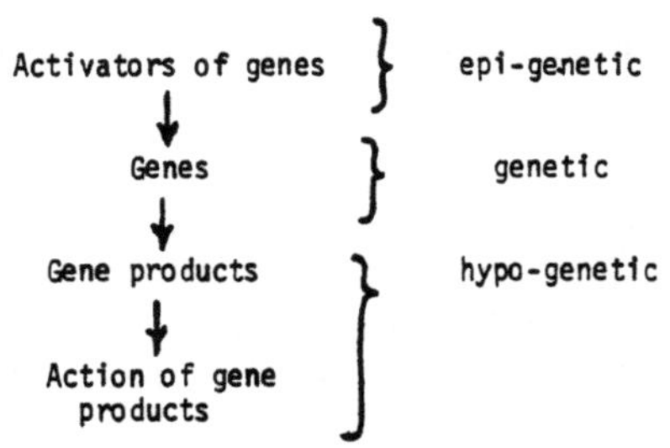

Fig. 3. Geneticist's taxonomy of gene action. The gene is the central focus. Those factors that control its activity are epigenetic in nature. Those factors that stem from its activity are hypogenetic in nature. The complex interactions of the many sets of epi-, hypo-, and genetic factors in the cell are commonly known as metabolism.

dering of "Make" into a three-dimensional response surface. All the parameters governing the gene action under consideration can be collapsed into 3 constructed variables, and the distribution of metabolic states in the cells in a flask or a tissue is represented as a three-dimensional response surface (Fig. 4). Waddington (35) has described such representations, calling them "epigenetic landscapes," and that is exactly what they are. They are the distribution of the cellular metabolic states (the epi-genes) which control the action, on-off, of the genes.

The 3 axes describe a volume containing all the unencumbered combinations of levels of metabolites. The response surface through that volume describes all the combinations of levels of metabolites that can actually occur. This does not say that a certain level of an amino acid may not occur. It does say that a certain level of an amino acid may occur if and only if other metabolites are present in certain combinations of levels. Just as there are permitted energy levels for electrons in atomic or molecular orbitals, there are permitted "constellations" for the levels of a set of metabolites in a complex, regulated pathway. While the concentrations of the metabolites are the variables in a system of metabolic equations, if some are to be independent variables, others must become dependent variables. Change in the level of one metabolite compels changes in the levels of others.

This is certainly not the world-view of most plant physiologists, I realize, but it is a respected tradition among students of

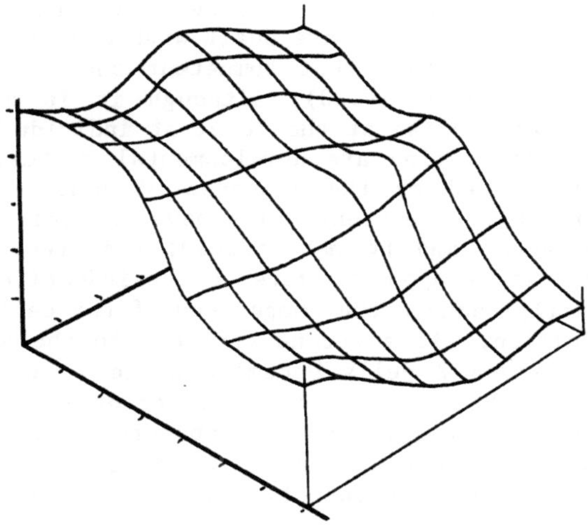

Fig. 4. An epigenetic landscape. The 3 axes are summations of the controlling aspects of plant metabolism. The response surface is those values that plant cells actually can take. Development or metabolic change can be visualized as movement along this surface.

the developmental biology of animals. It is also exactly the procedure our colleagues, the plant systematists, use to reduce the huge volumes of data gathered by numerical taxonomy down to plots showing the clustering of taxa. Later, we will return to examine the details of their methodology and how it can be applied to plant development.

The construction of such response surfaces and an examination of their properties has led to some remarkable and useful insights into the dynamics of plant metabolism. They clarify the nature of some epigenetic events, for example. Writing out the chemical equilibrium equations for all the reactions in a reasonably well-described metabolic pathway leads to a system of simultaneous equations. The concentrations of the metabolites are the variables and the rate constants are the known quantities. These equations describe all the biologically permitted combinations of levels of the intermediates and end-products. The solutions to such sets of equations may be represented as a response-surface in 3 dimensions. And these can be shown to have Eigen values, maxima and minima. Some minima are "potential wells." The surface in the immediate vicinity is so steep that cells in the vicinity "fall" to the bottom and are locked in a particular metabolic state. Such self-perpetuating states of biosynthesis are well known in tissue-cultures, and are predicted from a mathematical consideration of plant metabolism. Interested readers should refer to the original papers for further details (24,27,34).

Considering response surfaces might clarify the nature of cellular totipotency in higher plants, and its coexistence with cellular commitment and determination in those very same higher plants. Previous papers in this volume have dealt with competence and determination. There is no doubt that competent and determined states exist in higher plants (5,19,20), although it is not completely clear that such states exist at the level of individual cells (14). Competence and determination are developmental states, and as such are defined and assayed by the response of cells or tissues to external stimuli (20). Competent cells have the ability to respond to an inductive agent and become induced; competent cells change from state 1 to state 2 upon exposure to the inductive agent. This response necessarily changes the competence of the cells, if only by a trick of definition. If a second exposure to the same inductive agent results in state 2 cells becoming state 2 cells, there has been no observable change and the cells cannot be judged to have been competent. If, however, a second exposure to that agent, or an exposure to a different agent, causes the cells to change to state 3, then they can be judged competent, for the $2 \rightarrow 3$ induction.

The end result of a series of inductions is a cell determined for a particular developmental fate. That cell and its mitotic derivatives are pre-set and can go through the processes of morphological growth and differentiation to become a particular type of specialized cell, tissue, or organ. All they need is the trigger that says "express your cellular phenotype." Developmental geneticists working with animal systems call this latter a "permissive induction," and the former, the linear enchainment of inductions of competent cells, "directive inductions" (17). These, I believe, are very important distinctions to be able to make.

Waddington (35) maintained that determination always precedes morphological change; advances in molecular biology let us think of proteins and mRNA molecules as structures and let us demonstrate their presence during the process leading to a determined state. We can refine Waddington's contention and say that determination is the result of directive induction, that it precedes structural changes, and that it is the direct consequence of underlying molecular changes. Morphological expression and growth is the result of a permissive induction which causes cells to express their current developmental fate. This distinction is shown schematically in Fig. 5.

What then is totipotency? We have already noted that the whole set of permitted biochemical or physiological states for any cell can be summarized in a simple three-dimensional response surface, an epigenetic landscape sensu Waddington. Directive inductions are movements along this surface. Waddington very aptly describes this in terms of the path a ball would take rolling along the surface. The whole enchainment of sequential inductive steps in a developmental process traces a path from point A to point B; Waddington (35) uses the Greek word for path, Chreod, to describe this. There are always several, perhaps an infinite number of, paths from point A to point B; some direct, some meandering. While there are exact mathematical ways of coming to this same conclusion, it should be obvious to any of us who have rolled balls along curved surfaces in games of chance that certain paths or path segments will be favored and have an inherent stability. Such paths are ones that follow troughs in the surface.

Such a description is perhaps the best physical metaphor for the phenomenon "canalization," the tendency for a developmental process to resist perturbation by outside forces. Possible paths or path segments from A to B that go over many ridges, saddles, or extensive stretches of flat surface, or that go "uphill," will have a high failure rate. These paths are not well canalized. Natural selection will, over time, preserve those genes that shunt cells along well canalized paths and eliminate others.

Each plant cell, in vivo or in vitro, whether differentiated, undifferentiated, or dedifferentiated, has a particular constellation of biochemical components at any given moment; this assigns its position on the response surface (Fig. 4). External factors, such

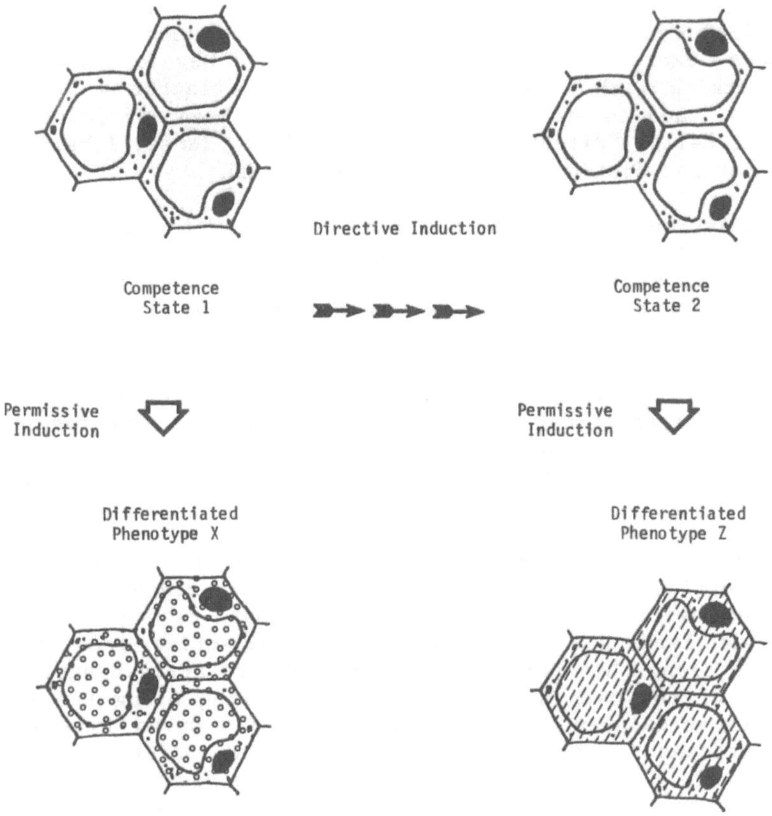

Fig. 5. Directive and permissive inductions. Under the influence of directive induction, competent cells will change from State 1 to State 2. This change is accompanied by a change in competence, but not by any structural change in the cells. Permissive induction allows competent cells to express a differentiated phenotype. The particular phenotype is determined by the current state of competence of the cell. This morphological expression of differentiated phenotype necessarily involves structural change in the cells.

as explantation onto a defined chemical medium, initiate changes in that biochemistry, and each cell moves across the response surface with a certain initial momentum. Some physiological states will collect cells; these are the low spots on the response surface. Some physiological states will trap cells; these are the potential wells, the steep-sided holes in the response surface, and are responsible for self-perpetuating, epigenetic, cellular, biochemical phenotypes.

All this theory translates directly into practical manipulation of plant cell and tissue cultures. Papers in this volume have already dealt with almost magical formulations for media and equally arcane preconditioning procedures. The magnitude and direction of the initial momentum of the cell on the response surface is a function of the composition of the medium; the starting point of the cell is a function of its physiological state in planto. Movement across the response surface, barring local steep-sided hills or ridges, is possible in all directions under the influences of stochastic changes in the biochemistry of individual cells and of additional nudges given to cells by changes in the external medium, be they depletion or replacement. Temperature changes, we should note, can change the shape of the response surface in many ways. This free movement along the response surface is cellular totipotency.

Upon receipt of a permissive inductive signal, cells will express their current developmental potential. This is cellular determination. You can imagine this as a topological mapping or projection of a three-dimensional surface onto a flat surface (Fig. 6). Certain kinds of cells will be caught determined for embryogenesis, others, for rhizogenesis; the developmental potential of many cells will be simply for continued growth as "undifferentiated" cells. A different permissive inductive signal on the same collection of cells could realize very different morphological changes; imagine the projection of the three-dimensional surface onto a flat surface at right angles to the first one (Fig. 6). This permissive induction might let certain cells become tracheids, others become highly vacuolate, and still others, shoot apices, in addition to the large numbers who simply continue growth as "undifferentiated" cells.

Permissive inductions are by nature passive; they let or allow the expression of the developmental state of those cells competent to respond to that particular permissive inductor. Directive inductions are not passive; they change competent cells from state 1 to state 2.

AN EMBRYOGENIC CULTURE OF SOYBEAN

The development of a set of manipulations to produce somatic embryoids in a given species, then, hinges on cells competent to respond to a permissive inductive stimulus by the production of somatic embryoids. This description of an embryogenic culture may not seem very different at all from the one that opened this paper. It does, however, let us separate the trigger from the state of the culture, and to work on each separately.

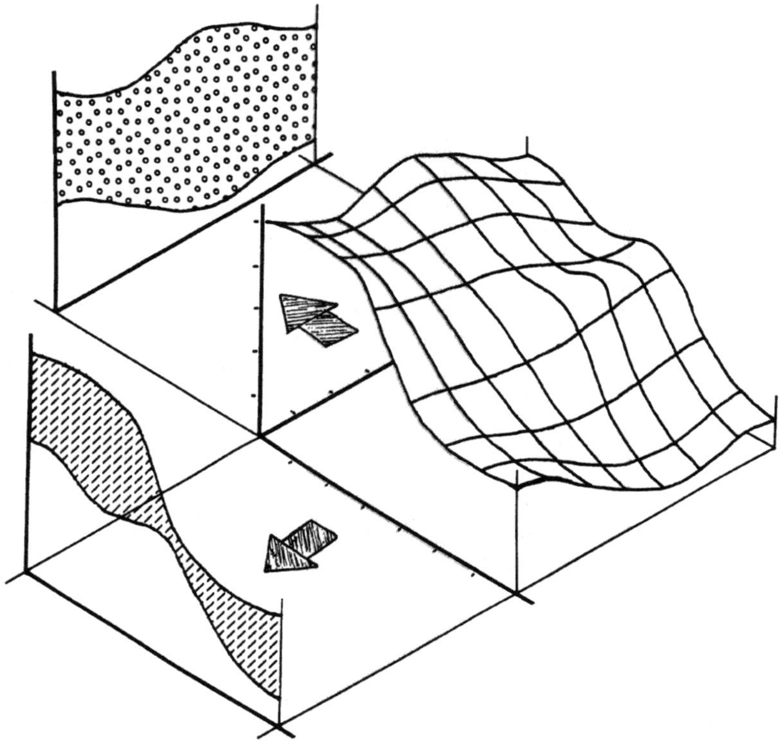

Fig. 6. Topological projections of epigenetic landscapes by permissive inductions. The effects of a permissive induction on a set of cellular physiologies can be visualized as the projection of the three-dimensional response surface onto a flat surface. Projection from one angle gives one shape, while projection along a different axis gives a different shape. This corresponds to the different end results of 2 different permissive inductions of a single plant cell culture: one kind of transfer of an alfalfa culture results in embryoids, another kind of transfer results in masses of roots.

In many cases, callus tissue grown in vitro develops distinct phenotypes, distinct at least to the eye of a practiced observer. These phenotypes, "wet," "rough," "hard," "dense," and "glossy," arise and can be subcultured on the same medium. They can reflect different developmental potentials. Tom Orton (22), for example, distinguished 5 callus phenotypes in cultures derived from immature ovules of barley and observed that one of them had significant potential for plantlet regeneration. Visual selection and subcultures of this pheno-variant callus produced cultures where plantlet regeneration was a routine procedure and highly efficient.

The cultivated soybean, Glycine max, remained without a report of plant regeneration from cell culture for a long period of time. Despite a growing number of successes in achieving shoot or plantlet formation from explants, callus, or suspension cultures of various legumes, published reports in the genus Glycine were limited to shoot production from hypocotyl slices, multiplication of cotyledonary buds, and an incomplete form of somatic embryogenesis (3,15,16,23). This last report (23) and its confirmation from another laboratory (10) were particularly encouraging; soybean cultures could be manipulated to form somatic embryoids with regular development as far as the late torpedo stage. More recently Jack Widholm's group has reported shoot formation from callus cultures of G. canescens (40). Although rumors from highly reliable sources and press releases suggest the existence of additional successful attempts, currently, our report is, unfortunately, the only published report of successful plant regeneration from a culture of G. max (7).

We initiated callus from young embryos of soybean taken from 2.5-3.0 cm pods of G. max cv. Mitchell. This is a maturity group VI variety grown in the southern United States. The cotyledons were removed and discarded; the embryonic axes were cut into 1-2 mm segments and placed on a Murashige and Skoog (MS) salt-based medium containing 5 mg/l 2,4-D. [The complete details of this and the other media we used can be found in the original report (7).] Callus development ensued, and a few weeks later, the friable parts were discarded and the non-friable, slow-growing parts were placed in fresh medium, again with 5 mg/l 2,4-D. This material was transferred a month later to medium with 2 mg/l indoleacetic acid (IAA) and 0.2 mg/l N_6-(2-isopentenyl)aminopurine (2-iP), and incubated in low light for some 3 months. Inspection revealed a number of callus phenotypes, from a white friable type (Fig. 7a) to a smooth, green, glossy type (Fig. 7b). These green glossy pieces resembled highly abnormal embryo axes; such structures are termed "neomorphs" by Krikorian and Kann (18). Submitting neomorphs of daylily to additional rounds of 2,4-D exposure and withdrawal was successful in developing a highly morphogenetic daylily culture. This, of course, is what Steward did with his original wild carrot cultures: take the embryos from the first round, slice them up, and repeat the process (30). A similar scheme of recurrent selection was used to develop a highly rhizogenic line of barley (4), and was also used in the recent report on the development of a morphogenetic culture of cotton (9). Our recurrently selected cultures of soybean would reproducibly form neomorphs in response to the withdrawal of 2,4-D (Fig. 7c). They would not make real embryos. We had gotten a culture with morphogenetic competence, but the trigger, the permissive inductor, was somehow not correct. To extend the physical metaphor we used previously, it was as if the three-dimensional response surface was being projected not onto a flat surface, but onto a slanted one: the outlines were wrong although the shape was almost there.

(a)

(b)

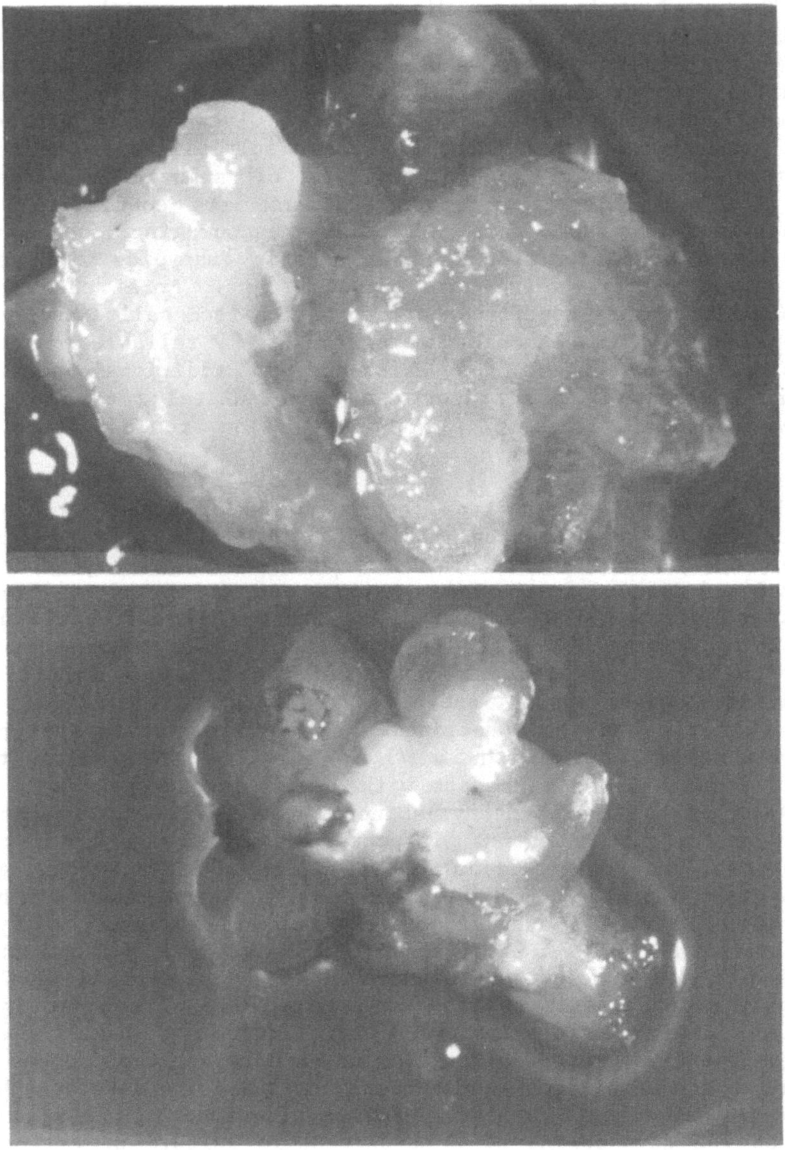

Fig. 7. Somatic embryogenesis in soybean. (a) White friable callus from embryonic axes explanted to 5 mg/l 2,4-D. This phenovariant callus is fast-growing and has no embryogenic potential. (b) Green, glossy callus from embryonic axes explanted to 5 mg/l 2,4-D. This phenovariant callus is slow-growing and was used to develop the morphogenetic culture.

(c)

(d)

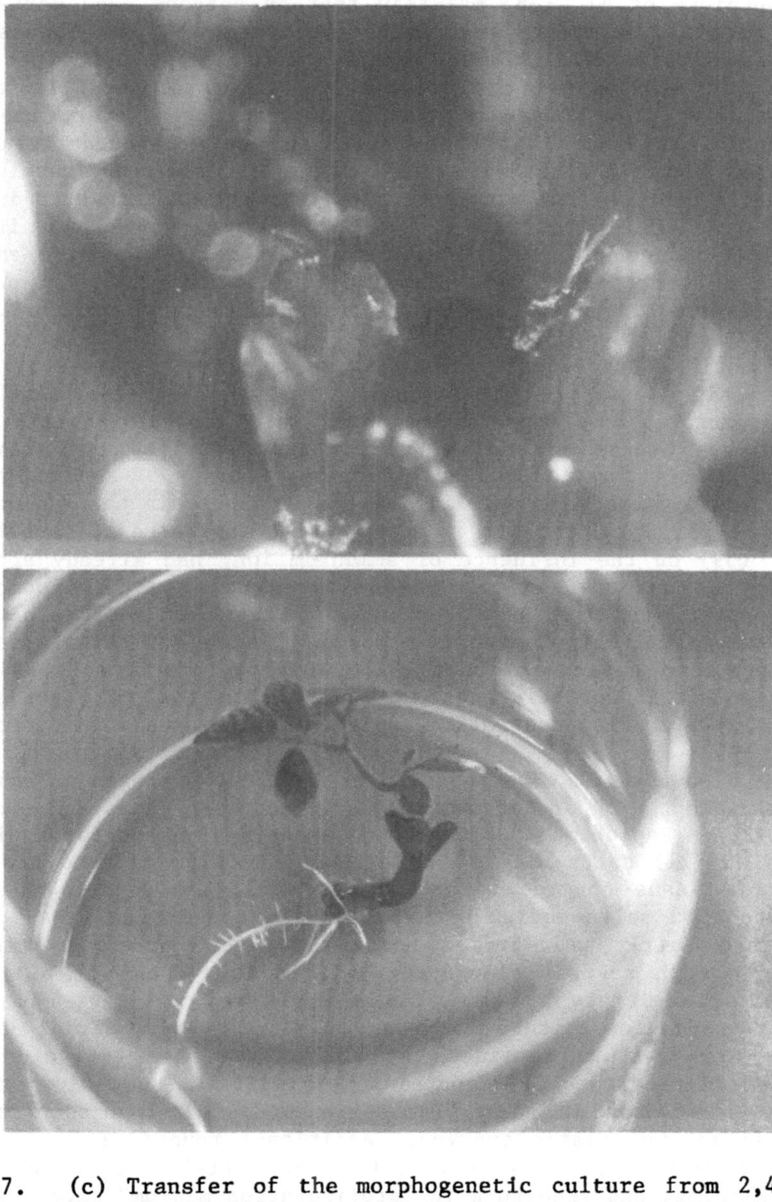

Fig. 7. (c) Transfer of the morphogenetic culture from 2,4-D to Cheng's hormones resulted in the production of neomorphs. This example appears to be an attempt at a leaf; notice the eglandular hairs characteristic of soybean leaves. (d) After 3 years in culture, the cell line continues to make embryoids. Continued rounds of selection have solved the problem noted in our original report: the culture now produces embryoids with balanced growth of the root and shoot.

Wernicke et al. (38) maintain that in sorghum, and perhaps in other species, embryogenic cultures proliferate as small masses of suppressed primordia. They, myself, and others feel that high levels of 2,4-D suppress differentiation or morphological expression of the underlying competent state. In carrot, however, there is substantial evidence that the suppression of morphogenetic competence can occur in other ways, and, conversely, that the development of somatic embryos from a competent culture can be triggered in several ways. While withdrawal of 2,4-D does work, changes in nitrogen level and changes in cell density can also trigger somatic embryogenesis in carrot cultures. These alternate triggers will produce embryoids even in the presence of significant amounts of 2,4-D.

Somatic embryogenesis from alfalfa cultures requires an exposure to high levels of 2,4-D and is triggered by the transfer of cell aggregates to "expression media" which do not contain 2,4-D (28). Dave Stuart and Keith Walker have described the promotive effects of ammonium ion and certain other sources of reduced nitrogen on somatic embryogenesis in alfalfa in papers in this volume and elsewhere (31,37). The first papers, extending Wetherall's original report (39) on the influence of ammonium ion, had already appeared at the time our soybean culture was reproducibly making neomorphs. Those papers said high levels of ammonium ion promoted the development of somatic embryos even in a hormone regime appropriate for the development of masses of roots (37). Our own experiments on shoot organogenesis in another genus had convinced us that nitrogen source could play a crucial controlling role in morphogenesis in vitro (6).

We also knew that we could supply all the nitrogen as ammonium and maintain the pH of the medium, if we supplied it as ammonium citrate (26). Replacement of the 40 mM ammonium ion and 20 mM nitrate ion in MS (21) salts with 20 mM dibasic ammonium citrate led to a piece of tissue covered with numerous small embryoids. The half moved to a MS salts medium with 5 µg/l IBA and 200 µg/l BAP developed into fairly well-formed somatic embryos, and gave rise to plantlets with several elongating internodes. The half placed in the ammonium citrate 2,4-D medium proliferated and gave rise to a nodular suspension culture which can be maintained by serial subculture. This culture retained the ability to form embryoids when transferred from the 2,4-D containing "proliferation medium" to the 2,4-D-less standard MS salts "expression medium." This same culture is still making somatic embryos. Additional rounds of 2,4-D exposure and withdrawal, visual selection of well-formed young embryoids, slicing, and reculture in "proliferation medium" have resulted in a culture that will make somatic embryos whose cotyledons have distinct petioles and almost exactly mimic the cotyledons of zyogtic embryos. The embryos from our current culture produce plantlets with balanced growth of the root and shoot (Fig. 7d). Although this is routine, it is still not as efficient as we would like. We now use an expression medium slightly changed from our original report:

sucrose at 40 g/l and replacement of Cheng's (3) hormones with no phytohormones at all. In our experience, treatments with ABA or gibberellic acid are not necessary or even helpful, nor is the inclusion of proline in either medium.

DESCRIPTION OF THE PERMISSIVE INDUCTION

Somatic embryogenesis in our soybean culture occurs after the coordinate removal of 2,4-D and change from 40 mM ammonium to 20mM ammonium and 40 mM nitrate. Experiments showed that a simple withdrawal of the 2,4-D was not enough to trigger complete somatic embryogenesis in our soybean culture: transfer of morphogenetic suspension from ammonium citrate 2,4-D medium to ammonium citrate 2,4-D-less medium did not result in embryoid formation and development.

The change in the nitrogen source was shown to be essential. One experiment established a subline of the morphogenetic suspension in 2,4-D medium but with standard MS salts. Transfers to expression medium after one week produced a few abnormal embryoids. After 4 weeks in standard MS salts, transfer to expression medium resulted in glossy green structures, but not a single embryoid. More recent experiments with an advanced morphogenetic culture suggest that visual selection of the waxy, yellow-green, "embryogenic pieces" at weekly intervals, and elimination of the rapidly growing friable white pieces that arise during culture on standard MS salts, can prolong the embryogenic competence of the culture.

Other experiments documented the adverse effects of a 40 mM nitrate (potassium salt) supplement added to the ammonium citrate containing proliferation medium, and also documented the adverse effect of 20 mM ammonium (as the citrate) to the MS salts-based expression medium. In contrast, 20 mM potassium citrate added to the expression medium did not stop the production of somatic embryos. The effective transfer is the change from 40 mM ammonium (as citrate) to 20 mM ammonium plus 40 mM nitrate.

The change of nitrogen source alone is not enough, however. The presence of 5 mg/l 2,4-D in the "expression medium" did not allow the development of somatic embryos. Our results show the coordinate changes in auxin level and nitrogen source to be the keys to triggering complete somatic embryogenesis in competent cultures of soybean (Tab. 1).

In carrot cultures, any one of a number of changes can be the permissive inductor and lead to the development of somatic embryos. In corn and alfalfa, 2,4-D withdrawal alone is a sufficient permissive induction (12,28,36). For both corn and alfalfa, however, particular kinds and combinations of reduced nitrogen can strongly and

Tab. 1. Effects of different permissive inductors on a component culture of soybean. The morphogenetically competent culture of Glycine max was manipulated in a variety of ways. These treatments were assayed for their effectiveness as permissive inductors of somatic embryogenesis by direct observation of the quality and extent of subsequent embryoid development.

Treatment Combination						Result
Proliferation Medium			Expression Medium			
hormones	NH_4^+	NO_3^-	hormones	NH_4^+	NO_3^-	
5 mg/l 2,4-D	40mM	-	TYC*	20mM	40mM	embryoids
5 mg/l 2,4-D	40mM	-	TYC	40mM	-	callus
5 mg/l 2,4-D	40mM	-	5 mg/l 2,4-D	20mM	40mM	callus
5 mg/l 2,4-D	40mM	40mM	TYC	20mM	40mM	callus
5 mg/l 2,4-D	40mM	-	TYC	40mM	40mM	callus
5 mg/l 2,4-D	20mM	40mM	TYC	20mM	40mM	neomorphs**

* TYC = 5 μg/l IBA plus 200 μg/l BAP

** neomorphs = abnormal embryoid-like structures, after Krikorian and Kann (18)

dramatically reinforce this trigger (31,37). In soybean, both effects, 2,4-D withdrawal and change in nitrogen source, are necessary.

An amenable starting material, immature embryos, combined with visual selection for non-friable, embryo-like structures and recurrent selection in a regime of 2,4-D exposure and withdrawal can be used to develop a soybean culture with embryogenic competence. This culture can be triggered to produce somatic embryoids capable of germination and growth as plants by the coordinate change of nitrogen source and 2,4-D level. This appears to be a general phenomenon; a culture of cultivated pea, Pisum satiuum, we are treating in this manner has produced occasional somatic embryos. More rounds of selection are needed to turn this extremely slow-growing culture into a reliable embryo producer.

Z. Renée Sung has used the various single triggers to somatic embryogenesis in carrot to describe a set of proteins intimately associated with callus proliferation and with embryoid development (32,33). We suspect the dual trigger to somatic embryogenesis in our soybean culture can be exploited to advance our understanding of the molecular controls on embryo development. But this work all focuses on the permissive induction and the enchainment of events we see as morphological differentiation and growth. What of the necessary precondition for the action of the permissive inductor: a culture with morphogenetic competence? How can we approach the molecular nature of "competence" and "suppressed differentiated states?"

APPROACHING COMPETENCE

One approach to the nature of "embryogenic competence" which seems to promise a certain degree of success is to map the three-dimensional response surfaces described in the first part of this paper. This is conceptually simple, will not be too difficult in practice, and has the distinct advantage of identifying those aspects of cellular biochemistry that are major contributors to embryogenic competence and those that are not.

Actual construction of a three-dimensional response surface requires 2 things: that we can measure a significant fraction of the biochemical status of small amounts of plant tissue, and that we can reduce this large array of data from each sample to a single data point defined by X, Y, and Z coordinates. Advances in analytical biochemistry have been such, that with a very small sample of plant tissue and a gas chromatograph, we can know the amounts of each of the free amino acids, sugars as well as organic acids. This embarrassment of data, hopefully stored in a file on a computer, is similar to the collections of data in the hands and computer files of our colleagues, the plant systematists. They have been in this predicament longer than we have, and, fortunately for us, they have discovered ways of ordering and extracting the information from such arrays. One technique is called "multivariate analysis," and, in particular, "principal components analysis." There are works describing this approach in great detail (29). Basically, what it does is take all the data, look at all the covariances, and distill the measured parameters down into 3 constructed variables, the "principal components."

Those "constructed variables" are, in fact, linear combinations of the original variables (29). The first one takes care of most of the variance, the second one takes care of most of what is left over, and the third one takes care of most of what is still left. The residual is the "error term." These are the 3 axes for our metabolic response surfaces. The "error term" may be thought of as the thickness of that surface. Such summary plots of metabolic data can be as useful to us as "Wagner trees" have been to plant systematists.

Imagine a mediocre embryogenic culture subcloned into small cultures, each assayed for the entire spectrum of small metabolites and assessed in a quantitative way for embryogenic competence (i.e., count the number of embryos each subculture can make). An experiment much along these lines has already been done. John Cross and coworkers at Pfizer Central Research have taken a marginally embryogenic carrot culture, subcloned it, and assayed each subclone for both the amount and type of phenolic compounds, as well as for embryogenic response (8). Unfortunately, this is a sparse set of data from secondary, not primary, plant metabolism.

Analysis of similar data from our as-yet-imagined experiment will result in a map of permitted states for plant cell biochemistry. We can superimpose the embryogenic response of each subclone on this response surface, and connect similar levels as if they were barometric pressures on a map of the earth (Fig. 8). The end result will be a very precise description of those intracellular states most compatible with embryo development in response to a permissive induction.

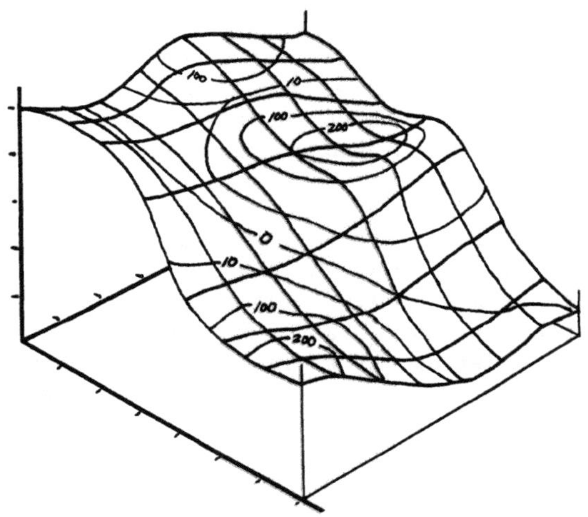

Fig. 8. Iso-embryonic map of states of cellular physiology. Small subcultures of widely varying embryogenic potential can be assayed for the actual number of embryos produced after permissive induction, as well as for their particular constellation of small metabolites. This data, reduced by principal components analysis to a three-dimensional surface, is a map of the possible metabolic states of such cultures. Superimposition of the actual numbers of embryos produced by subcultures can allow the construction of iso-embryonic contour lines on this surface, much as if they were barometric pressures superimposed on a map of the earth.

If there is one cultivar of a species which regenerates at all we should be able to construct a map. If we move the intracellular states of non-regenerating cultivars to match those of this map, can we achieve efficient regeneration for the entire species? If so, the constellation of small metabolites is what "determination" is at a molecular level, and gene action simply serves to get cells from metabolic state A to metabolic State Z. Will we find that embryogenic competence in a wide range of species or families is a similar constellation of small metabolites? Will we be able to regenerate any species, from any tissue of any age or physiological state, by matching its intracellular metabolic state to such a map? This may very well be true. It may even be the route to an answer to James Bonner's question, "How does a zygote know it's a zygote?" (2). And it has the distinct advantage that it is amenable to direct experimental test.

ACKNOWLEDGEMENTS

I would like to acknowledge the collaboration of D.A. Warnick and P.S. Carlson in some of the experiments reported here, and to acknowledge my debt to Leo Willis and Rae Phelps Mericle who introduced me to plant development and genetics. This work was facilitated by the friendship and encouragement of R. Gene Geisler.

REFERENCES

1. Ammirato, P.V. (1983) Embryogenesis. In Handbook of Plant Cell Culture, Vol. 1, D.A. Evans, W.R. Sharp, P.V. Ammirato, and Y. Yamada, eds. Macmillan, New York, pp. 82-123.
2. Bonner, J. (1965) The Molecular Biology of Development, Oxford Press.
3. Cheng, T.Y., H. Saka, and T.H. Voqui-Dinh (1980) Plant regeneration from soybean cotyledonary node segments in culture. Plant Sci. Lett. 19:91-99.
4. Chin, J.C., and K.J. Scott (1977) The isolation of a high-rooting cereal callus line by recurrent selection with 2,4-D. Z. Pflanzenphysiol. 85:117-124.
5. Christianson, M.L., and D.A. Warnick (1983) Competence and determination in the process of in vitro shoot organogenesis. Dev. Biol. 95:288-293.
6. Christianson, M.L., and D.A. Warnick (1984) Phenocritical times in the process of in vitro shoot organogenesis. Dev. Biol. 101:382-390.
7. Christianson, M.L., D.A. Warnick, and P.S. Carlson (1983) A morphogenetically competent soybean suspension culture. Science 222:632-634.
8. Cross, J.W., and W.R. Adams (1983) Phenolic levels and embryogenic potential in carrot cultures. Plant Physiol. 72:s46.

9. Davidonis, G.H., and R.H. Hamilton (1983) Plant regeneration from callus tissue of Gossypium nirsutum L. Plant Sci. Lett. 32:89-93.
10. Gamburg, O.L., B.P. Davis, and R.W. Stahlhut (1983) Somatic embryogenesis in cell cultures of Glycine species. Plant Cell Reports 2:209-212.
11. Goldschmidt, R. (1938) Physiological Genetics, McGraw-Hill, New York.
12. Green, C.C., and R.L. Phillips (1975) Plant regeneration from tissue cultures of maize. Crop Sci. 15:417-421.
13. Halperin, W. (1970) Embryos from somatic plant cells. In Control Mechanisms in the Expression of Cellular Phenotypes, A. Padykula, ed. Academic Press, New York, pp. 169-191.
14. Henshaw, G.G., J.F. O'Hara, and K.J. Webb (1982) Morphogenetic studies in plant tissue cultures. In Differentiation "in vitro", M.M. Yeoman and D.E.S. Truman, eds. Cambridge Press, pp. 231-251.
15. Kameya, T., and J. Widholm (1981) Plant regeneration from hypocotyl sections of Glycine species. Plant Sci. Lett. 21:289-294.
16. Kimball, S.L., and E.T. Bingham (1973) Adventitious bud development of soybean hypocotyl sections in culture. Crop Sci. 13:758-760.
17. Kratochwil, K. (1983) Embryonic induction. In Cellular Interactions and Development: Molecular Mechanisms, K.M. Yamada, ed. Wiley, New York, pp. 99-122.
18. Krikorian, A.D., and R.P. Kann (1981) Plantlet production from morphogenetically competent cell suspensions of daylily. Ann. Bot. 47:679-686.
19. McDaniel, C.N. (1978) Determination for growth pattern in axillary buds of Nicotiana tabacum L. Dev. Biol. 66:250-255.
20. Meins, Jr., F., and A.N. Binns (1979) Cell determination in plant development. Bioscience 29:221-225.
21. Murashige, T., and F. Skoog (1962) A revised medium for rapid growth and bioassays with tobacco tissue cultures. Physiol. Plant 15:473-497.
22. Orton, T.J. (1979) Quantitative analysis of growth and regeneration from tissue culture of Hordeum vulgare, H. jubatum and their interspecific hybrid. Env. Exp. Bot. 19:319-335.
23. Phillips, G.C., and G.B. Collins (1981) Induction and development of somatic embryos from cell suspension cultures of soybean. Plant Cell Tissue Organ Culture 1:123-129.
24. Ratner, V.A., and R.N. Tehuraev (1978) Simplest genetic systems controlling ontogenesis: Organization principles and models of their function. Prog. Theo. Biol. 5:81-127.
25. Rendel, J.M. (1967) Canalization and Gene Control, Logos Press, Ltd., London.
26. Rhodes, D., G. Jamieson, and M.L. Christianson (1982) Analysis of (15N)H_4^+ assimilation in suspension cultures of the field bindweed, Convolvulus arvensis L., using combined gas chromatography-mass spectrometry and computer simulation techniques. In Vitro 18:296.

27. Riggs, D.S. (1970) Control Theory and Physiological Feedback Mechanisms, Williams and Wilkins Co., Baltimore.
28. Saunders, J.W., and E.T. Bingham (1972) Production of alfalfa plants from callus tissue. Crop. Sci. 12:804-808.
29. Sneath, P.H.A., and R.R. Sokal (1973) Numerical Taxonomy: The Principles and Practice of Numerical Classifications, W.H. Freeman, San Francisco.
30. Steward, F.C., M.O. Mapes, A.E. Kent, and R.D. Holsten (1964) Growth and development of cultured plant cells. Science 143:20-27.
31. Stuart, D.A., and S.G. Strickland (1984) Somatic embryogenesis from cell cultures of Medicago sativa L. II. The interaction of amino acids with ammonium. Plant Sci. Lett. 34:175-181.
32. Sung, Z.R., and R. Okimoto (1981) Embryonic proteins in somatic embryos of carrot. Proc. Natl. Acad. Sci. 78:3683-3687.
33. Sung, Z.R., and R. Okimoto (1983) Coordinate gene regulation during somatic embryogenesis in carrots. Proc. Natl. Acad. Sci. 80:2661-2665.
34. Tyson, J.J., and H.G. Othmer (1978) The dynamics of feedback control circuits in biochemical pathways. Progress in Theo. Biol. 5:1-62.
35. Waddington, C.H. (1977) Tools for Thought, Basic Books, Inc., New York.
36. Walker, K.A., M.L. Wendeln, and E.G. Jaworski (1979) Organogenesis in callus tissue of Medicago sativa: The temporal separation of induction processes from differentiation processes. Plant Sci. Lett. 16:23-30.
37. Walker, K.A., and S.J. Sato (1981) Morphogenesis in callus tissue of Medicago sativa: The role of ammonium ion in somatic embryogenesis. Plant Cell Tissue Organ Culture 1:109-121.
38. Wernicke, W., I. Potrykus, and E. Thomas (1982) Morphogenesis from cultured leaf tissue of Sorghum bicolos - the morphogenetic pathways. Protoplasma 111:53-62.
39. Wetherall, D.F., and D.K. Dougall, (1976) Sources of nitrogen supporting growth and embryogenesis in cultured wild carrot tissue. Physiol. Plant. 37:97-103.
40. Widholm, J.M., and S. Rick (1983) Shoot regeneration from Glycine canescens tissue cultures. Plant Cell Reports 2:19-20.
41. Yusufov, A.G. (1982) Origin and evolution of the phenomenon of regeneration in plant (problem of evolution ontogenesis). Usp. Sovrem. Biol. 93:89-104. (Translated from the Russian by Leo Kanner Associates.)

SOMATIC EMBRYOGENESIS FOR MASS CLONING OF CROP PLANTS

Joseph D. Lutz, James R. Wong, Jan Rowe,
David M. Tricoli,* and Robert H. Lawrence, Jr.

Applied Genetics Laboratory
Agrigenetics Corporation
3375 Mitchell Lane
Boulder, Colorado 80301

INTRODUCTION

The large-scale cloning of crop plants potentially has broad applications in agriculture (23,28,45). The possibility of producing large numbers of plants of a single genotype evokes visions of applications to breeding programs, new seed production methods, and even an alternative to normal crop production from seeds. Over the last 10 to 15 years tissue culture techniques have found wide use in the commercial propagation of horticultural plants, particularly ornamentals, and in the elimination of specific pathogens for the production of pathogen-free plants. There has been little commercial use of tissue culture in the production of agricultural crops. The methods commonly used for the production of horticultural propagules are relatively labor-intensive, low volume, and have high unit costs compared with current agricultural seed practices. Highly mechanized culture systems able to efficiently produce large numbers of propagules must be developed if high-frequency cloning is to be a viable concept.

The generally accepted methods for multiplication and regeneration are diagrammed in Fig. 1 (23). These methods have been extensively reviewed in detail elsewhere (7,21,30,39). Here we will briefly outline the differences as they apply to mass cloning systems. The most widely used commercial method of regeneration is axillary bud multiplication (21,28,29,30). Clonal propagation gen-

*Monsanto Company, 800 N. Lindbergh Blvd., St. Louis, Missouri 63167

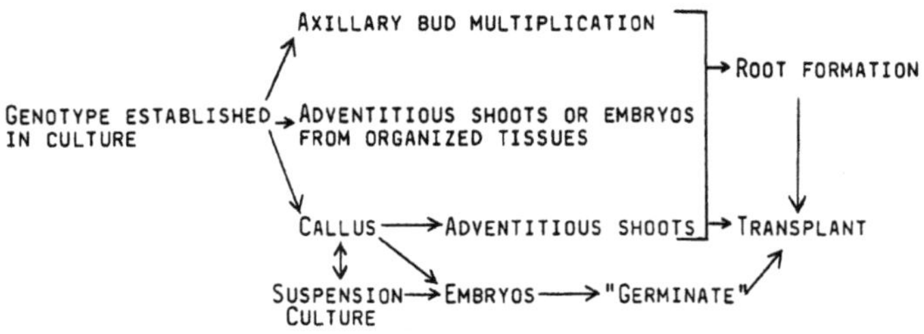

Fig. 1. Methods of multiplication and regeneration for in vitro cloning.

erally involves placing shoot tips and/or axillary buds on a multiplication medium to induce axillary shoot development. These axillary shoots are recycled on the multiplication medium until the desired level of production is achieved. Shoot tips are then rooted, transplanted to soil, and hardened off. Typically a multiplication rate of 5 to 10 propagules per culture cycle is considered reasonable for commercial systems. This method is very labor-intensive due to the number of individual manual manipulations involved and the low multiplication rates. In the best cases the cost is estimated to be 13 to 25 cents per microcutting (1,32). Phenotypic variability expressed among propagules generally is low, and the quality of the product has continued to gain acceptance as evidenced by a large and growing micropropagation industry.

Another method of regeneration is adventitious bud formation, a form of organogenesis. Clonal multiplication by this method includes 2 forms. In some systems adventitious shoots can be induced directly on the original explants. These adventitious shoots must then be rooted, transplanted to soil, and hardened off. This method requires fewer manipulations than shoot tip cultures, but potentially requires a large explant source depending upon the multiplication rate. In other systems adventitious buds form on callus cultures. Again these shoots have to be rooted, transplanted to soil, and hardened off. The multiplication rate of organogenic systems can be up to an order of magnitude higher than for axillary bud multiplication. In some cases the number of individual manipulations is less, bringing down the cost per unit. Some organogenic systems are amenable to mechanization such as mechanical chopping in blenders or food choppers (8). Bulk transfers may be possible, such that large numbers of tissue pieces could be transferred simultaneously using some carrier; however, each individual propagule still has to be isolated by hand.

Finally, the third method of regeneration outlined in Fig. 1 is somatic embryogenesis. Somatic embryogenesis involves the induction of bipolar structures bearing both root and shoot meristems, and has an obvious advantage over other forms of regeneration for use in the mass cloning of agricultural crops. That advantage is that the ability to produce free, individual propagules in liquid culture should be subject to labor-reducing mechanization. Clonal multiplication by this method involves the induction of embryos either on the surface of explants, callus, or in a suspension culture. In terms of their utility, the formation of embryos on callus or explants may present little advantage over adventitious shoots. Isolation of individual embryos still has to be done manually. However, in suspension cultures, embryos sometimes form free and separate from other tissues. Such embryos can be isolated to produce populations of single propagules which should be highly amenable to large-scale handling and mechanized planting systems.

Using the well-characterized carrot system as a model we have identified many of the major problems, and have begun to work on a few solutions involved with the use of somatic embryogenesis on a commercial scale. We have separated the problems into the following 3 areas: cell culture techniques, delivery systems, and whole plant biology.

CELL CULTURE TECHNIQUES

Somatic embryogenesis has been observed in the in vitro cultures of over 100 plant species (4,5,37,44,47). In the majority of these reports embryos develop attached to other tissues, and individual embryos must be physically separated from each other and surrounding tissues. In a few systems, such as some species of Umbelliferae, embryogenesis occurs in liquid culture producing free, individual embryos. Somatic embryogenesis was first recognized in tissue cultures derived from the tap root of carrot in the late 1950s (34,35,41). The standard morphological stages of zygotic embryogenesis, i.e., globular, heart, and torpedo, have been observed in regenerating carrot cultures (17,18,42). The process by which embryoids form on pro-embryonic masses has been described in detail (19,25).

When suspension cultures are transferred to auxin-free media, globular embryos form on small pro-embryonic masses. These then further develop and continue to break apart to form a population of embryos containing individual and attached globular, heart and torpedo embryos. The addition of abscisic acid (ABA) to the later stages of embryo development has been reported to normalize development, producing somatic embryos with a more normal balance between cotyledon and radicle growth (3,20).

Figure 2 is a flow diagram illustrating the general steps needed to establish a somatic embryogenesis cloning system. Since all of these steps are feasible with a carrot tissue culture system, we have used this system to identify potential problems encountered after the establishment of a good embryogenic system.

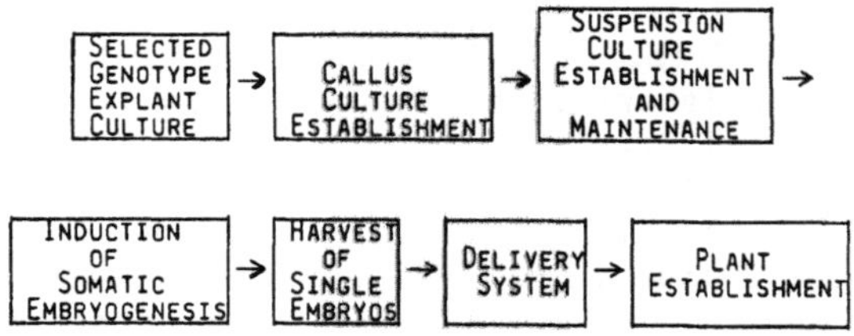

Fig. 2. Generalized somatic embryogenesis cloning scheme.

Explant Source

The severity of any given problem will depend on the end use of the product, i.e., applications to breeding programs will have a different set of requirements than applications to crop production. Young explant tissue, and preferably aseptically-grown seedling material, generally constitutes the better explant for tissue culture. However, plant breeders generally need to clone from mature plants after they have been evaluated in the field. The lack of juvenile tissue and difficulty in establishing sterile cultures can be a problem.

Variation Between Culture Lines

Tissue cultures and especially suspension cultures are dynamic populations of many cell types with different developmental fates. In the routine transfer of these suspensions, very different cell types can be selected for unknowingly (26). Quality control throughout the cloning process will be a major problem in the large-scale application of this process. In our lab, celery lines initiated from the same plants at the same time, and treated the same in culture, exhibit different degrees of regeneration when transferred to a standard regeneration medium. A similar case has been reported by Drew (12) with cell lines derived from carrot petioles.

Genotypic Variation Between Cultivars

Variation exists in the tissue culture response from one cultivar to another as well (15). Plant breeders generally deal with large numbers of cultivars at any one time. In our experience most genotypic variation in culture response can be corrected by small changes in the hormone regimes. Any mass cloning system would need to be fine-tuned for the cultivar being used. Broad tissue culture experience across cultivars, pilot productions, and close communication between breeders and tissue culturists can alleviate the impact of this problem.

Developmental Synchrony

Developmental synchrony is probably the major cell culture problem after establishing embryogenesis. Classically there are 3 ways to attain developmental synchrony: simultaneous induction, block and release, and selection of a synchronous portion of a population. We and others (14,46) have attempted to select a developmentally synchronous population of embryos using size and density separations. An initial sizing to less than 94 μ selects for a population of pro-embryonic masses and globular embryos to inoculate regeneration runs. Table 1 shows the results of a size separation experiment. Cultures were filtered through stainless steel screens of different size mesh 14 days after transfer to hormone-free medium. The greater-than-520 μ size class has the most homogenous population of embryos. It is also possible to separate different embryo stages by density. Table 2 shows the results of a density separation on Percoll gradients. We have found that if we combine density and size separations sequentially (Tab. 3), we can get a large and uniform population of torpedo embryos between 0.5 and 2 mm in length.

As reported by others (2,3,5) low concentrations of ABA inhibit precocious germination during embryo development. Table 4 illustrates that the addition of 1 μM ABA during the second week of regeneration increases the proportion of torpedo-stage embryos less than 2 mm. This confirms the results of Kamada and Harada (20). Figure 3 is a flow diagram depicting the carrot embryo production scheme used in our lab. One milliliter packed cell volume produces approximately 5000 embryos at the end of 14 days. Fifty percent of these are lost in the density and size separations producing a developmentally synchronous population of uniformly sized embryos.

Abnormal Development

The somatic embryos produced by our embryo production scheme appear to be normal, having gone through all the generally recognized stages of zygotic embryogenesis. However, when compared to

Tab. 1. Carrot embryo separation by SIZE.

Sieve Sizes (μ^2 Mesh)	Developmental Stages (% of Total) Globular	Heart	Torpedo <1	Torpedo 1-2	Torpedo >2 mm
> 520	0	0	24	76	0
280 - 520	11	7	66	16	0
190 - 280	34	32	32	2	0
< 190	90	9	1	0	0

Tab. 2. Carrot embryo separation by DENSITY.

Percoll	Developmental Stages (% of Total) Globular	Heart	Torpedo <1	Torpedo 1-2	Torpedo >2 mm
35%	64	17	16	3	0
45%	36	22	28	12	2

Tab. 3. Carrot embryo separation by DENSITY and SIZE.

Percoll/Sieve	Developmental Stages (% of Total) Globular	Heart	Torpedo <1	Torpedo 1-2	Torpedo >2 mm
35% / >520	0	0	57	43	0
35% / 280-520	8	17	33	42	0
45% / >520	0	0	32	68	0
45% / 280-520	0	17	60	23	0

Tab. 4. Carrot embryo development and timing of ABA treatments (10^{-6} M).

1st/2nd Week	Embryos per flask	Developmental Stages (% of Total) Globular	Heart	Torpedo <2 mm	Torpedo >2 mm
-ABA/-ABA	773	8	9	58	25
-ABA/+ABA	973	9	21	65	5
+ABA/-ABA	853	12	8	56	24
+ABA/+ABA	607	30	14	56	0

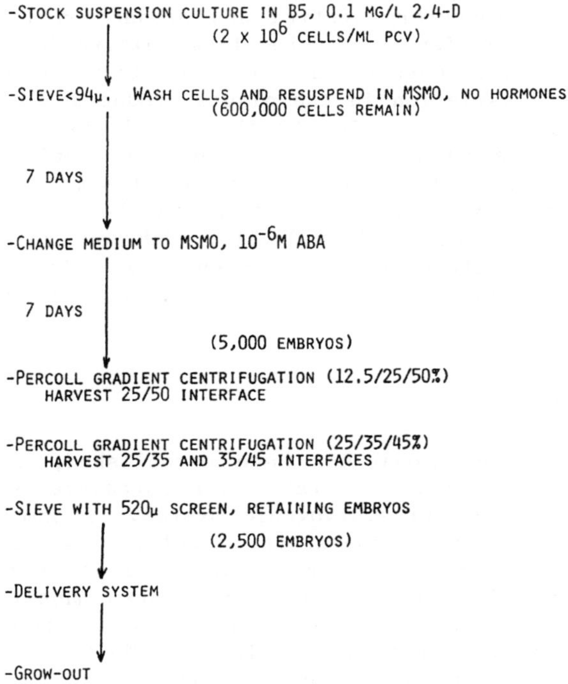

Fig. 3. Carrot regeneration protocol.

zygotic embryos excised from mature seeds, there are obvious morphological differences. The somatic embryos have reduced cotyledon growth and increased hypocotyl and/or radicle growth. These embryos do not develop further if maintained in 1 μM ABA, and if placed in hormone-free medium they germinate, never exhibiting a stage equivalent to zygotic embryos from seeds. Somatic embryos will develop into plantlets, but only if nurtured on media to an advanced stage. The larger cotyledons of zygotic embryos contain stored food reserves that could be important for the conversion of somatic embryos to plantlets.

A second problem of abnormal development is the formation of adventitious embryos along the sides of primary embryos. An ABA treatment in the second week of regeneration suppresses them in carrots (20).

Time Constraints

A commercial mass cloning operation would be under strict time constraints. The selection of genotypes to be cloned at one time must be matched to production schedules to meet planting schedules at other times. Inventory control and production levels are also time-dependent. The length of time from culture initiation to regeneration has been correlated with the loss of regenerability and/or increased variation in the plants produced (22,40,43). Culture storage may be a solution to these problems. The biological time-scale is presumably measured by cell generations. If growth rates in vitro can be slowed down, control of some of these parameters might be possible. Culture storage in the cold works well for some shoot tip cultures (6,11,24,27,31). Embryos can be made quiescent and stored in the cold also. In our lab we have stored carrot embryos for 120 days at 10°C with no deleterious effect on the subsequent conversion to plants.

DELIVERY SYSTEMS

Once a culture system has been established, a handling and delivery system is needed to take full advantage of the reduced labor associated with an embryogenic system. A somatic embryo is not equivalent to a seed, therefore, a delivery system is required which nourishes and protects the embryo until it is fully autotrophic. For commercial uses the delivery system must be highly amenable to mechanization. The need and benefits from such systems have been reported elsewhere (4,23,29). There are basically 2 choices: fluid drilling and encapsulation.

Fluid Drilling

Fluid drilling of somatic embryos has been pioneered by the group at Wellsbourne, England (12). Fluid drilling technology was developed for the planting of pre-germinated seeds (9,10,16) to provide more uniform germination and less time to emergence in adverse field conditions. The concept of fluid drilling is to suspend the propagules in a carrier gel supplemented with additives, which is then pumped into the soil. Drilling devices have been developed which separate and precision drill pre-germinated seeds with high efficiency (16). As a delivery system, fluid drilling should work well if the embryos have sufficient vigor to become autotrophic with the limited additives in the gel. The problem, to this point, in our lab and as reported by Drew (12), has been that carrot somatic embryos small enough to bulk handle and fluid drill require a carbon source under conditions of sterility and high humidity for an extended period of time to become autotrophic.

Encapsulation

The encapsulation of somatic embryos was first reported by Redenbaugh (33). We have been investigating an encapsulation system using carrot embryos as a model system. The encapsulation involves forming a gel matrix around individual embryos. The attraction as a delivery system is that isolation can be handled mechanically, and that seeding operations can use conventional vacuum seeders. Embryos do germinate and grow out of these beads, but only when supplemented with nutrients as described in the previous section. Embryos small enough for bulk handling and encapsulation still require an external carbon source in sterile conditions with high humidity for an extended period to become autotrophic. Currently, embryo quality is a limiting factor in developing a delivery system.

WHOLE PLANT BIOLOGY

Variation

Variation, either genotypic or phenotypic, exhibited in tissue culture regenerated plants is a limiting factor in the commercialization of these technologies. The existence of somaclonal variation in plants regenerated from a variety of tissue culture systems has been widely reported (13,22,36,38). Some systems and genotypes seem to be very stable, while other systems and genotypes are extremely unstable. Despite a wealth of circumstantial evidence and descriptive examples, a real lack of knowledge exists about the basis and mechanisms of this variation.

Field Performance

Again, there is a real lack of information about the field performance of agricultural plants regenerated from tissue culture. Only when mass cloning systems can produce large numbers of plants economically will field trials be feasible.

SUMMARY

In summary, somatic embryogenesis has several distinct advantages as an in vitro system for the mass cloning of agricultural crops. Several problem areas exist beyond the establishment of an embryogenic culture system. We feel the most noteworthy are embryo quality or development and genetic stability.

REFERENCES

1. Anderson, W.C., G.W. Meagher, and A.G. Nelson (1977) Cost of propagating broccoli plants through tissue culture. Hort. Sci. 12:543-544.
2. Ammirato, P.V. (1974) The effects of abscisic acid on the development of somatic embryos from cells of caraway (*Carum carvi*). Bot. Gaz. 135:328-337.
3. Ammirato, P.V. (1977) Hormonal control of somatic embryo development from cultured cells of caraway. Plant Physiol. 59:579-586.

4. Ammirato, P.V. (1983) Embryogenesis. In Handbook of Plant Cell Culture Vol. I: Techniques for Propagation and Breeding, D.A. Evans, W.R. Sharp, P.V. Ammirato, and Y. Yamada, eds. MacMillan Publ. Co., New York, pp. 82-123.
5. Ammirato, P.V. (1983) The regulation of somatic embryo development in plant cell cultures: Suspension culture techniques and hormone requirements. Bio/technology 3:68-74.
6. Boxus, Ph., M. Quoirin, and J.M. Laine (1977) Large scale propagation of strawberry plants from tissue culture. In Applied and Fundamental Aspects of Plant Cell, Tissue, and Organ Culture, J. Reinert and Y.P.S. Bajaj, eds. Springer-Verlag, Berlin, Heidelberg, New York, pp. 130-143.
7. Conger, B.V., ed. (1981) Cloning Agricultural Plants via in vitro Techniques, CRC Press, Inc., Boca Raton, Florida.
8. Cooke, R.C. (1979) Homogenization as an aid in tissue culture propagation of Platycerium and Davallia. Hort. Sci. 14:21-22.
9. Currah, I.E. (1977) Fluid drilling research. National Vegetable Research Station Report, Wellesbourne, Warwick, England.
10. Currah, I.E., D. Gray, and T.H. Thomas (1973) The sowing of germinated vegetable seeds using a fluid drill. Ann. Appl. Biol. 76:311-318.
11. Damiano, C. (1980) Strawberry micropropagation. In Proceedings of the Conference on Nursery Production of Fruit Plants Through Tissue Culture: Applications and Feasibility, Agricultural Research Science and Education Administration, USDA, Beltsville, Md., pp. 11-22.
12. Drew, R.K.L. (1979) The development of carrot (Daucus carota L.) embryoids (derived from cell suspension culture) into plantlets on a sugar-free basal medium. Hort. Res. 19:79-84.
13. Earle, E.D., and Y. Demarly, eds. (1982) Variability in Plants Regenerated from Tissue Culture, Praeger Publishers, New York.
14. Fujimura, T., and A. Konamine (1979) Synchronization of somatic embryogenesis in a carrot cell suspension culture. Plant Physiol. 64:162-164.
15. Gengenbach, B.G., C.E. Green, and C. Donovan (1977) Inheritance of selected pathotoxin resistance in maize plants regenerated from cell cultures. Proc. Natl. Acad. Sci. USA 74:5113-5117.
16. Gray, D. (1981) Fluid drilling of vegetable seeds. Hort. Rev. 3:1-27.
17. Halperin, W., and D.F. Wetherell (1964) Adventive embryony in tissue cultures of the wild carrot, Daucus carota. Amer. J. Bot. 51:274-283.
18. Halperin, W., and D.F. Wetherell (1965) Ontogeny of adventive embryos of wild carrot. Science 147:756-758.
19. Jones, L.H. (1974) Factors influencing embryogenesis in carrot cultures Daucus carota L. Ann. Bot. 38:1077-1088.
20. Kamada, H., and H. Harada (1981) Changes in the endogenous level and effects of abscisic acid during somatic embryogenesis of Daucus carota L. Plant and Cell Physiol. 22:1423-1429.

21. Krikorian, A.D. (1982) Cloning higher plants from aseptically cultured tissues and cells. Biol. Rev. 57:151-218.
22. Larkin, P.J., and W.R. Scowcroft (1981) Somaclonal variation--a novel source of variability from cell cultures for plant improvement. Theor. Appl. Genet. 60:197-214.
23. Lawrence, Jr., R.H. (1981) In vitro plant cloning systems. Environ. Expt. Bot. 21:289-300.
24. Lundergan, C., and J. Janick (1979) Low temperature storage of in vitro apple shoots (Malus domestica cv. Golden Delicious). Hort. Sci. 14:514-519.
25. McWilliam, A.A., S.W. Smith, and H.E. Street (1974) The origin and development of embryoids in suspension cultures of carrot (Daucus carota). Ann. Bot. 38:243-250.
26. Meins, Jr., F. (1982) The nature of cellular, heritable change in cytokinin habituation. In Variability in Plants Regenerated from Tissue Culture, E.D. Earle and Y. Demarly, eds. Praeger Press, New York, pp. 202-210.
27. Mullin, R.H., and D.E. Schlegel (1976) Cold storage maintenance of strawberry meristem plantlets. Hort. Sci. 11:100-107.
28. Murashige, T. (1974) Plant propagation through tissue cultures. In Annual Review of Plant Physiology, Vol. 25, W.R. Briggs, ed. Annual Reviews, Palo Alto, pp. 135-166.
29. Murashige, T. (1978) The impact of plant tissue culture on agriculture. In Frontiers of Plant Tissue Culture 1978, T.A. Thorpe, ed. The International Association for Plant Tissue Culture, Calgary, pp. 15-26.
30. Murashige, T. (1978) Principles of rapid propagation. In Propagation of Higher Plants Through Tissue Culture, A Bridge Between Research and Application, K.W. Hughes, R. Henke, and M. Constantin, eds. Technical Information Center, U.S. Dept. of Energy, Oak Ridge, pp. 14-24.
31. Nitzsche, W. (1983) Germplasm preservation. In Handbook of Plant Cell Culture. Vol. I: Techniques for Propagation and Breeding, D.A. Evans, W.R. Sharp, P.V. Ammirato, and Y. Yamada, eds. MacMillan Publishing Company, New York, pp. 782-805.
32. Oblesby, R.P. (1978) Tissue cultures of ornamentals and flowers: Problems and perspectives. In Propagation of Higher Plants Through Tissue Culture, A Bridge Between Research and Application, K.W. Hughes, R. Henke, and M. Constantin, eds. Technical Information Center, U.S. Dept. of Energy, Oak Ridge, pp. 59-61.
33. Redenbaugh, K., J. Nichol, M.E. Kossler, and B. Paasch (1984) Encapsulation of somatic embryos for artificial seed production. In Vitro 20:256.
34. Reinert, J. (1958) Untersuchungen uber die Morphogenese an Gewebekulturen. Ber. Dtsch. Bot. Ges. 71:15.
35. Reinert, J. (1959) Uber die Kontrolle der Morphogenese und die Induktion von Adventureembryonen an Gewebekulturen aus Karotten. Planta 53:318-333.
36. Scowcroft, W.R. (1977) Somatic cell genetics and plant improvement. Adv. Agron. 29:39-81.

37. Sharp, W.R., M.R. Sondahl, L. Caldas, and S.B. Maraffa (1980) The physiology of in vitro asexual embryogenesis. In Horticultural Reviews, Vol. 2, J. Janick, ed. AVI, Westport, Connecticut, pp. 268-310.
38. Skirvin, R.M. (1978) Natural and induced variation in tissue cultures. Euphytica 27:241-266.
39. Sharp, W.R., P.O. Larsen, E.F. Paddock, and V. Raghaven, eds. (1979) Plant Cell and Tissue Culture: Principles and Applications, Ohio State University Press, Columbus, Ohio, 892 pp.
40. Smith, S.M., and H.E. Street (1974) The decline of embryogenic potential as callus and suspension cultures of carrot (Daucus carota L.) are serially subcultured. Ann. Bot. 38:223-241.
41. Steward, F.C. (1958) Interpretations of the growth from free cells to carrot plants. Amer. J. Bot. 45:709-713.
42. Street, H.E., and L.A. Withers (1974) The anatomy of embryogenesis in culture. In Tissue Culture and Plant Science, H.E. Street, ed. Academic Press, London, pp. 71-100.
43. Syono, K. (1965) Changes in organ forming capacity of carrot root calluses during subculture. Plant and Cell Physiol. 6:403-419.
44. Tisserat, B., E.B. Esan, and T. Murashige (1979) Somatic embryogenesis in angiosperms. In Horticultural Reviews, Vol. I, J. Janick, ed. AVI, Westport, Connecticut, pp. 1-78.
45. Walker, K.A., M.L. Wendeln, and E.G. Jaworski (1979) Organogenesis in callus tissue of Medicago sativa. The temporal separation of induction processes from differentiation processes. Plant Sci. Lett. 16:23-30.
46. Warren, G.S., and M.W. Fowler (1977) A physical method for the separation of various stages in the embryogenesis of carrot cell cultures. Plant Sci. Lett. 9:71-76.
47. Wetherell, D.F. (1978) In vitro embryoid formation in cells derived from somatic plant tissues. In Propagation of Higher Plants Through Tissue Culture, A Bridge Between Research and Application, K.W. Hughes, R. Henke, and M. Constantin, eds. Technical Information Center, U.S. Dept. of Energy, Oak Ridge, pp. 102-124.

BIOREACTOR TECHNOLOGY FOR PLANT PROPAGATION

D.J. Styer

DNA Plant Technology Corporation
2611 Branch Pike
Cinnaminson, New Jersey 08077

INTRODUCTION

The application of tissue culture methods to the propagation of plants has dramatically altered the way many plants are routinely propagated. As tissue culture technology continues to expand and improve, these techniques will be applied to a growing list of plant species. Most of the species that are presently being propagated on a large scale through tissue culture are herbaceous ornamentals and fruits (Tab. 1). This technology provides the opportunity to produce uniform, disease-free plants where lack of uniformity and/or virus infestations had previously been a serious problem. While tissue culture propagation is not likely to be applied in the near future to crops that are easily and satisfactorily propagated through true seed, there are a large number of species, many of which are important agronomic and horticultural crops, that are reproduced asexually and could benefit from this technology. Most cultivars of crops such as banana, cassava, potato, pineapple, and sugarcane are propagated vegetatively by dividing stems, suckers, tubers, crowns, or other plant parts that sprout to form new plants. Unfortunately, conventional vegetative propagation practices can lead to the spread of plant pathogens, resulting in loss of yield and quality. Through tissue culture, pathogen-free stock plants can be generated, maintained, and increased for cultivation (29). These same benefits apply to fruit trees and ornamentals that produce phenotypically heterogeneous true seed and are propagated asexually in order to reproduce superior individuals.

In addition to plants that are normally vegetatively propagated, there are many other species for which high-frequency propagation methods would be valuable. For example, the ability to vegetatively reproduce superior forest trees, which do not produce supe-

Tab. 1. Plant genera that are mass-propagated by tissue culture. (Adapted from George and Sherrington, Ref. 8a.)

Aechmea	Episcia	Narcissus
Amaryllis	Ficus	Nephrolepsis
Ananas	Freesia	Nerine
Anthurium	Gerbera	Ophiopogon
Asparagus	Gladiolus	Pelargonium
Begonia	Gloxinia	Petunia
Browallia	Guzmania	Philodendron
Caladium	Haworthia	Pinguicula
Chrysanthemum	Hemerocallis	Platycerium
Clematis	Hippeastrum	Saitpaulia
Cordyline	Hosta	Sinningia
Cyclamen	Hyacinthus	Streptocarpus
Davallia	Iris	Torenia
Dicentra	Kalanchoe	
Dracaena	Lilium	

rior progeny through seed, would be extremely useful. Conventional methods are generally not appropriate, as cuttings fail to root successfully. Methods for the in vitro propagation of several forest trees, including Pseudotsuga menziesii, Sequoia sempervirens, and numerous Pinus spp., have been developed recently (31). Similarly, for crops such as coconut and oil palm for which there are no satisfactory methods of conventional vegetative propagation (23) and genetic variability in the seed crop is high, tissue culture methods allow propagation of superior phenotypes. Methods have been developed for the high-frequency in vitro propagation of oil palm (12). For seed crops, there are cases where rapid tissue culture propagation of a particular genotype is warranted. Hybrid plants that arise from crosses that are tedious or difficult to make may be quickly increased for evaluation of the hybrids' performance. The rapid production of parents for hybrid seed synthesis would allow large-scale hybrid seed production without the normal delay for parental seed increase (3). This could result in a time and cost savings for the hybrid seed producer. Another application of in vitro plant cloning is the reproduction of unique somatic hybrids or other plants developed in the laboratory. These plants may be sterile, or exhibit genetic segregation during the sexual cycle, eliminating the genetic combinations intentionally created in vitro. In vitro propagation could permit the reproduction of identical copies of these unique hybrids.

PLANT PROPAGATION IN VITRO

The three major pathways for producing plants in vitro are organogenesis, shoot-tip culture, and embryogenesis. Organogenesis is generally not suitable for cloning plants on a large scale because it usually occurs at low frequency, and plant production is a labor-intensive process. Callus tissue growth is induced from explant material (pieces of leaf, stem, roots, etc.), followed by shoot initiation and growth. Each shoot must be excised from the callus piece, placed on one medium for further shoot growth, then transferred again for root formation. At each step, individual plants are the units that are transferred. In addition to the labor-intensive nature of the process, plants resulting from organogenesis may exhibit a range of genetic changes (somaclonal variation) resulting in altered phenotypes (6).

Shoot-tip culture is used widely for propagating plants because moderately high multiplication rates are achieved. In addition, stock plants can be freed of systemic viruses and other pathogens by culturing very small meristem explants (30). Once the plants are pathogen-free, there is little opportunity for recontamination during the multiplication process. Shoot-tip culture propagation is sufficiently productive that millions of plants can be generated annually. Genetic variation among shoot tip-cultured plants occurs at very low frequencies (10). Unfortunately, shoot-tip culture is a highly labor-intensive and expensive means of propagation. Each new shoot must be excised from the mother plant and individually placed either on multiplication medium or on rooting medium. Thus, the procedure has generally had its application limited to plant species for which the economic return is high.

The production of plants via somatic embryogenesis has been recognized as a method of producing large numbers of plants efficiently (1). The propagation of oil palm through somatic embryos is currently the best example of such a propagation scheme (12). Somatic embryogenesis is attractive for scale-up because it occurs at a high frequency and many somatic embryos are formed per explant. In many species, both the growth of the embryogenic tissue and the subsequent development of the embryos occur in liquid medium. Thus, hundreds of individuals can be handled at once. This provides a significant labor savings compared to shoot-tip culture. In addition, the product of somatic embryogenesis is an embryo that is capable of developing into a regenerated plant; separate shoot growth and rooting steps are eliminated.

USE OF BIOREACTORS

The ability to grow embryogenic cell suspensions and to produce normal embryos on a large scale is mandatory if somatic embryogenesis is to be used as a propagation tool. Bioreactors have been used for growing plant cells on a large scale (16,19) and provide the basis for a cloning technology. However, scale-up of a cloning system does not necessarily imply the use of the large bioreactors (1-500,000 l) used in microbial fermentations. Plants are counted as individuals, and a bioreactor that could produce 10-100,000 plants would be practical. A typical program for propagating plants by somatic embryogenesis is outlined in Fig. 1. An elite plant is identified (Step 1), then the donor tissue is excised and placed in culture (Step 2). Callus is produced and is moved into liquid culture medium to generate a cell suspension (Step 3). This cell suspension serves as the inoculum for the first bioreactor (Step 4). In this bioreactor, the cells continue to proliferate in the auxin-containing culture medium. Once an appropriate cell density has been achieved, some cells are moved into a second bioreactor that contains auxin-free culture medium and that operates under a differ-

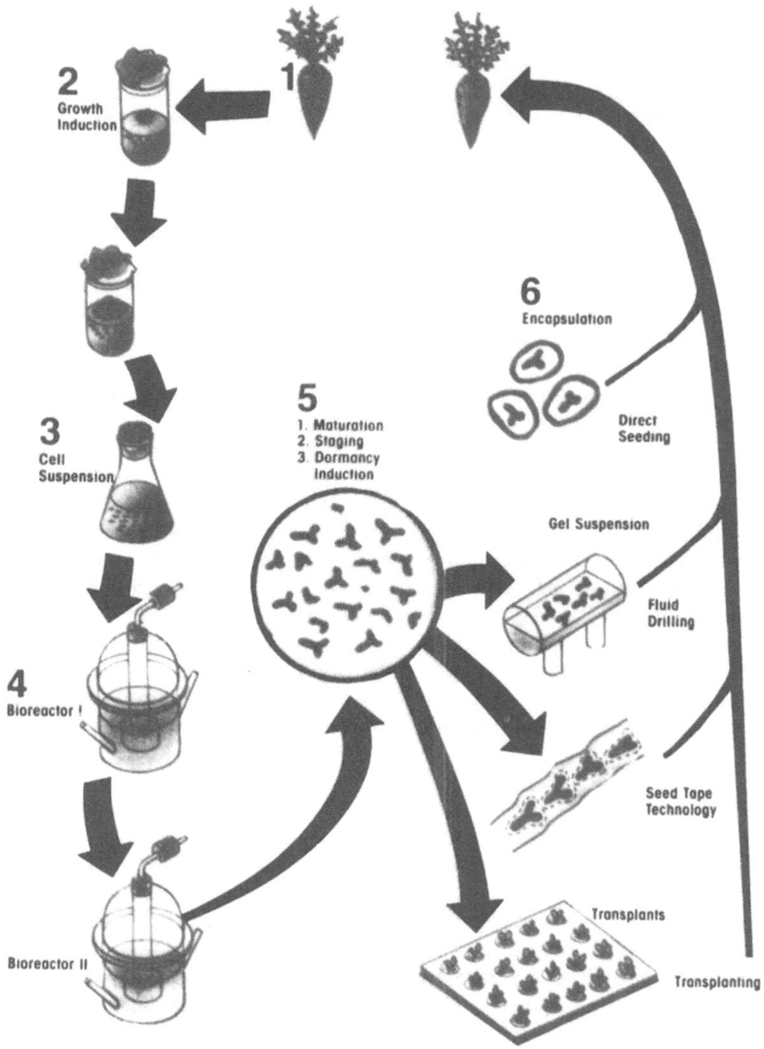

Fig. 1. Plant propagation using bioreactor technology (see text for description of steps).

ent set of environmental conditions. Under these new conditions the cells are induced to form somatic embryos (Step 5). After the embryos mature, they may enter any one of several delivery systems.

Allowing the embryos to grow directly into plantlets is the simplest delivery system. These plants can enter a nursery or go directly into the field as transplants. Alternatively, the somatic embryos could be planted by using fluid drilling systems (4,7,9). The gel matrix that surrounds each embryo could be supplemented with nutrients or protective compounds. For example, the addition of nutrients to the fluid drilling gel increases the growth rate of carrot seedlings (8). For fluid-drilled tomato seeds, Ohep et al. (22) reported that ethazol, ethazol plus thiophanate-methyl, and fenaminosulf controls seedling damping off. The development of seed tapes or encapsulated embryos requires the most sophisticated technology (5,21). In both cases, the embryos must be dormant for an appropriate time interval, and the material surrounding the embryo should be enriched with various compounds. Mechanisms to maintain dormancy of the embryo during storage and to break dormancy when the embryo is in the appropriate growing conditions are necessary. The matrix should include nutrients for embryo germination, antimicrobial compounds to protect both the embryo and the nutrients from decay organisms, and herbicides and insecticides to protect the emerging plant. Protection from physical damage and from desiccation are also important to maintain embryo viability. All these requirements must be fulfilled in a package small enough to be handled as a seed.

Bioreactor Designs

Several bioreactor designs have been employed to grow plant cells. Of at least 10 general types of novel bioreactors that have been identified (18), several are suitable for plant cells. The choice of design is determined by the plant cell type used and the purpose of the experiment. For example, the production of secondary plant products for industrial purposes has different requirements than the control of growth and development of cells into somatic embryos. Hollow fiber bioreactors or bioreactors that utilize immobilized cells are more suitable for secondary product synthesis than embryo development. Production of phenolic compounds by carrot (13) or soybean (28) cells trapped in hollow fiber bioreactors have been demonstrated. Brodelius and co-workers (2) have used immobilized cell lines of *Morinda citrifolia* for the de novo synthesis of anthroquines. The formation of the indole alkaloid ajmalicine from the distant precursors tryptamine and secologanin by *Catharanthus roseus*, and the biotransformation of the steroid digitoxin to digoxin by immobilized cells of *Digitalis lanata* have also been reported (2).

Air lift bioreactors are mechanically simple because both the aeration and agitation of the medium are provided by forced air entering the bottom of the vessel. An early design was simply a

20 l carboy that had an aeration tube extending to the bottom of the carboy (32). Kurz (15) designed a cylindrical vessel in which single, large air bubbles filled the entire cross-section of the bioreactor. More recent air lift designs employ draft tubes to control the flow of the culture medium (18,19).

Stirred tank designs require mechanical agitation by either paddles, impellers, or magnetic stir bars. The use of a magnetic coupling to agitate the culture eliminates the drive shaft and associated seal that blade stirrers require. Magnetic stirrers can operate at higher rpms than blades without shearing the cells.

Miller et al. (20) described and utilized a magnetically stirred bioreactor, termed a Phytostat, for continuous cultures of Glycine max. Constabel et al. (16) also used the bioreactor for Haplopappus gracilis. Steady state growth rates were achieved in both experiments. Wilson et al. (34) designed a bioreactor that could operate either as a chemostat or turbidostat. Constant values of cell numbers, packed cell volume, dry weight, and chemical composition were obtained for a culture of Acer pseudoplantanus (14,34).

Growth Kinetics

The growth of plant cells in bioreactors of any design can be described by the equations developed for microbial fermentation. Cells can be grown as batch, semi-continuous, or continuous cultures depending on the product. In batch cultures, the cells exhibit a lag phase, then enter the log phase of exponential growth. This is followed by the stationary phase, the period of secondary product accumulation. Throughout the growth period, the conditions of the culture are changing (Fig. 2). The cell mass increases, the concen-

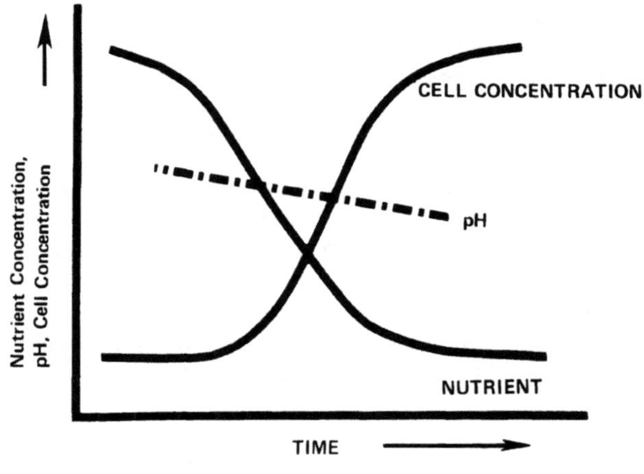

Fig. 2. Nutrient concentrations, pH, and cell concentration vs. time in an idealized growth curve.

tration of all the nutrients decreases, and often the pH value decreases. Batch cultures are useful if changing conditions are needed for product formation or, at least, if a continuously changing environment is not detrimental to the process of interest.

In most cultures, growth conditions as nearly optimal as possible are desired throughout the course of the culture in order to produce the maximum quantity of cells or of a compound produced by cells. These conditions will only occur for a short period in the growth curve of a batch culture. In addition, the time spent filling, draining, and re-sterilizing a large bioreactor can be substantial. For microbial fermentations, this time can amount to 8 of every 24 hours.

Semi-continuous cultures are operated as batch cultures until the growth cycle is in late log or early stationary phase. Then 50-80% of the culture is drained out and replaced with fresh medium. The cells remaining in the bioreactor are the inoculum for the next growth cycle. The kinetics of these cultures are cyclical with one log phase following another. Semi-continuous cultures require less time for draining and refilling the bioreactor than a batch culture, and have a reduced lag time because of the large inoculum. The growth characteristics of a semi-continuous culture of *Arachis hypogea* were reported by Verma and Van Huystee (33).

The efficiency of a bioreactor can be increased by continuous culture. The initial filling and inoculation of the bioreactor is identical to the process for a batch culture. The growth of the cells proceeds from lag phase into the log phase. Then fresh medium is introduced at a low rate while the same volume of spent medium and cells is removed. The cell population then reaches a steady-state level that is determined by the nutrient feed rate. Continuous culture provides more uniform growth conditions that are stable throughout the culture (34). Continuous cultures also are constantly productive since the draining and refilling of the bioreactor are not discrete steps but occur as ongoing processes. Besides the sustained growth of the culture and the continuous output of product, continuous culture has the advantage that almost any physiological state of the cells can be maintained. The nutrient feed rate determines the cell population and the state of the cells in the bioreactor. For example, high nutrient feed rates produce conditions similar to those of the early log phase of a batch culture; the high nutrient levels allow the cells to grow actively. However, the high feed rate must be balanced by a high rate of medium removal and, thus, a high cell washout rate. The cell population under these conditions is relatively low.

To produce a high cell population, the cell washout rate must be low. This limits the nutrient feed rate and the concentration of nutrients in the bioreactor. Thus at high cell populations, the cells are not actively growing, and an early stationary phase culture is established. The relationship between nutrient feed rate and cell populations imposes its own restrictions on the performance of a bioreactor, and in most cases high populations of rapidly dividing cells cannot be achieved. However, the use of a spin filter bioreactor (Fig. 3) is one means of uncoupling the strict relationship between cell density and feed rate. The spin filter bioreactor contains a filter in the bioreactor that spins via a magnetic coupling to a stir plate. The spinning of the filter provides agitation of the medium without generating the shear observed with blade stirrers. The spinning also prevents plugging of the filter. Filters of different pore sizes can be used to exclude cells of different sizes. The spin filter provides an added dimension to a bioreactor because medium can be removed through the filter without cell loss. For example, assume steady-state conditions exist in a continuous culture with a 1X rate of nutrient feed. When the feed rate is increased to 2X in a conventional bioreactor, the cell population drops because of the increased washout rate. In a spin filter bioreactor, the washout rate can be maintained at 1X, and the

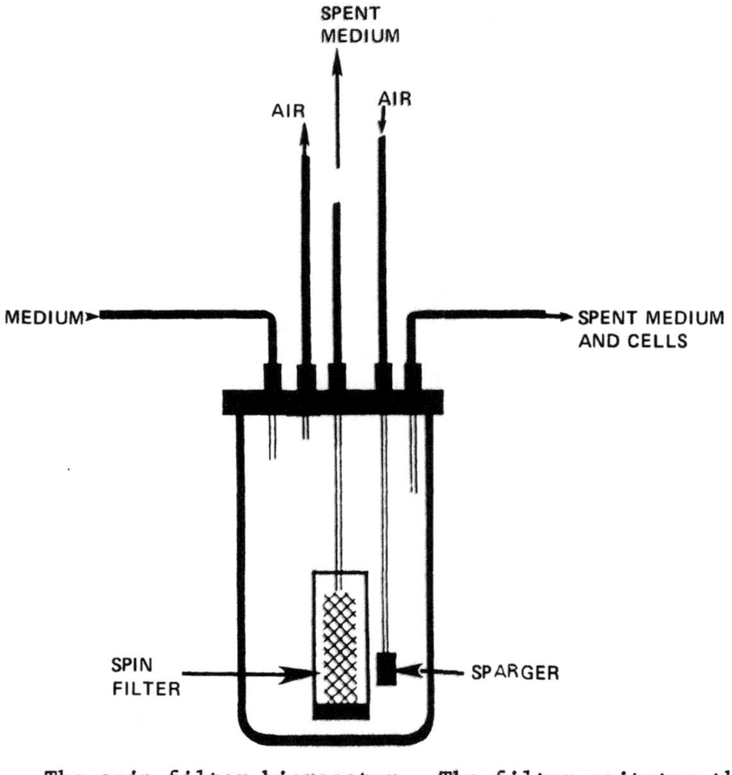

Fig. 3. The spin filter bioreactor. The filter agitates the medium and allows withdrawal of spent medium.

other half of the spent medium can be removed through the filter. This results in a cell population twice as dense as the population during a 1X rate of nutrient feed, but the cells are in the same physiological state. In the spin filter bioreactor, the physiological state of the cells is determined by the actual washout rate; the cell population is determined by the ratio of medium removed through the filter to the washout rate. In this manner, any cell state can be stabilized at high cell populations. The ability to produce increased cell concentrations of microbial and mammalian cells has been demonstrated (11).

Experimental Results

I chose carrot (*Daucus carota* L.) as a model species because of the well-characterized somatic embryogenesis in this crop (25). Batch cultures were conducted initially to provide experience with the bioreactors and to provide the kinetic data needed for continuous cultures. A typical growth curve of carrot cell suspension growing on 2,4-dichlorophenoxyacetic acid (2,4-D) medium is shown in Fig. 4. The shape of this curve is identical to microbial growth curves. However, an important difference is the longer time scale observed with carrot cells. The lag phase lasted 7 days and the log phase was 9 days in this experiment.

The length of the lag phase is influenced by the inoculum concentration (24). At a higher inoculum, the lag phase can be reduced to less than one day (Fig. 5). The experiments shown in Fig. 4 and

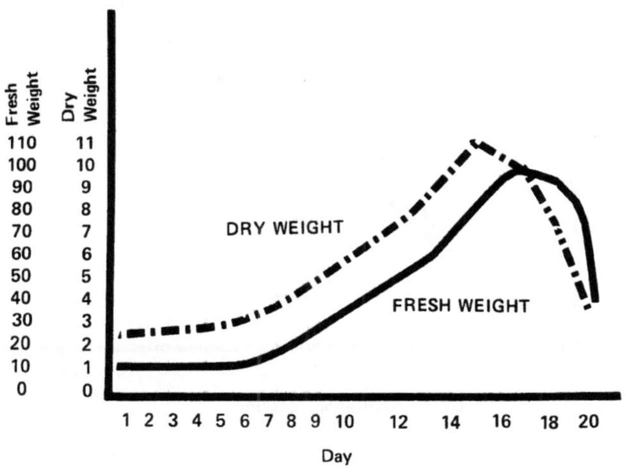

Fig. 4. Cell fresh weight (g/l) and dry weight (g/l) vs. days of a carrot cell line grown in a spin filter bioreactor. Inoculum was 7.7 g/l fresh weight.

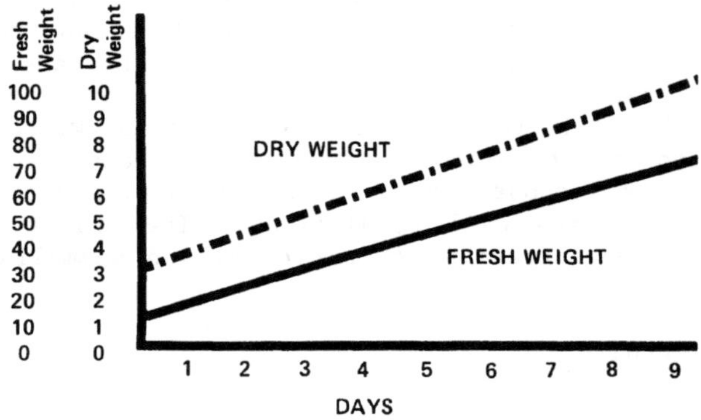

Fig. 5. Cell fresh weight (g/l) and dry weight (g/l) vs. days of a carrot cell line grown in a spin filter bioreactor. Inoculum was 16.0 g/l fresh weight.

5 employed the same cell line, and both reached the same maximum cell density (10.3 mg/ml dry weight). Batch growth data can be analyzed in several ways (26) to provide useful information for continuous cultures. Luedeking and Piret (17) described a graphical method of analysis in which the growth of the cells [d(gm/l)/dt] is plotted against cell concentration (g/l) (Fig. 6). This curve yields the maximum growth rate of the cells. In both experiments,

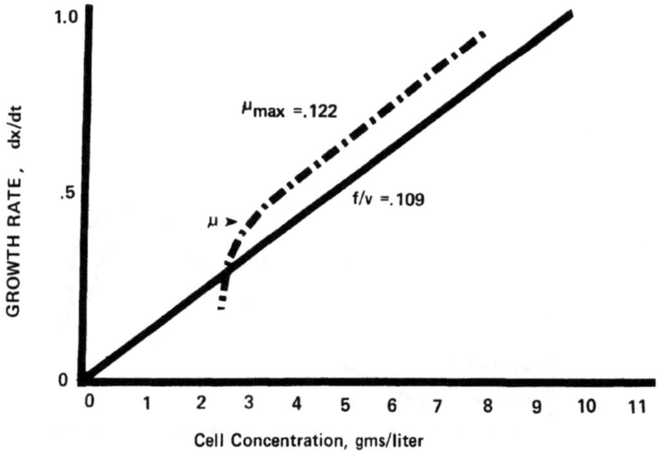

Fig. 6. Growth rate ($dx/dt=\mu$) vs. cell concentration (x) plotted for experiments graphed in Fig. 4 and 5. The maximum growth rate (μ max) of carrot cells in both experiments was 0.122 cells/cell/day. The nutrient feed rate (f/v) of 0.109 volumes/day was used in continuous cultures.

the maximum growth rate was identical (0.122 cells/cell/day). The identity of the 2 values suggests that neither experiment contained inhibiting conditions that introduced artifacts. This analysis is important because the maximum growth rate cannot be exceeded by the dilution rate in continuous culture. Using a dilution rate of 0.109 volumes/day, a continuous culture was maintained for 25 days in the spin filter bioreactor.

The spin filter can be used as the sole means of removing spent medium (Fig. 7). In this design, there is no cell washout and the cell density continuously increases. It is difficult to maintain a constant environment under these growth conditions, although such a system has been described by Wilson et al. (34). However, if culture medium that stimulates embryo differentiation rather than cell proliferation is used in the bioreactor, the number of cell clusters

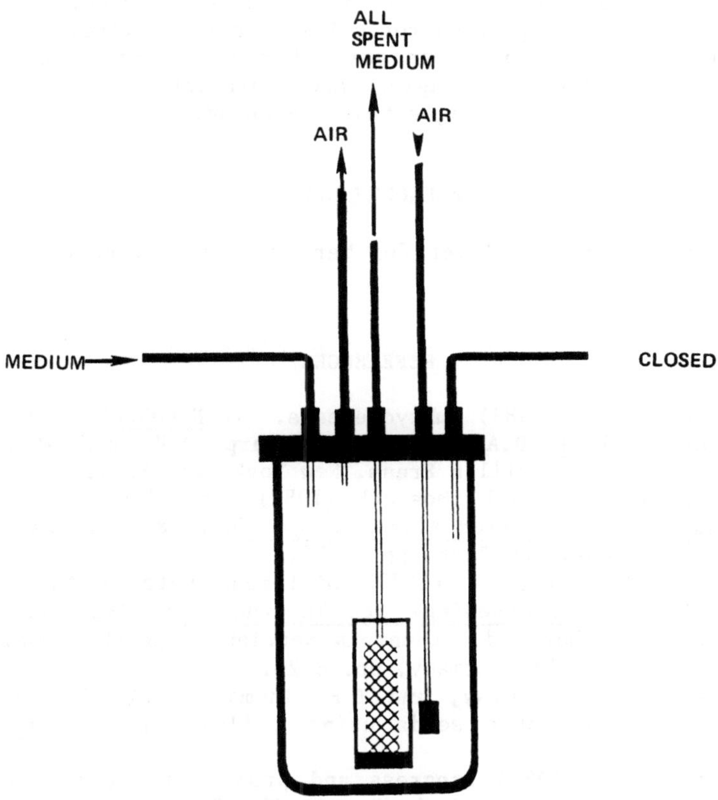

Fig. 7. The spin filter bioreactor designed for medium perfusion. All the spent medium is removed through the filter with no cell removal.

remains constant. The increases in cell number and cell mass that occur do so as part of organized structures rather than a proliferating cell population. This reduces the effect of the increasing cell mass on the turbidity and viscosity of the culture. It also allows complete exchange of the culture medium in a time frame that is independent of the cells' growth rate. This is in contrast with other bioreactors that require a dilution rate that is less than the maximum growth rate of the cells.

SUMMARY

The ability to produce somatic embryos on a large scale offers new opportunities for propagating many crops. A fully developed cloning system can be applied to the propagation of those crops for which high frequency somatic embryogenesis has been reported (1). It can be extended to other crops as the research on the induction and development of somatic embryos is conducted with more plant species. The ability to grow carrot cells in a spin filter bioreactor, as presented here, is only one part of a fully functional cloning system. The development, staging, and manipulation of embryos on a commercial scale are challenges that lie ahead.

ACKNOWLEDGEMENT

I thank Dr. Sally Miller for her critical review of the manuscript.

REFERENCES

1. Ammirato, P.V. (1983) Embryogenesis. In Handbook of Plant Cell Culture, Vol. 1, D,A. Evans, W.R. Sharp, P.V. Ammirato, and Y. Yamada, eds. Macmillan Press, New York, pp. 82-123.
2. Brodelius, P., and K. Mosbach (1982) Immobilized plant cells. In Advances in Applied Microbiology, Vol. 28, A.I. Laskin, ed. Academic Press, New York, pp. 1-26.
3. Corriols-Thevenin, L. (1979) Different methods in asparagus breeding. In Proceedings 5th International Asparagus Symposium, G. Reuther, ed. Eucarpia Section, Vegetables Geisenheim Forschungsanstalt, Germany, pp. 8-20.
4. Currah, I.E., D. Gray, and T.H. Thomas (1974) The sowing of germinated vegetable seeds using a fluid drill. Ann. Appl. Biol. 76:311-318.
5. Durzan, D.J. (1980) Progress and promise in forest genetics. In Paper Science and Technology - The Cutting Edge, Institute of Paper Chemistry, Appleton, Wisconsin, pp. 31-60.
6. Evans, D.A., W.R. Sharp, and H.P. Medina-Filho (1984) Somaclonal and gametoclonal variation. Amer. J. Bot. 71:756-774.

7. Finch-Savage, W.E. (1984) The effects of fluid drilling germinating seeds on the emergence and subsequent growth of carrots in the field. J. Hort. Sci. 59:411-417.
8. Finch-Savage, W.E., and C.J. Cox (1982) Effects of adding plant nutrients to the gel carrier used for fluid drilling early carrots. J. Agric. Sci. 99:295-303.
8a. George, E.F., and P.D. Sherrington (1984) Plant Propagation by Tissue Culture, Exegetics Ltd., Eversley, Basingstoke, Hants, England.
9. Gray, D. (1981) Fluid drilling of vegetable seeds. Hort. Rev. 1:1-27.
10. Hu, C.Y., and P.J. Wang (1983) Meristem, shoot tip and bud cultures. In Handbook of Plant Cell Culture, Vol. 1, D.A. Evans, W.R. Sharp, P.V. Ammirato, and Y. Yamada, eds. Macmillan Publishing Co., New York, pp. 177-227.
11. Johnson, D.E., L.R. Woodland, C.J. Kensler, and P. Himmelfarb (1972) Apparatus for cell culture. U.S. Patent No. 3,642,632.
12. Jones, L.H. (1983) The oil palm and its clonal propagation by tissue culture. Biologist 30:181-188.
13. Jose, W., H. Pedersen, and C.-K. Chin (1983) Immobilization of plant cells in a hollow fiber reactor. Annals N.Y. Acad. Sci. 413:409-412.
14. King, P., and H.E. Street (1973) Growth patterns in cell cultures. In Plant Cell and Tissue Culture, H.E. Street, ed. Blackwell Scientific, Oxford, pp. 269-337.
15. Kurz, W.G.W. (1971) A chemostat for growing higher plant cells in single cell suspension cultures. Exp. Cell Res. 64:477.
16. Kurz, W.G.W., and F. Constabel (1981) Continuous culture of plant cells. In Continuous Cultures of Cells, Vol. 2, P.H. Calcott, ed. CRC Press, Boca Raton, Florida, pp. 141-157.
17. Luedeking, R., and E.L. Piret (1959) Transient and steady states in continuous fermentation: Theory and experiment. J. Biochem. Microbiol. Technol. Eng. 1:431.
18. Margaritis, A., and J.B. Wallace (1984) Novel bioreactor systems and their applications. Bio/Technology 2:447-453.
19. Martin, S.M. (1980) Mass culture systems for plant cell suspensions. In Plant Tissue Culture as a Source of Biochemicals, E.J. Staba, ed. CRC Press, Boca Raton, Florida, pp. 149-166.
20. Miller, R.A., J.P. Shyluk, O.L. Gamborg, and J.W. Kirkpatrick (1968) Phytostat for continuous culture and automatic sampling of plant cell suspensions. Science 159:540.
21. Murashige, T. (1980) Plant growth substances in commercial uses of tissue culture. In Plant Growth Substances, F. Skoog, ed. Springer-Verlag, New York, pp. 426-434.
22. Ohep, J., R.T. McMillian, Jr., H.H. Bryan, and D.J. Cantliffe (1984) Control of damping-off of tomatoes by incorporation of fungicides in direct-seeding gel. Plant Disease 68:66-67.
23. Purseglove, J.W. (1972) Tropical Crops: Monocotyledons, Longman Group Ltd., Harlow, Essex, United Kingdom.
24. Puhan, Z., and S.M. Martin (1971) The industrial potential of plant cell culture. Prog. Ind. Microbiol. 9:13.

25. Raghavan, V. (1983) Biochemistry of somatic embryogenesis. In Handbook of Plant Cell Culture, Vol. 1, D.A. Evans, W.R. Sharp, P.V. Ammirato, and Y. Yamada, eds. Macmillan Press, New York, pp. 655-671.
26. Ricica, J., and P. Dobersky (1981) Complex systems. In Continuous Cultures of Cells, Vol. 1, P.H. Calcott, ed. CRC Press, Boca Raton, Florida, pp. 63-96.
27. Sheehan, T.J. (1983) Recent advances in botany, propagation, and physiology of orchids. In Horticultural Reviews, Vol. 5, J. Janick, ed. AVI Publishing Co., Westport, Connecticut, pp.279-315.
28. Shuler, M.L., O.P. Sahai, and G.A. Hallsby (1983) Entrapped plant cell tissue cultures. Annals N.Y. Acad. Sci. 413:373-382.
29. Slack, S.A. (1980) Pathogen-free plants by meristem-tip culture. Plant Disease 64:15-17.
30. Styer, D.J., and C.K. Chin (1983) Meristem and shoot tip culture for propagation, pathogen elimination, and germplasm preservation. In Horticultural Reviews, Vol. 5, J. Janick, ed. AVI Publishing Co., Westport, Connecticut, pp. 221-277.
31. Thorpe, T.A., and S. Biondi (1984) Conifers. In Handbook of Plant Cell Culture, Vol. 2, W.R. Sharp, D.A. Evans, P.V. Ammirato, and Y.Yamada, eds. Macmillan Publishing Co., New York, pp. 435-470.
32. Tulecke, W., and L.G. Nickell (1959) Production of large amounts of plant tissue by submerged culture. Science 130:863.
33. Verma, D.P.S., and R.B. Van Huystee (1971) Derivation, characteristics, and large scale culture of a cell line from Arachis hypogaea L. cotyledons. Exp. Cell Res. 69:402.
34. Wilson, S.B., P.J. King, and H.E. Street (1971) Studies of the growth in culture of plant cells. XII. A versatile system for the large scale batch or continuous culture of plant cell suspension. J. Exp. Bot. 22:177.

TREASURE YOUR EXCEPTIONS

Peter S. Carlson

Crop Genetics International N.V.
7170 Standard Drive
Dorsey, Maryland 21076

INTRODUCTION

In 1908, William Bateson admonished his fellow geneticists to "Treasure your exceptions! When there are none, the work gets so dull that no one cares to carry it further. Keep them always uncovered and in sight. Exceptions are like the rough brickwork of a growing building which tells that there is more to come and shows where the next construction is to be." From the time of Mendel, the recovery, characterization, and analysis of variability has played a central role in establishing the axioms and assumptions within which geneticists continue to view the world. Visually identifiable traits have been collected and named by geneticists for over a century. Among the characters which have been collected and named are included variants affecting plant color such as albina, xantha, and viridis types, and variants affecting plant morphology such as erectoides, speltoids, and dwarfism. Geneticists have developed methods to produce variants de novo, further characterizing them as macro, micro, or pseudoallic. By following the transmission of this variability, various modes of plant reproduction were defined including apomixis, allomixis, and automixis. Variant types were taxonomically differentiated in categories: the amorphs, hypomorphs, hypermorphs, antimorphs, and neomorphs. There was a realization that developmental processes eliminated certain classes of variants and skewed the spectrum of recovered types, resulting in haplontic and diplontic selection. The recovered phenotype of a variant is a multiple bioassay for the functioning of a gene, resulting in pleiotropy. The role of the environment on gene function was recognized, resulting in terms such as dauermodification.

As lovely as all these words and terms sound it is obvious that they describe the phenomonology of genetics; they do not imply an understanding of the underlying biological processes. They are descriptive, not analytical. They represent the recognition of an event or process and its characterization based on possible experimental manipulations. New experimental manipulations will continue to enlarge upon the phenomonology of the discipline and will increase the understanding of causal processes. In vitro manipulations, which are only a discrete subset of these new and developing techniques, are causing yet more lovely terms to be coined in attempts to describe an expanding data base. Consider somaclonal and gametoclonal variability as examples of this trend. In vitro manipulations present new ways to produce, recover, and analyze variability, and offer opportunities to develop methods beyond the traditional paradigms and procedures of plant and microbial genetics. These opportunities demand increased sophistication in the selection of variability. Our goal is not to find comfort in a descriptive new term or label, but to increase the understanding of plant biology.

VARIABILITY FROM CULTURE

During its early use as a genetic tool, plant cell and tissue culture was viewed as a technique for facilitating the recovery of variants in plant systems. Numerous variants have been recovered using tissue culture selection procedures, and a number of recent reviews in this area have been written (16,24). Isolation of variants resistant to normally toxic or lethal levels of a substance was relatively straightforward and resulted in mutants resistant to amino acid analogs (9,22,25,26), herbicides (5,11), salt and heavy metal tolerance (18), antibiotics (28), toxins (8), growth regulators (27), and other compounds. The majority of selection procedures involve exposing large numbers of cells to a lethal or toxic level of the selective agent and rescuing surviving cells. Although successful, this approach is not without its problems (14,21).

Early researchers in the area of in vitro mutation and selection were puzzled to find that mutant recovery seemed to be somewhat higher than expected. It rapidly became obvious that variants could be easily isolated from cell cultures, without any mutagenic treatment, and that plant cell cultures, and indeed plants themselves, were inherently genetically unstable (6,12,15). Several possible sources of genetic instability have been demonstrated in plant tissue culture systems (22) and others no doubt exist.

In spite of the numerous variants isolated from culture, few have been genetically characterized. Analysis of variability from culture, where genetically possible, has indicated that one or few genes are responsible for nuclear genetic traits (1-4,10,13,17,19). Unfortunately, traits of agronomic interest often are polygenic in nature, e.g., yield, disease resistance, salt tolerance, etc. It is apparent that new selection procedures which are <u>not</u> based on selective cell death in culture will need to be devised to obtain alterations in multigenic traits. In this presentation, I will attempt a brief consideration of new selection methodologies successfully utilized with microorganisms with potential importance for in vitro cultures.

SELECTION FOR A COMPLEX MULTIGENETIC CHARACTER

In cultures of yeast, ethanol is produced at low concentration because of its toxicity to the producing microorganism. The toxic levels (6-9% ethanol) are a biological limit which directly affects the economics of ethanol production. Increases in the tolerance of yeast to ethanol would have significant cost benefits. Oliver and his associates (20) have carefully examined this problem and have developed an elegant selection methodology to enrich yeast populations, over time, for genotypes able to accomplish this goal.

When ethanol is added to a log phase culture of yeast, the growth of the culture is immediately inhibited. This inhibitation is due to: a) a decrease in the growth rate of the living cells, and b) cell death. Ethanol production per se is dependent upon: a) growth rate, b) viability, and c) rate of fermentation; each with a different ethanol inhibition constant. Rate of fermentation is the most ethanol tolerant. In none of these 3 processes is the mechanism of ethanol inhibition known.

Defined single gene mutations which increase ethanol sensitivity of yeast can provide information which may help elucidate the mechanisms of alcohol toxicity. However, mutants have made no direct contribution to the goal of increasing ethanol tolerance of yeast and reducing the cost of alcohol production. The complexity of the toxic action of ethanol on yeast means that screening or selection for tolerant mutants by the conventional assay of growth or no growth in the presence of a selective agent is unlikely to be successful. Only small quantitative increases in ethanol tolerance are likely to be obtained by these techniques, and these variants will probably result from multiple, rather than single, mutations.

Continuous selection techniques could be used to obtain mutants with small selective advantages over the wild-type strain. However, the complexity of the inhibitory effects of ethanol on yeast makes it difficult to design a workable selection regime. Because of these difficulties, Oliver devised a workable selection system in which the intensity of selection was determined by the fermentation rate itself via a feedback circuit rather than cell death.

In this system, an infrared analyzer was used to continuously monitor the fermentation activity of the culture by determining the concentration of carbon dioxide in the exit gas. When the carbon dioxide concentration equalled or exceeded the preset value on the controller, a relay closed, switching on a peristaltic pump and introducing a concentrated ethanol solution into the culture vessel. When addition of the alcohol had reduced the carbon dioxide concentration of the exit gas to below the value set, the relay opened and the ethanol pump was switched off. The ethanol was then gradually diluted out of the fermenter vessel and the rate of carbon dioxide production increased. When the carbon dioxide concentration in the exit gas exceeded the value, the pump was switched on again. Hence, the system was functionally analogous to a turbidostat. However, it was the culture's rate of carbon dioxide production, rather than its turbidity, which was held constant and the supply of inhibitor, rather than nutrients, which was regulatory. Metabolism, not growth, controlled selection.

Improvement in the culture's ethanol tolerance is correlated with an increase in the frequency of the ethanol pump's activity. In one experiment the pump did not act on at all for the first 12 days of the run. After that, it was activated only occasionally until, after 27 days, there was a dramatic increase in the frequency with which the pump was activated. This increase in ethanol introduction coincided with a number of changes in the physiological state of the yeast culture. The specific rate of carbon dioxide production and the concentration of ethanol in the growth medium increased, as did the viability of the culture under increased ethanol concentration. These changes were accompanied by a fall in the concentration of yeast biomass in the culture vessel. The abruptness of these changes might be interpreted as being the result of some particular mutational events. This seems likely: variants with a range of phenotypes were recovered and genetic analysis showed that these carried mutations in more than one gene. Mutants were first isolated from the altered cultures by selecting cells which formed colonies on plates containing a normally lethal concentration of ethanol (12%). The fermentation performance of these mutants was investigated and some were found to ferment at twice the rate of the parental strain in the presence of 10% ethanol. Unexpectedly mutants were isolated which could survive and ferment at enhanced rates in the presence of high concentrations of alcohol, even though the selection regime never exposed the culture to more than 5% ethanol. This illustrates the advantage of allowing physiological aspects of the culture to determine the intensity of selection via the feedback system. If direct selection had been imposed, ethanol concentrations higher than 5% would have been used.

The problems encountered in the production of ethanol, the toxicity of the product, and its consequent low concentration in the fermentation liquor, are common to most bulk chemical fermentations. Hence, the lessons learned from their example should be widely applicable. This is particularly true of the feedback continuous selection system which was used in this example to isolate (located in more than one gene tolerant mutants more) and to push back the biological limits of ethanol production. This system could be utilized to isolate thermotolerant plate cells or cells able to withstand high concentrations of heavy metals or some toxic substrate. The method is potentially applicable to the selection of mutants for improved tolerance to any inhibitory condition of either its physical or chemical environment.

SELECTION FOR PERFORMANCE OF CELLS IN VITRO

Affinity chromatography is a well-established technique for separating small and large molecules as well as in cell fractionation. In cell fractionation work, separation is usually accomplished by the recognition of specific membrane surface molecules by immobilized antibodies or lectins. Recently affinity chromatography has been used to separate bacterial populations, and used to isolate mutants with altered surface components and surface properties (7). Affinity chromatography has the potential of recovering cellular genotypes with altered protein location and/or function. Enrichments of variants via affinity chromatography requires: a) populations of single cells which do not adhere to nonselection surfaces or to one another, b) characters to be altered must be accessible on the cell surface, and c) the immobilized antibody or ligand be accessible to the cell surface. Possible applications of affinity chromatography include: a) engineering of a surface receptor for a formerly non-interactive antibody or ligand, b) regulation of a cell surface character, and c) relocation of proteins to the cell surface.

The attractiveness of affinity chromatography is increased in view of advances in monoclonal antibody production. Selection by affinity chromatography has potential advantages over the currently accepted strategy for protein engineering involving in vitro modification of genes. In the affinity selections there is no need to know a great deal about the protein structure or its active site or sequence, or to clone the structural gene. Since the changes in affinity to be obtained are directly selected, there is no need for the predictions of the effects of sequence changes required for in vitro modification. The selection methodology can recognize and enrich whatever functional configuration the cell itself designs. It is particularly difficult to see how multiple mutations giving specific changes in activity may be predicted a priori for in vitro modifications. More speculatively, affinity selections may also result in changes in surface composition and protein relocation.

It would appear that plant protoplasts or evaculated protoplasts offer an excellent opportunity for work with affinity chromatography selections. As a first step, perhaps selection via immunoprecipitation from protoplast or suspension cultures with defined antiserum against a particular surface antigen would prove beneficial. In the longer term this technique may prove useful in the isolation of plant cell cultures able to produce high levels of secondary compounds under continuous flow conditions.

CONCLUSIONS

Over the last century plant breeders have been bombarded by techniques which offer more rapid progress in variety production. Many of these techniques have not performed as well as initially anticipated. Consequently, the plant breeder has developed a healthy skepticism about the utility of new technologies for varietal production. I feel that in vitro techniques will, in time, overcome that skepticism, providing routine techniques for plant improvement. Critical to our success is continuing effort to better understand the biological mechanisms underlying plant biology, and the utilization of this understanding to design better selection systems. We must learn to ask our genetic questions in a language in which our plants can clearly respond.

REFERENCES

1. Bourgin, J.P. (1978) Valine-resistant plants from in vitro selected tobacco cells. Molec. Gen. Genet. 161:225-230.
2. Carlson, P.S. (1973) Methonine-sulfoximine resistant mutants of tobacco. Science 180:1366-1368.
3. Chaleff, R.S., and M.F. Parson (1978) Direct selection in vitro for herbicide resistant mutants of Nicotiana tabacum. Proc. Natl. Acad. Sci. USA 75:5104-5107.
4. Chaleff, R.S., and M.F. Parson (1978) Isolation of a glycerol utilizing mutant of Nicotiana tabacum. Genetics 89:723-728.
5. Chaleff, R.S., and T.B. Ray (1984) Herbicide-resistant mutants from tobacco cell cultures. Science 223:1148-1151.
6. Evans, D.A., and W.R. Sharp (1983) Single gene mutations in tomato plants regenerated from tissue culture. Science 221: 949-951.
7. Forenso, T. (1984) Genetic manipulation of bacterial surfaces through affinity-chromatographic selection. Trans. Biol. Sci. 98:44.
8. Gengenbach, B.G., C.E. Green, and C.M. Donovan (1977) Inheritance of selected pathotoxin resistance in maize plants regenerated from cell cultures. Proc. Natl. Acad. Sci. USA 74:5113-5117.
9. Harms, C.T., J.J. Oerli, and J.M. Widholm (1982) Characterization of amino acid analog resistant somatic hybrid cell lines of Daucus carota L.Z. Pflanzenphysiol. 106:239-249.
10. Hibbard, K.A., and C.E. Green (1982) Inheritance and expression of lysine plus threonine resistance selected in maize tissue culture. Proc. Natl. Acad. Sci. USA 79:559-563.

11. Hughes, K.W., D. Negrotto, M.E. Daub, and R.L. Meeusen (1984) Free radical stress response in paraquat-sensitive and resistance tobacco plants. Env. and Exp. Bot. 24:151-157.
12. Larkin, P.J., S.A. Ryan, R.I.S. Brettell, and W.R. Scowcroft (1984) Heritable somaclonal variation in wheat. Theor. Appl. Genet. 67:443-455.
13. Maliga, P. (1981) Streptomycin resistance is inherited as a recessive nuclear trait in a Nicotiana sylvestris line. Theor. Appl. Genet. 60:1-3.
14. Malmberg, R.L., P.J. Koivuniem, and P.S. Carlson (1980) Plant cell genetics - Stuck between a phene and its genes. In Plant Cell Cultures: Results and Perspectives, F. Sala, B. Parisi, R. Cella, and O. Ciferr, eds. Elsevier/North Holland Press.
15. Marx, J.L. (1984) Instability in plants and the ghost of Lamarck. Science 224:1415-1416.
16. Meredith, C.P. (1983) On being selective: Trends in biochemical sciences, mutants from cultured cells. Plant Molecular Biology Reporter, S.G. Oliver, ed., 1:05.
17. Muller, A.J. (1983) Genetic analysis of nitrate reductase deficient tobacco plants regenerated from mutant cells: Evidence for duplicate structural genes. Molec. Gen. Genet. 192:275-281.
18. Nabors, M.W., A. Daniels, L. Nadolny, and C. Brown (1975) Sodium chloride-tolerant lines of tobacco cells. Plant Sci. Lett. 4:155-159.
19. Negrutiu, I., R. Dirks, and M. Jacobs (1983) Regeneration of fully nitrate-reductase deficient mutants from protoplast culture of Nicotiana plumbaginifolia (Viviani). Theor. Appl. Genet. 66:341-347.
20. Oliver, S.G. (1984) Economic possibilities for fuel alcohol chemistry and industry. 12:42S.
21. Parke, D., and P.S. Carlson (1979) Somatic cell genetics of higher plants appraising the application of bacterial systems to higher plant cells cultured in vitro. In Physiological Genetics, J.G. Scandalios, ed. Academic Press.
22. Ramulu, K.S., P. Dijkhuis, S. Roest, G.S. Bokelmann, and B. DeGroot (1984) Early occurrence of genetic instability in protoplast cultures of potato. Plant Sci. Lett. 36:79-86.
23. Ranch, J.P., S. Rick, J.E. Brotherton, and J.M. Widholm (1983) The expression of 5-methyltryptophan-resistance in plants regenerated from resistant cell lines of Datura innoxia. Plant Physiol. 71:136-140.
24. Thomas, B.R. (1984) Selection of phenotypes in plant cell cultures. Plant Molecular Biology Reporter 2:46.
25. Wakasa, K., and J.M. Widholm (1982) Regeneration from resistance cells of tobacco and rice to amino acids and amino acid analogs. In Plant Tissue Culture, A. Fujiwara, ed. Maruzen Co., Ltd., Tokyo, pp. 455-456.
26. Widholm, J.M., and J.P. Ranch (1983) Expression of 5-methyltryptophan resistance in regenerated Datura innoxia plants. In Plant Cell Culture in Crop Improvement, S.K. Sen and K.L. Giles, eds. Plenum Press, New York, pp. 381-386.

27. Wong, J.R., and I.M. Sussex (1980) Isolation of abscisic acid-resistant variants from tobacco cell cultures. Planta 148:103-107.
28. Yurina, N.P., M.S. Odintsova, and P. Maliga (1978) An altered chloroplast ribosomal protein in a streptomycin-resistant tobacco mutant. Theor. Appl. Genet. 52:125-128.

SOMACLONAL VARIATION IN PROGENY OF PLANTS FROM CORN TISSUE CULTURES

Elizabeth D. Earle and Vernon E. Gracen

Department of Plant Breeding and Biometry
Cornell University
Ithaca, New York 14853

INTRODUCTION

Spontaneous variation in plants recovered from tissue culture has been documented for many species (1). This variation, recently termed "somaclonal variation" (1), has been seen in many types of culture systems, including ones involving protoplasts, long-term callus cultures, and fresh explants. The occurrence of somaclonal variation has both positive and negative aspects. For those concerned with in vitro propagation, it is undesirable; progeny not true-to-type are usually of little value. Spontaneous changes may also be a problem in attempts to transform plant cells; a high frequency of change not related to the experimental manipulations can complicate interpretation of results and can yield material with alterations other than the desired specific gene transfer. Although variability among cultured plant cells may reduce or obviate the need for mutagenesis prior to in vitro selection, it can again result in variants other than those being specifically selected.

On the other hand, the occurrence of unexpected changes following culture raises important questions about the nature and plasticity of the plant genome. Moreover, new sources of genetic variation are attractive to plant breeders. The prospect of obtaining limited changes in otherwise well-adapted plant types without resorting to hybridization is particularly appealing. However, tissue culture-derived variation can be taken seriously as a component of breeding programs only if a number of conditions are satisfied:

a. Changes seen must be stable, either heritable through repeated sexual crosses, or transmitted asexually through cycles of vegetative propagation. Changes seen only in the regenerated plants themselves and not in their progeny are of little practical interest.

b. Changes must be in traits of agricultural importance such as vigor, yield, maturity, plant type, fertility, etc. Clearcut, positive changes in characteristics such as yield or resistance to stresses are most desirable, but alterations not of direct field value may be useful in clarifying the genetic control of the traits involved. Variation limited to deleterious seedling mutations, etc., will not be very enthusiastically received.

c. The most desirable somaclonal variants are those which provide positive traits not already available in the lines of interest to breeders, or not easily transferred to them. Heritable shifts to existing plant types may be of value in studies of genetic mechanisms but do not directly advance breeding programs.

d. The effort involved in identifying and characterizing somaclonal variants must not exceed that required by conventional breeding approaches. Recourse to somaclonal variation usually necessitates regeneration and analysis of large populations of plants. This already sacrifices the frequently cited benefits of manipulations at the cell and protoplast level. A positive feature of screening at the plant level is that such screening can reveal variants for which no in vitro selection schemes or gene transfer approaches have yet been devised. However, if the frequency of useful variants is much smaller than that from conventional methods, the enterprise will have little appeal for plant breeders.

e. The significance of somaclonal variation as a resource for plant breeders also hinges on its occurrence in important lines of major crop plants. It is encouraging that most of the documentation of somaclonal variation comes from crop species (1). However, breeders are concerned with finer distinctions: not just with species, but with lines suitable for their specific environments, crossing programs, or markets. They will be most attracted by results showing that desirable new traits can be obtained in the lines of real importance to them. Thus, the extent to which regenerability is a genetic trait, high in some lines and low or absent in others, will dictate the potential of somaclonal variation in breeding programs for specific crops.

The requirements for a really significant contribution of somaclonal variation to a crop breeding program are thus quite stringent and have only rarely been met. This is hardly surprising since rather few extensive analyses of progeny from regenerated plants of major crops have been done (but see Ref. 2,3).

This report describes current work at Cornell University on somaclonal variation in corn which addresses the concerns listed above. General procedures and some of the results to date are presented. Changes in the expression of male-fertility are given particular emphasis.

PLANT REGENERATION PROCEDURES

Corn inbred W182BN, an early maturity inbred used as a male parent in a number of successful hybrids in the Northeast, has been the material in most of our initial studies. We have cultured W182BN with male-fertile "N" cytoplasm and each of the 3 major groups of male-sterile cytoplasms: T, C, and S (S, CA, and L subgroups).

Our first cultures, started in 1980, were established from immature embryos between 1 mm and 2 mm long, placed scutellum up on Murashige-Skoog (MS) medium containing 1 to 5 mg/l 2,4-dichlorophenoxyacetic acid (2,4-D). Some embryos then developed organized proliferations from the scutellum. We have subcultured this type of tissue in semiorganized form at 4- to 6-week intervals for the last 4 years, usually on medium containing 5 mg/l 2,4-D. Tissue grew more vigorously on medium solidified with 0.2% Gelrite (Kelco Corp.) than on agar-solidified media.

Plants were recovered from these cultures by one of 2 routes. Occasionally plantlets appeared on tissue in plates of maintenance medium not subcultured for several months. More often, development of partially organized areas into plants occurred after one or more transfers to fresh medium with little or no 2,4-D. A shift of the basal medium from MS medium with 3-4% sucrose to N_6 medium with 10% sucrose was often helpful but not essential. Cultures on regeneration medium usually produced firm white organized areas before shoots were seen.

The pathway by which plant regeneration occurs is not completely clear. The organized structures seen on W182BN embryo explants and subcultured callus are less clearly embryo-like than those we have seen on other genotypes, and we have not yet done histological studies. However, the general appearance of the regenerating tissue is rather similar to that of other embryogenic cereal callus. We have observed the trichome-bearing leafy green structures generally interpreted as scutellar development in cereal cultures. Some plantlets also had a mesocotyl-like structure between the callus and the leaves, again suggesting an embryogenic pathway. We are probably dealing with "mixed" cultures in which both organogenesis and somatic embryogenesis occur.

Dark-grown cultures with shoots or plantlets were transferred to lab light where leafy structures usually turned green and continued to grow. Plantlets were then moved onto filter paper moistened with liquid MS medium (0 2,4-D) in deep Petri dishes in a culture room (16 hr light/day). The next transfer was to Magenta GA-7 boxes containing filter paper and liquid half-strength MS medium (0 2,4-D; 1% sucrose). Plantlets with several well-developed leaves and a good root system were removed from sterile culture, potted in sterile vermiculite watered with 1/4x MS salts, and maintained in a growth chamber (12 hr light/day). Well-established plantlets were moved to soil-less mix and grown to maturity in a greenhouse.

Following these general procedures, we have regenerated hundreds of W182BN plants. In the process, we learned more about the experimental system. From additional embryo cultures established in 1981, 1982, and 1983, we confirmed that W182BN is favorable material. Up to 40% of the W182BN embryos cultured in 1983 showed initially promising growth while most other inbreds tested under our conditions did less well. Co410, a 3-way hybrid with W182BN as the male parent, was also very responsive, suggesting that transfer of genes for "regenerability" from W182BN may be possible.

Although nuclear genotype clearly affects response in culture, the cytoplasm does not have major effects, at least not in W182BN. Embryos from the various cytoplasms we tested all responded similarly.

Addition of 20 mM proline to MS medium enhanced production of regenerable tissue from the scutellum of excised embryos. Although many of the W182BN embryos initially produce promising material, the tissue tended to shift to the undesirable type of tan rooty callus familiar to many corn tissue culture workers. This tendency was counteracted by rigorous subculture procedures selecting for firm, pale semiorganized tissue, by culture in the dark, and by subculture on higher 2,4-D levels (2-5 mg/l) which suppress root development.

As our culture procedures became more efficient, losses of plants before maturity, especially in vermiculite, decreased. The main source of loss remaining is poor continued development of some small plantlets on 2,4-D-free medium. In some cases, the shoots of such plants become dark red rather than green, have a husk-like appearance rather than separated leaves, and eventually brown without elongating more than a few centimeters.

To summarize, we have recovered many mature W182BN plants from embryo-derived tissue cultures, even though the experimental system would profit from further refinement. The plants came from cultures established in 1980, 1981, 1982, and 1983. Most of the plants used for field studies came from cultures over a year old, and even 4-year-old cultures were still capable of regeneration.

CHARACTERISTICS OF REGENERATED PLANTS

It is difficult to make useful evaluations of greenhouse-grown regenerated corn plants. Traits such as days to silking, used as an index of maturity, cannot be realistically measured since there is no clearcut starting point such as date of seed planting. Characters such as yield require populations and cannot be assessed on a single plant basis. Ratings for resistance to most common diseases are difficult, both because of the greenhouse environment and because our highest priority was to take as many plants as possible to seed production without adding additional stresses. Nevertheless, several general features of the regenerated plants were easily apparent; reduced size, altered morphology, and reduced male-fertility.

The W182BN plants grown from seed in the greenhouse are typically over 200 cm high. Regenerated plants varied considerably in height, but were only very rarely over 100 cm high. Many plants were less than 50 cm high, and some were only 5 cm.

Another striking feature was reduced male-fertility, seen even in plants from cultures of embryos with fertile cytoplasms. This included a general trend to feminization of the plants, not just appearance of sterile tassels. Plants with tassels (either fertile or sterile) often had silks emerging from the tassels as well as from ear shoots. Other plants completely lacked tassels and had only terminal silks. These terminal silks occurred both on plants that had elongated to several internodes and on ones that showed very little vertical growth. The latter had the appearance of a corn ear set vertically into the soil. The husk-like plants that fail to develop well in vitro (p. 142) may represent an advanced form of this syndrome.

The reasons for the reduced stature and feminization of many of the regenerated plants are not yet clear. Similar observations have been made by Brettell et al. (4) and others working with corn cultures. The changes are not genetic ones; progeny of most regenerated plants have normal stature and tassels without silks. Long-term exposure of callus to 2,4-D and other components of the culture medium before plant regeneration are not responsible: we have seen these characteristics in some of the first plants recovered from new cultures. Moreover, we have recovered about 30 plants directly from 1.5-2.0 mm long excised embryos germinated on MS medium containing no 2,4-D and then taken through our standard procedures. These plants too were considerably shorter than typical seed-grown plants, and, unlike seed-grown W182BN plants in our greenhouse, many had terminal silks. Some had no tassels at all. Possibly precocious germination of immature embryos is involved. Perhaps nonhormonal components of the MS medium or environmental factors in vitro alter plantlet development. Another possible factor is genotype. Some genotypes other than W182BN tended to be taller with more normal tassels.

Regenerated plants in the greenhouse were pollinated to obtain progeny for further study. Whenever possible we selfed the plants. When the regenerated plant produced no pollen or had no pollen when silks appeared, we used pollen from seed-grown W182BN plants grown in the greenhouse for that purpose. By pollinating all available silks, we obtained at least a few seeds, even from many of the very aberrant plants. The seeds came either from ear shoots, from ovules at the base of terminal silks, or both. However the morphological peculiarities of many of the regenerated plants had 2 immediate consequences. It was not possible to self many of the regenerated plants, even those from lines with male-fertile cytoplasms. This delayed later identification of any recessive alterations. Also it was not possible to do meaningful ratings of male-fertility on many of the plants. Absence of a tassel obviates the possibility of a rating, and under the circumstances, even the occurrence of a sterile tassel is of questionable significance.

In view of the rather poor fertility of many regenerated plants from N cytoplasm cultures, it was striking that some plants from initially male-sterile lines were fertile. Plants from several CMS-S lines had fertile tassels with anthers exserted and viable pollen. (See p. 146.) Such reversions to fertility were never seen in W182BN CMS-S plants grown from seed in our greenhouse.

PROGENY OF REGENERATED PLANTS

We have grown seeds from some 88 regenerated W182BN plants for evaluation of their progeny. Some seeds were planted in the greenhouse or growth chamber for increase (if only a few seeds were initially available) and for ratings of male-fertility (if results were sought before field plantings were feasible). The plants obtained in the greenhouse looked much more like the inbred parent than the regenerated plants themselves. In greenhouse tests, progeny of plants which were abnormally short (down to 15 cm) grew to the typical 200 cm range. These plants had normal tassels without silks.

Most of the seed progeny of regenerated plants were grown in the field, either in summer plantings in New York in 1982, 1983, and 1984, or in winter nurseries in Homestead, Florida in 1982 to 1983 and 1983 to 1984. The field plantings permitted large-scale comparisons to the original inbred W182BN and ratings of agronomic traits. Crosses testing combining ability, fertility restoration, etc., have also been made. Some of our regenerated material had its first field tests last summer, but progeny of the first few plants regenerated are in their 6th generation of selfs or crosses.

In the field we saw considerable variability, both within rows consisting of progeny from a single regenerated plant and among progeny from different regenerates. Several categories of plants were seen. Some were similar to the inbred parent in all obvious characteristics. Some seeds gave no viable plants, either because of seedling lethality, or because of poor plantlet survival under stress. Other plants grew but were morphologically abnormal, e.g., dwarf, brachytic, lacking a tassel, etc. At least some of these are recessive mutations, probably similar to those recovered in several other corn tissue culture programs (5,6,7). (Most of the latter mutants involved seedling, leaf, or endosperm abnormalities or other deleterious traits, perhaps because these were most easily detected and analyzed.) The genetic basis of other aberrations has not yet been determined. Problems include small initial populations (usually no more than 10 seeds/row), and poor fertility of many of the abnormal variants which hampers further tests. It is interesting that some of the morphological abnormalities appeared in progeny of plants that were not selfed, but sibbed.

Some progeny of regenerated W182BN plants also showed heritable variability in important agronomic traits. Among the changes seen were increased vigor, later maturity, increased or decreased height, darker leaf color, morphological variation, and shifts to and from male-sterility.

Selected plants with altered characteristics were taken through several cycles of selfs and/or crosses. Not all selections showed their novel traits in further generations, but many did. Most of these did not segregate further after the first few generations of selfings. Some have been stable for at least 6 selfed generations. In this way, we have obtained W182BN lines with a range of new phenotypes. In the field last summer were some 60 different stable and uniform advanced families derived from our first few regenerated plants.

Some of the novel W182BN types could have real agricultural value. For example, a uniform line which sheds pollen several weeks later than the inbred has been selected. This new line and others like it could add flexibility to planting for hybrid seed production. Culture-derived inbreds with increased vigor could either replace the standard line or give stronger material through sister-line crosses with the standard line. A stronger W182BN line would find application as a female parent for crosses; at present W182BN can only be used as a male parent. An altered W182BN could also contribute to improved versions of the hybrids in which it is used, provided combinability is not affected. Seven of our best selections were being tested in hybrid combinations in 1984.

Detailed information about the most recent field tests, including disease ratings, hybrid combinations, and yield will be available only late in 1984. While we are encouraged by the appearance of heritable variation in several agriculturally significant traits in progeny of the very first regenerated plants examined, further comments about most of the agronomic variants are probably premature. The remainder of this paper will focus on one specific trait, male-fertility, about which we have already accumulated more information. Several different types of changes were observed.

Background Information on Male-Sterility in Corn

Cytoplasmic male-sterility (CMS), coupled with the use of nuclear genes that restore pollen fertility and permit seed set, has contributed greatly to efficient production of hybrid corn (8,9). Three major groups of CMS have been recognized; these are usually designated C (Charrua), T (Texas), and S (USDA). They differ in a variety of ways, including stability in different inbreds and environments, cytology of pollen abortion, fertility restoration after crosses with inbreds carrying specific restorer genes, and degree of restoration in plants heterozygous for restorer genes ("sporophytic" vs. "gametophytic" restoration). The sensitivity of the widely used T cytoplasm to *Helminthosporium maydis* Race T led to an epiphytotic of Southern Corn Leaf Blight in 1970 and to the prompt elimination of CMS-T from hybrid production (10). The remaining CMS types, C and S, have various drawbacks, so new or improved versions of CMS are of great interest to corn breeders.

It is also possible to distinguish the major CMS groups and sometimes even subgroups by gel banding patterns of mitochondrial (mt) DNA treated with restriction endonucleases (11,12). CMS-S can be recognized by the presence of 2 linear episomal DNA molecules (S-1 and S-2) even in gels of unrestricted mtDNA (13). Relatively few clearcut phenotypic traits are encoded by plant mtDNA, and of these CMS is the most significant. The different CMS lines available provide needed variants for study. Thus, CMS is a trait of interest to molecular biologists as well as to breeders. To date, corn is the species in which the most elaborate analyses of the molecular basis of CMS have been done.

For these reasons, we were particularly interested in possible somaclonal variation in CMS and/or nuclear restorer genes that influence the expression of sterility. Two such genes (Rf-1 and Rf-2) restore CMS-T. The gene Rf-3 restores CMS-S, and Rf-4 restores CMS-C. Additional modifier genes can also affect sterility. W182BN carries no major restorer genes; that is, crosses of W182BN with all the different CMS lines result in sterile plants.

Changes from CMS to Fertility

Reversion from CMS-S to fertile cytoplasm(s) occurred in our cultures. This was first seen in plants from our earliest CMS-S cultures, established in 1981. Four of 26 regenerated plants had fertile tassels with viable pollen. Because of the generally reduced male-fertility among the regenerates, this was a particularly noteworthy result. We obtained seed from 10 of the 26 plants, including 2 of the fertile ones, and tested 7 of these seed lots. All of the plants that were recovered were fertile, including ones from regenerated plants with missing or sterile tassels. Further crosses showed that the reversion to fertility is a cytoplasmic change. All progeny obtained when the revertants were used as females were fertile. The revertants used as males did not restore fertility to other CMS-S lines (or to CMS-C or T), so changes in nuclear restorer genes are not involved.

Further reversions to fertility were seen in plants from additional cultures of W182BN in CMS-S. Several cultures each of 3 different subgroups of CMS-S (S, CA, and L), were established in August, 1983. Four fertile regenerated plants were seen in a total of 45. These 4 were distributed among the subgroups (2 in 2 different cultures of S, one each in CA and L) in a way that precluded possibility of labeling error. While 6 other cultures have produced no fertile plants, the frequency of reversion seems quite high. Fertile plants were recovered from callus shifted to 2,4-D-free regeneration medium only 3 months after initiation of the cultures. Progeny tests of the newer material have been less extensive, but initial results again show fertile plants in the next generation.

Shifts from CMS-S to fertility in plants recovered from tissue cultures have not previously been reported. Such shifts have occasionally been seen in field-grown material of several genotypes (14) though not in W182BN. The field-derived cytoplasmic reversions to fertility have been correlated with loss of the S-1 and S-2 episomes from mtDNA and their integration into high molecular weight mtDNA (15). We therefore checked some of our callus cultures, regenerated plants, and progeny for these episomes using agarose gel electrophoresis. Callus cultures that yielded plants with sterile progeny contain the S-bands, as expected. Most callus cultures from which fertile progeny were obtained contained no S-bands. In these cases, the callus cultures themselves were altered, not just individual plants recovered from them. One culture gave rise to a fertile plant as well as to plants with sterile progeny. This callus retained the S-1 and S-2 bands, suggesting that sectoring for the cytoplasmic change occurs. The fertile revertant plants tested also lacked the bands. We are now doing Southern hybridization with radioactive probes of S-1 and S-2 sequences to confirm loss of the episomes.

Thus, we have seen a fairly rapid, high-frequency change from CMS-S to fertility not previously reported from cultured material or from this inbred in the field. The change appears to be correlated with a specific alteration in mtDNA. While the revertant material has no immediate agricultural value, it offers a promising system for study of the molecular biology of CMS. An obvious next step is to study if and where S-1 and S-2 are integrated into the main mtDNA of our revertants and what types of rearrangements occur (16). It should also be possible to study the loss of the S-bands from CMS-S callus from cultures. Freshly established callus could be monitored with time and under different culture conditions to see what affects the change in mtDNA. Loss of free S-1 and S-2 DNAs and integration into high molecular weight mtDNA has been observed in cultures of inbred Wf-9 carrying CMS-S (17). This alteration was correlated with a shift from organized to friable callus morphology. Plant regeneration from the friable callus or the compact callus from which it arose was no longer possible, so the relationship of the mtDNA changes seen in Wf-9 cultures to CMS are not clear. All of our revertants from CMS-S W182BN cultures arose from organized callus similar to that which yielded CMS-S plants.

In contrast to the results with CMS-S material, we have seen no reversions to fertility from CMS-C. Over 100 regenerated plants and progeny from over 20 of them all remained sterile. CMS-C is more stable than CMS-S in the field; the same appears true in vitro.

No reversions from CMS-T to fertility were seen in 26 regenerated plants and progeny from 4 of them. Such reversions have been noted with other materials. Brettell et al. (4) obtained over 50% fertile revertants from cultures of a hybrid involving Wf9T, W22, and A188. Of 9 plants we recovered from cultures of inbred Wf-9 in CMS-T, 8 were also fertile (Earle, unpub. results). However, Umbeck and Gengenbach (18) saw much lower rates of reversion (5% and 0%) from A188-T cultures. It is possible that nuclear genotype influences the frequency of reversion of CMS-T in vitro and that revertants from W182BN-T would be seen if larger populations were studied.

Changes from Male-fertility to CMS

Although many plants regenerated from fertile N cytoplasm cultures had poor male-fertility, most of their progeny had fully fertile tassels. However 7 plants, all derived from one culture of W182BN, had progeny with sterile tassels. Several generations of crosses indicated that this sterility was cytoplasmic rather than genic.

We are attempting to determine whether this represents an in vitro shift from fertility to one of the known CMS groups or to a new type of cytosterility. The failure of the sterile plants to show sensitivity to Helminthosporium maydis Race T toxin (19) suggests that they are not CMS-T. No S-1,S-2 episomes are present so they are not typical CMS-S types.

Crosses with 3 inbreds, each known to carry Rf genes for only one of the male-sterile cytoplasms (20), gave unexpected results. Both in Florida in 1983-1984 and in New York this summer, the culture-derived steriles were fully restored by the restorer for CMS-C (Co150) and by the one for CMS-S (MS64-7). Some partial restoration was also seen with the restorer for CMS-T (RD4515). Pollen counts showed that the pattern of restoration is sporophytic; that is, virtually all the pollen grains in the anthers of the heterozygous restored plants are fertile. Pollen abortion in the nonrestored plants occurred early, after the young microspore stage.

Taken together, the results indicate that our male-sterile cytoplasm is neither CMS-T, CMS-S, nor a typical CMS-C. Restoration with Co150, sporophytic restoration, and early abortion are all consistent with a CMS-C, but MS64-7 does not normally restore W182BN-C cytoplasms. Thus, it appears that we may have recovered either a novel variant of CMS-C or a new type of CMS for which restorers are present across several inbreds that distinguish known sterile cytoplasms.

Crosses with additional inbreds have been made to help resolve this issue. Comparisons of mtDNA of the culture-derived steriles with that of known cytoplasms are also needed. Either of the possible results would be significant. A new cytoplasm would be of most agricultural interest, but even a spontaneous shift to a known CMS type would aid in the study of the origin and molecular nature of CMS.

Nuclear Gene Changes Involving Male-fertility

Shifts to and from male-sterility can involve changes in nuclear as well as cytoplasmic genes. We have some evidence for the appearance of genic male-sterility in one selection from a plant regenerated from an N cytoplasm culture. Unlike the maternally inherited culture-derived steriles described above, this sterile line segregates for fertility and sterility in the F_2 generation of crosses with seed-grown W182BN.

Finally, we have evidence for appearance of new restorer genes in some selections among the progeny of our first regenerated plant. W182BN does not normally restore any of the 3 groups of cytoplasms to fertility. However, in 2 seasons of tests, one selection from selfed progeny of this plant restored both CMS-C (Pr subgroup) and CMS-S (CA and LBN subgroups); it did not restore CMS-T. Whether these results represent shifts to the dominant alleles of the known restorers for both CMS-C and CMS-S or shifts to novel restorers remains to be determined.

CONCLUSIONS

Several laboratories have recovered somaclonal variants from corn tissue cultures. These include maternally inherited shifts from sensitivity to *H. maydis* Race T toxin accompanied by reversion from CMS-T to fertility (4,18). Changes controlled by recessive nuclear genes also appeared often (5,6,7), but most of these involved deleterious traits. Progeny of plants from our cultures of corn inbred W182BN show some such abnormalities, but also have multiple heritable changes in important agronomic traits, including vigor, maturity, plant health, and male-fertility. Our stable selections with increased vigor and/or altered maturity offer useful modifications of an already desirable inbred, which may speed development of improved hybrid corn lines. Introducing such traits by conventional hybridization would alter the inbred significantly, and many years would be required to develop a new improved inbred. In contrast, the changes obtained by somaclonal variation appear to stabilize within the first few generations of selfings. Lines with maternally inherited shifts in male-fertility provide new material for molecular biologists studying CMS and perhaps for breeders as well.

Lines that differed from the standard inbred were obtained at a fairly high frequency. Progeny of the first few regenerated plants yielded many distinct, promising selections. Several different CMS-S cultures produced male-fertile plants after only a few months of culture.

Some of the changes (e.g., acquisition of new restorer genes) can be detected only by making the appropriate crosses with a range of inbreds or cytoplasms. Large-scale analysis of regenerated material requires links with a field breeding program, preferably one with much experience with the line under study. A substantial commitment of time and space is needed. At Cornell this commitment has been made; over 600 rows of the 1984 summer nursery contained material from regenerated plants, and over 60 experimental hybrids involving regenerated material were being evaluated in replicated trials.

Many questions about these studies are still unanswered. The relationship between variation and such factors as time in culture, culture conditions, and regeneration pathways must be analyzed. The origin of the somaclonal variation in this, as in other plant systems, requires clarification. Shifts to and from cytoplasmic male-sterility probably involve alterations of mitochondrial DNA, but the genetic basis of the changes controlled by nuclear genes is not yet well understood. Some of these changes are in phenotypes usually considered to be conditioned by quantitative genes. Alterations in repetitive DNA sequences have recently been implicated in diverse examples of unexpected variability in higher plants (21). The puzzling, nongenetic, reduced fertility of many regenerated plants prevents selfing and adds a generation to crosses required for identification of recessive mutants. Whether progeny of additional regenerates from newer cultures or different inbreds will also yield stable, useful variants remains to be seen. However, our results to date suggest that continued attempts to exploit somaclonal variation in corn are worthwhile and that somaclonal variants may soon have a place in hybrid corn production.

ACKNOWLEDGEMENTS

We thank Lou Ann Batts, Violet Best, and Sherrie Norman for their valuable contributions to these studies. The work was supported by the Agrigenetics Research Corporation.

REFERENCES

1. Larkin, P.J., and W.R. Scowcroft (1981) Somaclonal variation - A novel source of variability from cell cultures for plant improvement. Theor. Appl. Genet. 60:197-214.
2. Larkin, P.J., S.A. Ryan, R.I.S. Brettell, and W.R. Scowcroft (1984) Heritable somaclonal variation in wheat. Theor. Appl. Genet. 67:443-455.
3. Evans, D.A., and W.R. Sharp (1983) Single gene mutations in tomato plants regenerated from tissue culture. Science 221:949-951.
4. Brettell, R.I.S., R. Thomas, and D.S. Ingram (1980) Reversion of Texas male-sterile cytoplasm maize in culture to give fertile T-toxin resistant plants. Theor. Appl. Genet. 58:55-58.
5. Edallo, S., C. Zucchinali, M. Perenzin, and F. Salamini (1981) Chromosomal variation and frequency of spontaneous mutation associated with in vitro culture and plant regeneration in maize. Maydica 26:39-56.
6. Rice, T.B. (1982) Tissue culture induced genetic variation in regenerated maize inbreds. Proc. Annual Corn and Sorghum Research Conf. 37:148-162.
7. McCoy, T.J., and R.L. Phillips (1982) Chromosomal instability in maize (Zea mays L.) tissue cultures and sectoring in some regenerated plants. Can. J. Genet. Cytol. 24:559-565.
8. Duvick, D.N. (1965) Cytoplasmic pollen sterility in corn. Adv. Genet. 13:1-56.
9. Laughnan, J.R., S. Gabay-Laughnan, and J.E. Carlson (1983) Cytoplasmic male sterility in maize. Ann Rev. Genet. 17:27-48.
10. Ullstrup, A.J. (1972) The impacts of the southern corn leaf blight epidemic of 1970-71. Ann. Rev. Phytopath. 10:37-50.
11. Pring, D.R., and C.S. Levings, III (1978) Heterogeneity of maize cytoplasmic genomes among male-sterile cytoplasms. Genetics 89:121-136.
12. Leaver, C.J., and M.W. Gray (1982) Mitochondrial genome organization and expression in higher plants. Ann. Rev. Plant Physiol. 33:373-402.
13. Pring, D.R., C.S. Levings, III, W.W.L. Hu, and D.H. Timothy (1977) Unique DNA associated with mitochondria in "S"-type cytoplasm of male-sterile maize. Proc. Nat. Acad. Sci., USA 74:2904-2908.
14. Laughnan, J.R., and S.J. Gabay (1978) Nuclear and cytoplasmic mutations to fertility in S male-sterile maize. In Maize Breeding and Genetics, D.B. Walden, ed. John Wiley, New York, pp. 427-447.
15. Levings, III, C.S., B.D. Kim, D.R. Pring, M.F. Conde, R.J. Mans, J.R. Laughnan, and S.J. Gabay-Laughnan (1980) Cytoplasmic reversion of CMS-S in maize: Association with a transpositional event. Science 209:1021-1023.
16. Schardl, C.L., D.M. Lonsdale, D.R. Pring, and K.R. Rose (1984) Linearization of maize mitochondrial chromosomes by recombination with linear episomes. Nature 30:291-296.

17. Chourey, P.S., and R.J. Kemble (1982) Transposition event in tissue cultured cells of maize. In Plant Tissue Culture 1982, A. Fujiwara, ed. Japanese Association for Plant Tissue Culture, Tokyo, pp. 425-426.
18. Umbeck, P.F., and B.G. Gengenbach (1983) Reversion of male-sterile T-cytoplasm maize to male fertility in tissue culture. Crop Sci. 23:584-588.
19. Gracen, V.E., M.L. Forster, K.D. Sayre, and C.O. Grogan (1971) Rapid method for selecting resistant plants for control of southern corn leaf blight. Plant Dis. Rep. 55:469-470.
20. Gracen, V.E. (1982) Types and availability of male-sterile cytoplasms. In Maize for Biological Research, W.F. Sheridan, ed. University Press, Grand Forks, North Dakota, pp. 221-224.
21. Marx, J.L. (1984) Instability in plants and the ghost of Lamarck. Science 224:1415-1416.

WHOLE PLANT GENETICS IN THE STUDY OF PLANT ONTOGENY

Peter M. Gresshoff

Department of Botany/Genetics
Australian National University
Canberra ACT 2600
Australia

INTRODUCTION

One of the major factors stimulating the expansion of cell culture research and the growth of somatic cell genetics in plants was the assumption that many genetical processes were infrequent at the whole plant level. Thus, cellular approaches were needed to isolate products of events such as recombination, transformation, and somatic variation. Additionally, plant cell culture expanded as it offered a technique by which one could investigate the factors controlling differentiation and dedifferentiation in plant tissue. Another impetus came from the view that somatic cell and organ culture could be used to propagate clonally advantageous material.

Several things failed to come to reality. In several cell culture systems, it was illustrated that passage through the callus stage increased the amount of genetic variation. The term "somaclonal variation" describes this phenomenon, although it may be neither "soma" nor "clonal." However, the concept seems correct although the general applicability and the underlying mechanism(s) still need to be resolved.

In terms of the analysis of differentiation, cell culture technologies have yielded some insights, such as the discovery of cytokinins, analysis of somatic embryogenesis, and the development of the totipotency concept (for review, see Ref. 8). Plant cell cultures have also been used to facilitate the transfer of bacterial and eukaryotic genes to plants. This approach, especially when coupled with molecular biological techniques, provides a new way of looking at plant development and gene regulation. Sometimes, however, it is interesting to realize that while one understands the promoter function of, for example, a plant storage protein gene, the insights into fundamental processes (even at a biological or physiological level), such as mass transport, flowering, photoperiodicity, and growth regulators, are not advanced enough to permit molecular model building.

Here I do not want to discuss the value or shortcomings of plant cell culture technology. More importantly, I want to point out, by example, that other approaches, even "classical plant breeding and selection," can produce valuable results which can serve to foster the analysis of differentiation and "pure" gene regulation. Additionally, such approaches may also provide the economic and agronomic reward that also often motivates the cell culture approach.

GENETIC ANALYSIS OF GENE REGULATION IN SYMBIOTIC NITROGEN FIXATION

Symbiotic nitrogen fixation, as illustrated by the legume-*Rhizobium* symbiosis, involves the concerted interaction of eukaryotic and prokaryotic organisms. It results in a specialized structure, the root nodule (or stem nodule as in the case of *Sesbania*), which provides the milieu involving oxygen supply, nitrogen assimilation, and photosynthate supply to optimize the bacterial nitrogenase function. Of interest here is the observation that although many ontogenic responses during nodulation are based on cellular phenomena, it has been impossible to reliably repeat these responses under in vitro cell culture conditions. For example, *Rhizobium* does not invade in vitro cultured cells by infection threads, leghemoglobin synthesis is not derepressed, and it is impossible to establish a continued symbiosis between plant and bacterial cells. Clearly the cell culture state does not coincide with the state that nodule cells are in during and after nodule initiation and development.

There are some aspects of the symbiosis, however, which are investigated using cell culture. These include (a) in vitro derepression of nitrogenase by co-culture with plant cells, (b) attachment and "invasion" studies, and (c) phytohormone-related cell stimulation to enter cell division (10).

My original motivation to study plants via cell culture came from the wish to study development. The experimental analysis of the development of a supernumerary organ such as the root nodule has several advantages. One can isolate mutants without affecting the fitness of the whole organism. Likewise, physiological manipulations are less plagued by pleiotrophic effects. Additionally, this ontogenetic process is induced by the prokaryote, *Rhizobium*, and, as such, is open to analysis by genetic techniques such as insertion mutagenesis. The application of molecular techniques to this problem has resulted in the cloning and characterization of numerous *nod* and *nif* (nodulation and nitrogen fixation) genes in *Rhizobium* (for example, see Ref. 16).

So what do we know about the plant? Clearly there exists a large body of physiological/biological/biochemical information. This is highly useful, especially from a genetical viewpoint, as that body of knowledge defines the phenotype of nodulation and fixation. But what are the plant genes which control this process? How are they regulated? Are they only present in legumes? Are they only active in the nodule? Do they vary between legumes? How do the plant genes control host susceptibility? Big questions indeed, but what approach to take?

The molecular biologist pursues the problem by the production of nodule-specific cDNA libraries, the sequencing and Northern-blotting. Often such approaches yield sophisticated molecular data, but fail to elucidate the function of a clone or "spot" on the 2-D polyacrylamide gel.

The approach taken by my colleagues David McNeil, Bernard Carroll, and myself was naive, simple, and based on a transfer of thinking from the cellular technologies of microbial genetics. If one wants to study a process, one must get mutants. These then can be characterized in terms of their genetics, physiology, biochemistry, and molecular biology. To get a mutant, one mutagenizes, then selects. It is straight forward, but labor-intensive.

We chose soybean (*Glycine max* var. Bragg) because it is self-fertile, agronomically important, and has a substantial background as an experimental organism. Seeds were pregerminated in aerated solution, then exposed for 6 hr to 0.5% ethyl-methyl-sulphonate (EMS). Further detail on the selection procedure can be found in Carroll et al. (6). The resultant wet seeds were planted directly into soil and allowed to mature. The progeny seeds from each plant were pooled to make M2 families. A total of 12 seeds per M2 family was screened in sand-filled pots watered daily with 5.5 mM KNO_3. This concentration was chosen to select soybean mutants which nodulate heavily in the presence of nitrate, because it defined the steep portion of a nitrate vs. nodule mass sensitivity curve. After 5 to 7 weeks of growth, pots were emptied and the root systems were monitored for heavy nodulation. Genetics predicts that since the apical meristem inside a soybean embryo is made up of numerous cells, of which only 4 cells give rise to the reproductive tissue of the plant, that one should expect recessive mutations to segregate in the M2 family at a ratio of 15:1 (4:0 plus 4:0 plus 4:0 plus 3:1). We selected a total of 15 *nts* (nitrate tolerant symbiosis) mutants. Additionally, using slightly altered selection schemes, we isolated soybean mutants which either fail to nodulate, have delayed nodulation, or alter root morphology. These mutants breed true, behave in all tested cases as mendelian loci, and show either recessive or co-dominant characteristics. A population of efficiently-mutagenized M2 seeds provides a resource which can be screened for a variety of mutant phenotypes. For example, using a multititre plate assay on leaf discs, we have isolated 2 mutants of soybean which lack the constitutive nitrate reductase activity (5). Additionally, such treated populations can be screened for other variations such as herbicide resistance or disease resistance. Induced mutation was very much out of fashion, as the green revolution brought successes through wide crosses. For some crops, however, the genetic base is not sufficient, thus new variation has to be induced.

Before elaborating on the nts mutants, one needs to understand the biology of the nodulation response. The soybean nodulation response involves an initial attachment of Rhizobium cells to the root surface in the zone of emerging root hairs. Only this region is infectable. The presence of Rhizobium stimulates subepidermal divisions of cortical cells. At the same time, the proliferating Rhizobium colony on the root surface induces curling of the emerging root hair. Upon enclosure between 2 adjacent plant cell walls (being either root hair walls or even a hair wall and an epidermal wall), the Rhizobium causes the invagination of the root hair cell wall, thus forming the early infection thread. This thread (made up of plant cell wall components with proliferating bacteria in them) penetrates the back wall of the epidermal cell and invades the proliferating plant tissues of the hypodermal region. Lytic functions (either plant- or bacterial-derived) are clearly necessary for such penetration, as they also seem to be for the initial entry into the root hair. The infection thread ramifies, invading an increasing number of plant cells. The developing nodule meristem induces a secondary meristem in the pericycle tissues (which normally are the source of lateral roots). The secondary meristem enlarges and eventually breaks the endodermis and grows towards the nodule meristem. By this time the nodule is visible (7 days after inoculation). As the nodule meristematic region increases by division and as the infection proceeds into further cells, the pericycle-derived tissues progressively induce tissue surrounding the infected zone into division. This gives rise to the final nodule structure of a determinate type (as compared to the meristematic types as seen in clovers, alfalfa, and peas), in which the infected zone is surrounded by a boundary cell layer, then an inner cortex with few airspaces between cells, then a region of peripheral vascular bundles, and finally an outer cortex (with airspaces). The outer cortex of soybean nodules develops large cells with thickened cell walls (scleroid type). Additionally, one finds "puffings" on the surface, called lenticels, which may be involved with gas exchange of the nodule.

During these plant cellular developments several other symbiotically important steps occur. The infection thread-contained bacteria are released into plant membrane vescicles called the peribacteroid membrane. Division proceeds for 1 to 2 cycles, so that 2 to 4 bacteria are found per vescicle. The bacteria then stop dividing, although they still retain their ability to form colonies if reisolated (9). The bacteria differentiate into bacteroids, which are the nitrogen-fixing forms inside nodules. The enzyme nitrogenase and related symbiotic functions are derepressed about 12 days after inoculation. Prior to this the plant produces an oxygen carrier protein called leghemoglobin, which is located diffusely in the plant cytoplasm and provides the bacteroid with a low, but steady, supply of oxygen (1). The leghemoglobin genes of soybean have been cloned and analyzed (13). Leghemoglobins are found in all legumes, and recently hemoglobin proteins were also discovered in non-legumes, such as Parasponia, which nodulate either with Rhizobium (2,11,15) or Alnus, Myrica, Datisca, and Casurina species, which nodulate with the fungal-like (but prokaryotic) bacterium Frankia. Figure 1 illustrates some of the structural components of the legume and the non-legume (Parasponia) nodule as formed by the same bacterial inoculant (Rhizobium strain ANU289). At this stage it is also worthwhile to stress that the above general description of nodule ontogeny is based in part on weak evidence. More detailed studies, especially with attention to the fate of cell lineages, are needed. Additionally, one finds great variability between legume species in regard to their precise mechanism of nodule initiation. (For example, peanuts fail to show root hair infection threads, yet they nodulate.)

The nitrogen-fixing bacteroid converts N_2 gas to ammonia which is assimilated by plant-derived mechanisms, such as glutamine synthetase, to form nitrogenous translocation compounds like ureides or amines. The overall nitrogenase activity found in soybean nodules seems to be regulated by the amount of photosynthetate supplied, and, in the short term, by the amount of oxygen which is permitted to diffuse into the nodule interior. For a detailed review of nitrogen fixation, the reader is directed to the reviews and books (4,17-20).

There exists a nexus in legumes between nitrate utilization and symbiotic nitrogen fixation. This is best illustrated by field data of 7 independent crops (Tab. 1). The percentage of nitrogen fixation has a negative relationship to the initial nitrate content of the soil. Additionally, whether the nitrogen was obtained from nitrate or the atmosphere had no influence on the final grain yield. Nitrate inhibits nodule initiation and nodule growth, as well as the functioning of the established nitrogenase system. Mutants isolated in our laboratory indicate that the early events of nodulation initiation and growth are controlled via nitrate by a different mechanism than that which acts to control the later (i.e., fixation) stages. Recent experiments by Bernie Carroll and David McNeil in our laboratory suggest that, in soybean nodules, nitrate-induced nodule senescence is initially induced by an oxygen supply-related mechanism (7). It is not known whether the mode of action works through an alteration of the oxygen diffusion to the bacteroid, or through the increased demand by the bacteroid for oxygen as an alternative carbon source is supplied. Agriculturally, this nitrate

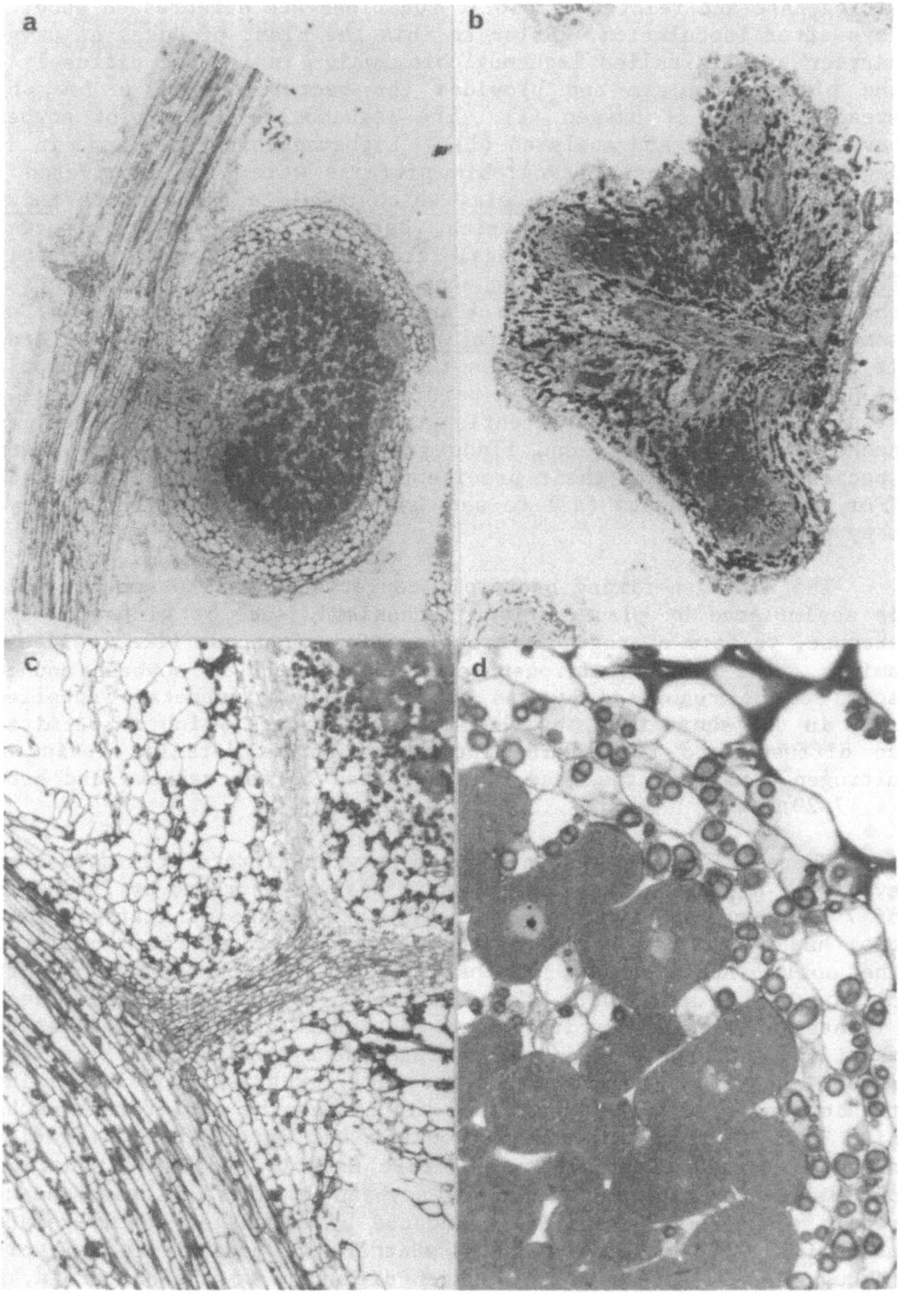
a
b
c
d

Fig. 1. Nodule structure in a legume (Macroptilium atropurpureum) and a non-legume (Parasponia rigida). (a) Young nodule of the determinate nodule type (diameter about 1 mm) as also found in soybean. Note the central invaded zone, which also contains cells without bacteroids. These interstitial cells are involved in ammonia assimilation. On the left side of the root one also notes the emergence of a lateral root. (b) A young Parasponia rigida nodule (length about 1 mm) illustrating non-legume nodule morphology. The vascular system is central, meristematic, and multilobed. Rhizobium bacteria remain in the infection thread and fix nitrogen there. (c) Close-up of the vascular junction of the legume nodule. Note starch grains in cortical cells of the nodule. (d) Close-up of the cell types in the legume nodule. Bacteroid and interstitial cells are encased by a boundary cell layer and the inner cortex (upper right). Note the hypertrophy of bacteroid-filled cells and the lack of a large visible vacuole. Clear space is the plant cell nucleus.

Tab. 1. Field data for 7 soybean crops grown in Australia: Nitrate effects on nitrogen fixation.

Crop	Soil NO_3 at sowing	N_2 fixed	Crop N uptake	% fixation	Grain yield
1	234	189	350	64	3930
2	369	76	287	26	3030
3	98	288	393	73	4550
4	117	248	568	44	4180
5	308	111	298	37	3370
6	83	268	450	60	2540
7	201	73	290	25	3150

All units in kg per ha. All data were obtained by Dr. David Herridge (N.S.W. Department of Agriculture, Tamworth).

inhibition lowers the amount of nitrogen fixed by a crop, and thus increases the amount of nitrate withdrawn from the soil (Tab. 1). Thus, grain legumes, in general, lower the fertility of the soil as the assimilated nitrogen is removed with the grain crop. This means that the subsequent crop, for example a cereal like corn or wheat, requires a substantial amount of nitrogenous fertilizer to crop well. Additional advantages from a nitrate-tolerant symbiosis could be achieved through an optimization of infection by the seed-carried inoculant. In general, one finds that soil nitrate delays the onset of nodulation, thereby increasing the chance for non-inoculant Rhizobium (such as strain USDA 123), which fix nitrogen less efficiently, to establish the majority of nodules throughout the soil profile. Mutant soybeans which have the nts-phenotype nodulate quickly and thus optimize the inoculant's chances.

The induced mutation program yielded a number of useful mutants. These not only included those which affect the symbiosis negatively, but also those which can be viewed as "up" mutations. Mutant nts 382 has been characterized most extensively (14). It nodulates (up to 1000 nodules of 1-3 mm diameter per 4-5 week-old root system) on 5 mM KNO_3, supplied daily, in glasshouse trials. The parent Bragg, in contrast, shows 30-50 nodules. The specific nitrogenase activity is identical in the nitrate-treated mutant and wild-type. However, because of the increased (35-fold) nodule mass, total nitrogen fixation per plant on nitrate is substantially higher in the mutant than in the wild-type. The nts phenotype does not appear to be strain-specific and can be demonstrated in a variety of growth media, such as sand, vermiculite, Leonard jars, potting mix, and black vulcanic soil [high nitrate content; obtained from a soybean field at the Breeza (N.S.W.) experimental station]. Mutant nts 382 is stable and inherited in a recessive fashion giving a 3:1 segregation in the M3 out of wild-type M2 sibling plants of selected M2 nts plants. In contrast, mutant nts 733 segregates from an intermediate M2 phenotype 3 (!) phenotypic classes (i.e., nts:intermediate nts:wild-type) at a ratio of 1:2:2, which because of low sample numbers is not statistically different from the expected 1:2:1.

The nitrate tolerance of mutant nts 382 is also expressed as urea or ammonia tolerance. In the absence of exogenously supplied, nitrogenous compounds, nts 382 still exhibits supernodulation by a factor of 10 compared to Bragg controls. Nitrogen fixation of such plants is between 50 to 100% higher than with wild-type controls, suggesting that although the nodule mass is enormously larger, restrictions (most likely in terms of photosynthetic supply or ammonia sink relations) exist on the whole plant level. Mutant plants seem to be set back during the first 4 weeks because of heavy nodulation. The nodule mass on nts 382 may be as much as 69% of the whole root mass. This large mass of tissue, if left in the soil after harvest, is expected to contribute significantly to the nitrogen status of the soil.

Complementation analysis of the nts mutants is presently in progress. In broad phenotypic terms we can distinguish 3, or possibly 4, classes, suggesting that at least 3 genetic elements are involved. A preliminary description of mutant nts 382 is to be published by McNeil et al. (14).

Several key questions remain. What are the nature and function of affected proteins? What regulates their expression? Are they conserved in other legumes or even other dicots? Is the mutant nts 382 altered in the infection process?

One can build some ontogenetic models which can be tested. For example, as a soybean root develops from a seed (under ideal nodulation conditions) one finds excessive nodulation in the "crown" (i.e., hypocotyl) region. This heavy nodulation subsides as the root develops. The overall process is labelled autoregulation and functions prior to the onset of nitrogen fixation as illustrated by recent published work by Kosslak and Bohlool (12) and Bauer (3). It is thus conceivable that nts 382 is altered in the autoregulation response, and that nitrate tolerance is seen because the nitrate (ammonia) regulation of nodulation normally uses part of the autoregulation control circuit. Thus, one can visualize a set of "symbiotic" signals, being either small metabolites, polysaccharides, proteins, or hormones, which permits the early "supernodulation." Whichever is involved, it will be the product of a protein function.

As the root grows, however, autoregulation sets in, and these symbiotic signals (or one critical component) either are actively suppressed by a "non-symbiotic" suppressor or are no longer manufactured at all. The first alternative of a positive suppressor function in autoregulation is supported by the nts 382 molecular phenotype.

In the foregoing, I have tried to demonstrate that for the genetic-biochemical-physiological analysis of an ontogenetic pathway and the genetic modification of a crop plant, one does not necessarily require plant cell culture approaches. This is not to say that cell culture has no application. There are tasks which are best approached through in vitro cellular technology. However, many aspects of plant biology still are solvable at the whole seed level. What is affected is the scale; thus, rather than using an Erlenmeyer flask, we used home-beer-making containers, and instead of a set of petriplates, we needed one hectare of land and, later, several glasshouses. However, this is within normal scales and the time and labor involvements are similar to those of present day cell culture technology.

It will be the combination of molecular biology, classical genetics, and cell biology, with their technologies appropriately applied and their experimental perspective aiding one another, that will manage to find out more about the development of plants.

REFERENCES

1. Appleby, C.A. (1984) Leghemoglobin and Rhizobium respiration. Ann. Rev. Plant Physiol. 35:443-478.
2. Appleby, C.A., J.D. Tjepkema, and M.J. Trinick (1983) Hemoglobin in a nonleguminous plant, Parasponia: Possible genetic origin and function in nitrogen fixation. Science 220:951-953.
3. Bauer, W.D. (1981) Infection in leguminous plants. Ann. Rev. Plant Physiol. 32:407-447.
4. Bergersen, F.J. (1980) Methods for Evaluating Biological Nitrogen Fixation, John Wiley and Sons, Chichester, New York, Brisbane, Toronto, 702 pp.
5. Carroll, B.J. (1984) Symbiotically altered mutants of soybean. Ph.D. Thesis, Australian National University, Canberra.
6. Carroll, B.J., D.L. McNeil, and P.M. Gresshoff (1985a) Isolation and properties of novel soybean mutants that nodulate in the presence of high nitrate concentrations. Proc. Nat. Acad. Sci. (USA) (submitted for publication).
7. Carroll, B.J., D.L. McNeil, and P.M. Gresshoff (1985b) Oxygen supply to nodules limits nitrogenase activity in nitrate-inhibited soybeans. Plant Physiol. (submitted for publication).
8. Gresshoff, P.M. (1978) Phytohormones and their role in differentiation and growth of plant cells in vitro. In Phytohormones and Related Compounds - A Comprehensive Treatise, D.S. Letham, T.J. Higgins, and P.J. Goodwin, eds. Elsevier/North Holland Press, pp. 1-29.
9. Gresshoff, P.M., and B.G. Rolfe (1978) Viability of Rhizobium bacteroids isolated from soybean nodule protoplasts. Planta 142:329-333.
10. Gresshoff, P.M., and S.S. Mohapatra (1982) Legume cell and tissue culture. In Tissue Culture of Economically Important Plants, A.N. Rao, ed. Costed/ANBS Publ., Singapore, pp. 11-24.
11. Gresshoff, P.M. et al. (1984) The Parasponia/Rhizobium nitrogen fixing symbiosis: Genetics, biochemistry, and molecular biology of a plant and bacterium. In Proceedings of 15th International Congress of Genetics, New Delhi, India (in press).
12. Kosslak, R.M., and B.B. Bohlool (1984) Suppression of nodule development on one side of a split-root system of soybean caused by prior inoculation of the other side. Plant Physiol. 75:125-130.
13. Marcker, K.A., K. Bojsen, E.O. Jensen, and K. Paludan (1984) The soybean leghemoglobin genes. In Advances in Nitrogen Fixation Research, C. Veeger and W.E. Newton, eds. Martinus Nijhoff/Dr. W. Junk Publishers, The Hague, Wageningen, pp. 573-578.
14. McNeil, D.L., B.J. Carroll, and P.M. Gresshoff (1985) A supernodulation and nitrate tolerant symbiotic (nts) of soybean mutant. Plant Physiol. (submitted for publication).
15. Price, G.D., S.S. Mohapatra, and P.M. Gresshoff (1984) Structure of nodules formed by Rhizobium strain ANU289 in the nonlegume Parasponia and the legume siratro (Macroptilium atropurpureum). Bot. Gazette (in press).

16. Scott, K.F., J. Hughes, P.M. Gresshoff, J. Beringer, B.G. Rolfe, and J. Shine (1982) Molecular cloning of symbiotic genes from Rhizobium trifolii. J. Molec. Appl. Genet. 1:315-326.
17. Szalay, A.A., and F. Ausubel (1984) Proceedings of the Second International Symposium on the Molecular Genetics of the Bacteria-Plant Interaction, Ithaca, N.Y. (in press).
18. Veeger, C., and W.E. Newton (1984) Advances in Nitrogen Fixation Research, Martinus Nijhoff/Dr. W. Junk Publishers, The Hague, Wageningen, 760 pp.
19. Verma, D.P.S., and S. Long (1983) The molecular biology of Rhizobium-legume symbiosis. Int. Rev. Cytol., Suppl. 14 (Intracell Symbiosis):211-245.
20. Vincent, J.M. (1982) Nitrogen Fixation in Legumes, Academic Press, Sydney, New York, London, 288.

APPLICATION OF TISSUE CULTURE PROPAGATION TO WOODY PLANTS

Richard H. Zimmerman

U.S. Department of Agriculture
Agricultural Research Service
Horticultural Science Institute
Fruit Laboratory
Beltsville, Maryland 20705

INTRODUCTION

During the past decade, commercial production of woody plants by tissue culture techniques has progressed from a future possibility to a rapidly expanding reality. The magnitude of this change is truly remarkable, both in terms of the number of plants produced and the number of species and cultivars now in production. Much of this expansion is now well documented in either the scientific or popular literature but must be obtained from other sources such as nursery catalogs and personal contacts.

If one reviews the status of woody plant propagation as reported in the proceedings of the first 2 symposia in this series (16,25), one is struck by the total lack of reference to actual production of woody plants (9,37,42). However, by 1978, commercial production of some woody plants had already begun, notably the production of rootstocks for apples and peaches in Europe (mainly Italy, France, and England). Interest in the possibilities of using the technique commercially was so high that 2 conferences on this topic were held within 6 months in late 1979 and early 1980 (7,52). Since then, production has expanded to include numerous species and cultivars representing at least 50 genera of ornamental, forest, and fruit plants (Tab. 1). Some of these genera represent numerous species and cultivars. For example, the catalog for one United States nursery now lists 150 azalea and rhododendron species and cultivars available from tissue culture with more than 30 additional types scheduled for introduction next year.

Tab. 1. Genera of woody plants now being commercially produced in tissue culture.

Acer	Elaeis	Potentilla
Actinidia	Escallonia	Prunus
Amelanchier	Eucalyptus	Pyrus
Arctostaphylos	Ficus	Rhododendron
Betula	Forsythia	Ribes
Buddleia	Garrya	Rosa
Camellia	Hydrangea	Rubus
Campsis	Hypericum	Salix
Castanea	Kalmia	Sequoia
Celtis	Lagerstroemia	Sequoiadendron
Clematis	Lapageria	Simmondsia
Corylopsis	Leucothoe	Spiraea
Corylus	Magnolia	Syringa
Cotinus	Malus	Vaccinium
Crataegus	Nandina	Viburnum
Daphne	Pinus	Vitis
Deutzia	Populus	Weigela

Such rapid changes in production techniques have not come without problems, however, and some of the problems are directly related to the fact of rapid change. Many of the current problems have an economic basis and the need to reduce the cost of producing woody plants in vitro has become evident. Another concern is the field performance of tissue culture-propagated plants, which must be evaluated for phenotypic stability as well as general growth performance in relation to the use for which the plant is grown.

This paper will survey the current status of tissue culture propagation of woody plants, examine recent developments in techniques, and consider problems that need to be resolved in order to increase efficiency and validate the method for large-scale plant production.

DEVELOPMENT AND APPLICATION OF WOODY PLANT TISSUE CULTURE

Historical Development

The development of tissue culture as a method for propagating plants has been reviewed numerous times with the review of Murashige (38) being perhaps the most comprehensive. In that review, only a few woody plants are cited as having been propagated, or displaying a potential for propagation, by tissue culture methods. However, the successful propagation of some tree species, e.g., Populus tremuloides (49), foreshadowed the rapid expansion in application of tissue culture techniques to propagation of woody plants (27,31,36, 37,53).

Applications

Fruit crops. Most of the applications of tissue culture propagation of woody plants in the past decade have been with horticultural crops. Of these, the fruit crops were the first for which large-scale production was initiated. Production of apple and peach rootstocks in vitro was perhaps the first major application of tissue culture methods to woody horticultural crops. For apple, the research on which this development was based was conducted at several centers, but the work done at Gembloux, Belgium (41), and at East Malling (26) and Long Ashton (1) in England, was particularly relevant to this application. The report of Jones (26) on the efficacy of phloroglucinol in enhancing shoot production and rooting of the apple rootstocks M.7 and M.26 clearly stimulated both research and development efforts in this field. Within 3 years of the publication of Jones' report, large-scale commercial production of apple rootstocks was underway in England and Italy (51). This production has continued in Italy (32) and has expanded to production facilities in other countries of Europe and North America.

During the same period, meristem-tip culture techniques were developed for a number of Prunus species and cultivars (10). These techniques were used and adapted for mass propagation of Prunus rootstocks suitable for peach (56,57), cherry (28), and plum (28) simultaneously with the work on apple rootstocks (51). Subsequent production of rootstocks for peach has probably exceeded that for the other fruit crops, particularly in Italy where nearly 9 million of these rootstocks had been produced by mid-1982 (32).

Production of raspberry, blackberry, and related cane fruits began about the same time as, or perhaps slightly earlier than, that on apple and peach rootstocks. Commercial cane fruit propagation in vitro was based on research undertaken independently in several countries (5,13,15,22). The methodology developed for mass propagation of strawberry (11) served as one guide in the development of these techniques.

More recently, cultivars of various fruit trees (apple, peach, pear, plum) growing on their own roots have been produced in limited quantities by several commercial laboratories for evaluation of the usefulness of self-rooted cultivars. Since the testing required for these trees will take a number of years, it is unlikely that self-rooted cultivars will be in mass production until the end of this decade or later.

Other fruit and nut crops, such as blueberry, grape, and filbert, are now commercially propagated in vitro on a limited scale, but blueberry propagation by this method seems to be increasing rapidly. Recently, a method for producing walnut rootstocks in vitro has been developed and is being introduced for commercial production at the present time (19).

Ornamental plants. With woody ornamental plants, the adoption of tissue culture propagation has been even more rapid than with fruit crops. Two examples will illustrate where research on a particular crop was introduced quickly into a commercial production system. In both cases, close cooperation between research workers and the commercial laboratory operator was instrumental to success of the commercial venture.

Use of tissue culture to produce rhododendrons and azaleas is perhaps the most striking example because the method is now used by laboratories throughout the world. The results on rhododendron tissue culture reported by Anderson (2-4) were the culmination of research begun in 1968 (12). A close liaison between the researcher and commercial growers resulted in the rapid adoption of these techniques by at least 2 large growers (4,30). These commercial growers have gone on to refine and modify the methods to meet their own needs and have now reached the stage at which each are producing approximately 1,000,000 plants per year of numerous cultivars and species of rhododendron and azalea. Research on tissue culture of azaleas was underway simultaneously at other research institutions (e.g., Ref. 33, 40) but this other research had little, if any, influence on the commercial laboratories described above, although it may have had an impact in other parts of the world.

A second example is that of roses for which research on in vitro propagation was underway simultaneously at several locations (23,24,34,43). A large French nursery, specializing in rose production, became interested in using tissue culture techniques (18) and established a cooperative working arrangement with the I.N.R.A. station in Dijon (34). As a result, the nursery established a large tissue culture laboratory starting in 1980 and has more recently built a second, similar laboratory in California (H. Delbard, pers. comm.). Production of roses at the first of these laboratories reached approximately 500,000 plants by 1983 with nearly as many plants of other woody species also produced in tissue culture.

Forest trees and plantation crops. Applications of tissue culture propagation to forestry have come more slowly but trees of *Pinus radiata* are now being micropropagated on a modest scale in New Zealand (R.L. Mott, pers. comm.). The starting material for this production is seed obtained from controlled pollinations (44,47). Other forest species may be propagated in this way once more is learned about the characteristics of the trees produced and a determination is made of the economic viability of the method.

Similarly, plantation crops, such as oil palm (*Elaeis guineensis*), can be produced on a commercial scale and it now appears that this is being done in England (21). Successful application of tissue culture to the propagation of such crops will depend upon the firms that use this method developing a viable and reliable market for their product.

Other Considerations

Several points need to be made regarding the development of the industry outlined above. First, the method of propagating plants has been almost exclusively by production of axillary and/or adventitious shoots from shoot tips, or in the case of Pinus radiata, embryos, and cotyledons. This technique was derived mainly from that used so successfully with numerous herbaceous ornamental plants and with strawberry. It has also been the simplest method and undoubtedly will continue to be the main one for woody plants for some time to come. Somatic embryogenesis, theoretically a less expensive and more productive technique, has not been possible with most woody plants. However, it has been used for the production of oil palm (8,48) and one cultivar of grape (29). This method looks promising for these crops, but extended testing for phenotypic stability is necessary before the method will be widely accepted by growers.

Second, laboratories established to produce specific crops have sometimes found it necessary to diversify in order to utilize their facilities throughout the year and because the cost of producing certain types of plants was too high. Thus some laboratories, particularly in North America, started with production of fruit tree rootstocks as the main goal but found that production costs were so high that it was difficult to compete with rootstocks propagated conventionally in stool beds. Nevertheless, some production of these rootstocks in vitro continues in North America to fill specific needs, but the method is unlikely to supplant conventional techniques unless the costs can be reduced significantly. In Europe, however, production of fruit tree rootstocks in vitro will continue to be important for some time because the rootstocks being propagated are new, in short supply, difficult to propagate conventionally, or a combination of these factors.

When the laboratories begin to diversify their production, they shift to crops that have a higher unit value, e.g., shade and ornamental trees such as maple, birch, and crabapple, rather than rootstocks. In doing so, they conduct in-house research on a number of new crops, the results of which are often not reported in the literature.

PROBLEMS FACING WOODY PLANT TISSUE CULTURE

Navatel (39) divided the problems facing applications of large-scale tissue culture propagation into 2 types: economic and technical. The problems currently facing woody plant tissue culture and its application on a large scale can be similarly categorized, although certain of the problems have components in each category.

Economic Problems

Labor costs. Clearly the most serious obstacle to greater utilization of in vitro plant propagation is the fact that this method is so labor intensive. Labor costs account for 60 to 80% of the total costs of producing plants in vitro, whether herbaceous (6,45) or woody (14). Thus any significant progress in cost reduction must be done by reducing labor input at all stages of plant production.

Reducing labor input can be done by eliminating steps in the production process or by simplifying the procedures used. A good example of eliminating a step in the process is the consolidation of rooting and acclimatization into a single step. To do this, in vitro techniques are used only for explant establishment and proliferation. After shoots are harvested, aseptic techniques are not used for rooting. Shoots are rooted, with or without auxin pretreatment, by inserting them into various types of media or preformed rooting plugs or blocks and placing them under mist or conditions of high humidity provided by a fog system or a plastic tent. Rooting occurs as rapidly under these conditions as it does in vitro, but shoot growth begins sooner than in vitro. Woody plants for which this method is now used commercially include *Rhododendron*, *Kalmia*, *Amelanchier*, *Betula*, *Malus* (apple rootstocks), *Vaccinium*, and *Syringa*.

Simplification of procedures can also save labor. One reported technique successful with ferns involves homogenizing a culture briefly in a blender and then incorporating the slurry of plant tissue in partly cooled medium before dispensing it into containers, or by pouring the slurry over already solidified agar medium (17). Harvesting shoots from proliferating cultures can be speeded by removing all the shoots from a large cluster with a single stroke of the scalpel and grading them for size later at the time they are inserted into rooting medium. This is usually done when rooting is done under greenhouse conditions. With this technique, the cluster of shoot bases is recultured for several passages creating in effect a small stool bed in vitro. For plants that are a bit more difficult to root, it is sometimes necessary to place them on a rooting medium in vitro for a few days in order to get root initiation, and then to transfer the shoots to an appropriate rooting medium where root development takes place simultaneously with acclimatization. This is used for some *Rhododendron* cultivars (B. Briggs, pers. comm.) and for apple cultivars (55).

Scaling up. Reports abound in the literature and popular articles regarding the tens of thousands or more plants that can be produced from a single bud, meristem, or shoot tip within a year and, in theory, these reports are correct. However, reality is quite different because the facilities and labor required to handle the cultures after the fifth or sixth subculture are so great that the situation becomes impractical if not physically impossible. The data available in published research reports that deal with the production of a few dozen or even a few hundred plants are not adequate when production schedules must be developed for thousands of plants. Planning is required for utilization of transfer and growth room facilities just as with herbaceous plants. In addition, provision must be made for moving woody plants outdoors where they will be grown to larger size before ultimate distribution. Although the plants can be produced in the laboratory throughout the year, they can be planted outside for approximately half that time, depending upon climatic conditions and location of the laboratory. Furthermore, customers purchasing young tissue-cultured woody plants impose further restrictions since they want to obtain the plants during an even shorter time in spring and early summer. This requires the development of different production strategies and facilities. For example, it may be necessary to proliferate the cultures necessary for the required number of shoots and then store either the cultures or the harvested shoots prior to rooting during the winter. Alternatively, once plants are rooted, they may be stored at low temperature in vitro or after acclimatization and establishment in soil.

An example of the scale required for testing a new crop in tissue culture is the procedure used for some new peach-almond hybrid rootstocks used for peaches. Nearly one year of research effort by the commercial laboratory was required before it was possible to have a trial run of 5,000 plants per week for 3 weeks (A. Martinelli, pers. comm.). After this trial proved the efficacy of the method, production trials were started with a goal of 20,000 plants per week until more than 100,000 plants had been produced; a final scaling up to 60,000 plants per week enabled the production of 300,000 plants in the summer of 1984. Production was stopped by mid-August in order to allow time for the plants to become fully established in the nursery or in protected growing beds before autumn. With this experience, production schedules for future years can be set depending upon the demand for this particular rootstock.

Technical Problems

Culture establishment. Selecting appropriate material for establishing cultures of some types of woody plants continues to be a problem, especially with conifers and certain genera of dicotyledonous trees, e.g., *Juglans*, *Castanea*, *Quercus*, and *Pistacia*. One of the greatest of these problems results from the fact that trees are not selected for propagation until they are mature at which time their establishment in culture is much more difficult. One solution proposed by Franclet (20), which has been used effectively on some conifers, is to graft scions from the selected mature tree onto a seedling rootstock. Once the graft is successful and the scion has grown, the tip of the new shoot is grafted on another seedling. This procedure is repeated until the scion exhibits juvenile characteristics or responds like a juvenile plant when placed in culture.

A second problem is browning or discoloration of the explants when placed in culture from the production of phenols or polyphenols. This has been overcome by including antioxidants in the medium, by using liquid medium for explanting before transferring to solid medium, and by transferring explants to fresh medium daily for periods up to several months when necessary.

Culture contamination. A continuing problem with woody plants is bacterial contamination of the cultures at the time of explanting or later during proliferation. With plants that have pubescent or villous shoots and buds, the problem may arise from inadequate disinfestation procedures. In many, if not most, cases however, the problem seems to be bacteria growing internally in the tissue. The problem was summarized recently for apple (54).

Culture indexing at the time of explant establishment is advisable, but some contaminants appear only after a number of subcultures and are not always detected by common indexing procedures. For these cases, the problem can be overcome in some cases by growing proliferating cultures under conditions favoring rapid shoot elongation, e.g., low cytokinin concentration in the medium and/or dark, then culturing shoot tips (5 mm or less) from these shoots. Cultures from such shoot tips will often be found to be free of bacterial contaminants. Recently a technique using 4 antibiotics simultaneously has been reported to control bacterial infections in a number of different plant genera (50).

Field performance and phenotypic stability. Any propagation system is useful only to the extent that the regenerated plants are faithful copies of the clone being increased. However, variation in plant type is probably less of a problem with most ornamental trees and shrubs so long as the regenerated plants are not deformed, have no other obvious defects, and resemble the parent clone in most, if not all, characteristics so that the plant fulfills the general purpose for which it is to be used. With fruit crops, forest trees, and plantation crops, however, phenotypic stability is essential and the way that this must be measured ultimately is through an evaluation of the field performance of the regenerated plants. In the case of most of these crops, that means long-term field studies of regenerated plants. Such time-consuming, expensive evaluations are necessary, nonetheless, to prove the usefulness of in vitro propagation. Once substantial field experience has been obtained with such plants, it will probably be feasible to eliminate or greatly curtail this type of testing. Plants produced from tissue culture are generally more vigorous vegetatively than ones produced by conventional methods [(46); Zimmerman, unpub. data]. Nevertheless, they can still fruit heavily at an early age and have a higher early yield than conventionally propagated controls (35,46). The expression of this potential for early heavy fruiting will vary with the crop and with the cultural methods used on the plants once they are established in the field.

FUTURE PROSPECTS

One can anticipate that tissue culture of woody plants as a propagation method will continue to grow rapidly in the next few years, at least for horticultural crops. This is clear from the number of new laboratories being built and existing laboratories being enlarged. Nevertheless, the number of plants propagated by this method will still be only a small percentage of the total annual production of woody plants. This expansion will be mainly with those crops for which the method has proved to be particularly well suited. Production of additional crops will depend upon development of suitable methods, and then upon 2 factors: the cost of production relative to the conventional propagation methods for the same crop, and the characteristics of plants produced by in vitro culture methods in comparison to those of plants produced by the conventional methods.

Development of somatic embryogenesis as a propagation method will continue and will expand as more experience is gained with those crops, such as oil palm, for which it is now used. One can anticipate that it will become feasible for some forest tree species; once this happens, large-scale production of forest trees in vitro will become a reality. Somatic embryogenesis will be little used for woody plants important in horticulture for some time to come despite the work of Krul and coworkers with 'Seyval' grape (29). Less effort is being put into research on somatic embryogenesis of woody horticultural crops and the requirement for clonal fidelity means many years of field testing will be required once methods are developed.

Production of plants by stimulation of axillary, and in some cases, adventitious shoot production will continue to be the main propagation method for woody plants. The method is well developed, it has proved to be effective in commercial production, and it produces plants with acceptably low percentages of aberrant types.

REFERENCES

1. Abbott, A.J., and E. Whitely (1976) Culture of _Malus_ tissues in vitro. I. Multiplication of apple plants from isolated shoot apices. _Scientia Hort._ 4:183-189.
2. Anderson, W.C. (1975) Propagation of rhododendrons by tissue culture: Part 1. Development of a culture medium for multiplication of shoots. _Comb. Proc. Int. Plant Prop. Soc._ 25:129-135.
3. Anderson, W.C. (1978a) Tissue culture propagation of rhododendrons. _In Vitro_ 14:334 (Abstr.).
4. Anderson, W.C. (1978b) Rooting of tissue cultured rhododendrons. _Comb. Proc. Int. Plant Prop. Soc._ 28:135-139.

5. Anderson, W.C. (1980) Tissue culture propagation of red raspberries. In Proceedings of the Conference on Nursery Production of Fruit Plants Through Tissue Culture - Applications and Feasibility. U.S. Dept. Agr., SEA, ARR-NE-11, Beltsville, Maryland, pp. 27-34.
6. Anderson, W.C., G.W. Meagher, and A.G. Nelson (1977) Cost of propagating broccoli plants through tissue culture. HortScience 12:543-544.
7. Bellini, E., F. Loreti, P. Rosati, F. Scaramuzzi, R. Tesi, and F. Tognoni (1979) Tecniche di colture "in vitro" per la propagazione su vasta scala delle specie ortoflorofrutticole. S.O.I., Firenze, 236 pp.
8. Blake, J. (1983) Tissue culture propagation of coconut, date and oil palm. In Tissue Culture of Trees, J.H. Dodds, ed. Croom Helm, London, pp. 29-50.
9. Boxus, P. (1978) The production of fruit and vegetable plants by in vitro culture. Actual possibilities and perspectives. In Propagation of Higher Plants through Tissue Culture, A Bridge Between Research and Application. K.W. Hughes, R. Henke, and M. Constantin, eds. Tech. Inf. Cent., U.S. Dept. Energy, Oak Ridge, Tennessee, pp. 44-58.
10. Boxus, P., and M. Quoirin (1974) La culture de méristèmes apicaux de quelques espèces de Prunus. Bull. Soc. Roy. Bot. Belg. 107:91-101.
11. Boxus, P., M. Quoirin, and J.M. Laine (1977) Large scale propagation of strawberry plants from tissue culture. In Applied and Fundamental Aspects of Plant Cell, Tissue and Organ Culture, J. Reinert and Y.P.S. Bajaj, eds. Springer-Verlag, Berlin, Heidelberg, New York, pp. 130-143.
12. Briggs, B.A., and S.M. McCulloch (1983) Progress of micropropagation of woody plants in the United States and western Canada. Comb. Proc. Int. Plant Prop. Soc. 33:239-248.
13. Broome, O.C., and R.H. Zimmerman (1978) In vitro propagation of blackberry. HortScience 13:151-153.
14. Brown, C.L., and H.E. Sommer (1982) Vegetative propagation of dicotyledonous trees. In Tissue Culture in Forestry, J.M. Bonga and D.J. Durzan, eds. Martinus Nijhoff/Dr. W. Junk Publishers, The Hague, pp. 109-149.
15. Carré, M., J. Martin-Tanguy, P. Mussillon, and C. Martin (1979) La culture de méristèmes et la multiplication végétative "in vitro" au service de la pépinière. Bull. Petits Fruits 14:7-65.
16. Constantin, M.J., R.R. Henke, K.W. Hughes, and B.V. Conger, eds. (1981) Propagation of higher plants through tissue culture: Emerging technologies and strategies. Environ. Exp. Bot. 21:269-452.
17. Cooke, R.C. (1979) Homogenization as an aid in tissue culture propagation of Platycerium and Davallia. HortScience 14:21-22.
18. Delbard, H. (1982) Micropropagation of roses at Delbard Nurseries. The Plant Propagator 28(3):7-8.

19. Driver, J.A., and A.H. Kuniyuki (1984) In vitro propagation of Paradox walnut rootstock. HortScience 19:507-509.
20. Franclet, A. (1980) Rajeunissement et propagation végétative des ligneux. Ann. AFOCEL 1980:12-41.
21. George, E.F., and P.D. Sherrington (1984) Plant Propagation by Tissue Culture, Exegetics Ltd., Eversley, Basingstoke, 709 pp.
22. Harper, P.C. (1978) Tissue culture propagation of blackberry and tayberry. Hort. Res. 18:141-143.
23. Hasegawa, P.M. (1979) In vitro propagation of rose. HortScience 14:610-612.
24. Hasegawa, P.M. (1980) Factors affecting shoot and root initiation from cultured rose shoot tips. J. Amer. Soc. Hort. Sci. 105:216-220.
25. Hughes, K.W., R. Henke, and M. Constantin (1978) Propagation of Higher Plants through Tissue Culture, A Bridge Between Research and Application. Tech. Inf. Center, U.S. Dept. Energy, Oak Ridge, Tennessee, 305 pp.
26. Jones, O.P. (1976) Effect of phloridzin and phloroglucinol on apple shoots. Nature 262:392-393.
27. Jones, O.P. (1979) Propagation in vitro of apple trees and other woody fruit plants: Methods and applications. Scientific Hort. 30:44-48.
28. Jones, O.P. and M.E. Hopgood (1979) The successful propagation in vitro of two rootstocks of Prunus: The plum rootstock Pixy (P. insititia) and the cherry rootstock F12/1 (P. avium). J. Hort. Sci. 54:63-66.
29. Krul, W.R., and G.H. Mowbray (1984) Grapes. In Handbook of Plant Cell Culture; Vol. 2, Crop Sciences, W.R. Sharp, D.A. Evans, P.V. Ammirato, and Y. Yamada, eds. Macmillan Publ. Co., New York, pp. 396-434.
30. Kyte, L., and B. Briggs (1979) A simplified entry into tissue culture propagation of rhododendrons. Comb. Proc. Int. Plant Prop. Soc. 29:90-95.
31. Lane, W.D. (1982) Tissue culture and in vitro propagation of deciduous fruit and nut species. In Application of Plant Cell and Tissue Culture to Agriculture and Industry, D.T. Tomes, B.E. Ellis, P.M. Harney, K.J. Kasha, and R.L. Peterson, eds. University of Guelph, Ontario, pp. 163-186.
32. Loreti, F., and S. Morini (1982) Mass propagation of fruit trees in Italy by tissue culture: Present status and perspectives. Comb. Proc. Int. Plant Prop. Soc. 32:283-291.
33. Ma, S.S., and S.O. Wang (1977) Clonal multiplication of azaleas through tissue culture. Acta Hort. 78:209-215.
34. Martin, C., M. Carré, and R. Vernoy (1981) La multiplication végétative in vitro des végétaux ligneux cultivés: Cas de rosiers. C.R. Acad. Sci. Paris, Ser. III, 293:175-177.
35. Martin, C., M. Carré, and R. Vernoy (1983) La multiplication végétative in vitro des végétaux ligneux cultivés: Cas de arbres fruitiers et discussion generale. Rev. Agron. 3:303-306.

36. Mott, R.L. (1981) Trees. In Principles and Practices of Cloning Agricultural Plants Via In Vitro Techniques, B.V. Conger, ed. CRC Press, Boca Raton, Florida, pp. 217-254.
37. Mott, R.L., and R.H. Zimmerman (1981) Trees: Round table summary. Environ. Exp. Bot. 21:415-420.
38. Murashige, T. (1974) Plant propagation through tissue cultures. Annu. Rev. Plant Physiol. 25:135-166.
39. Navatel, J.C. (1980) L'utilisation des cultures in vitro pour la multiplication de quelques espèces légumières et fruitières. C.R. Acad. Agr. France 66:681-691.
40. Preil, W., and M. Engelhardt (1977) Meristem culture of azaleas (Rhododendron simsii). Acta Hort. 78:203-208.
41. Quoirin, M. (1974) Premiers résultats obtenus dans la culture in vitro du méristème apical de sujets porte-greffe de pommier. Bull. Rech. Agron. Gembloux 9:189-192.
42. Rediske, J.H. (1978) Vegetative propagation in forestry. In Propagation of Higher Plants through Tissue Culture, A Bridge Between Research and Application, K.W. Hughes, R. Henke, and M. Constantin, eds. Tech. Inf. Cent., U.S. Dept. Energy, Oak Ridge, Tennessee, pp. 35-43.
43. Skirvin, R.M., and M.C. Chu (1979) In vitro propagation of 'Forever Yours' rose. HortScience 14:608-610.
44. Smith, D.R., K.J. Horgan, and J. Aitken-Christie (1982) Micropropagation of Pinus radiata for afforestation. In Proc. 5th Int. Cong. Plant Tissue and Cell Culture, A. Fujiwara, ed. Japanese Assoc. for Plant Tissue Culture, Tokyo, pp. 723-724.
45. Strain, J.R. (1980) Analyzing costs in tissue culture laboratories. Staff Paper 167, Food and Resource Economics Department, University of Florida, Gainesville. 22 pp.
46. Swartz, H.J., G.J. Galletta, and R.H. Zimmerman (1983) Field performance and phenotypic stability of tissue culture-propagated thornless blackberries. J. Amer. Soc. Hort. Sci. 108:285-290.
47. Thorpe, T.A., and S. Biondi (1984) Conifers. In Handbook of Plant Cell Culture; Vol. 2, Crop Sciences, W.R. Sharp, D.A. Evans, P.V. Ammirato, and Y. Yamada, eds. Macmillan Publ. Co., New York, pp. 435-470.
48. Tisserat, B. (1984) Date palm. In Handbook of Plant Cell Culture; Vol. 2, Crop Species, W.R. Sharp, D.A. Evans, P.V. Ammirato, and Y. Yamada, eds. Macmillan Publ. Co., New York, pp. 505-545.
49. Winton, L.L. (1968) Plantlets from aspen tissue cultures. Science 160:1234-1235.
50. Young, P.M., A.S. Hutchins, and M.L. Canfield (1984) Use of antibiotics to control bacteria in shoot cultures of woody plants. Plant Sci. Lett. 34:203-209.
51. Zimmerman, R.H. (1979) The Laboratory of Micropropagation at Cesena, Italy. Comb. Proc. Int. Plant Prop. Soc. 29:398-400.
52. Zimmerman, R.H., ed. (1980) Proceedings of the Conference on Nursery Production of Fruit Plants through Tissue Culture - Applications and Feasibility. U.S. Dept. Agr., SEA, ARR-NE-11, Beltsville, Maryland, 119 pp.

53. Zimmerman, R.H. (1983) Tissue culture. In Methods in Fruit Breeding, J.N. Moore and J. Janick, eds. Purdue Univ. Press, West Lafayette, Indiana, pp. 124-135.
54. Zimmerman, R.H. (1984) Apple. In Handbook of Plant Cell Culture; Vol. 2, Crop Species, W.R. Sharp, D.A. Evans, P.V. Ammirato, and Y. Yamada, eds. Macmillan Publ. Co., New York, pp. 369-395.
55. Zimmerman, R.H., and I. Fordham (1983) Simplified method for rooting apple cultivars in vitro. HortScience 18:618 (Abstr.)
56. Zuccherelli, G. (1979) Moltiplicazione in vitro dei portainnesti clonali del pesco. Frutticoltura 41(2):15-20.
57. Zuccherelli, G., V. Venturi, and C. Damiano (1978) Rapid propagation on a vast scale of Damasco 1869 rootstock by in vitro culture. Ann. Ist. Sper. Frutt., Roma 9:21-23.

SOMATIC EMBRYOGENESIS IN TROPICAL FRUIT TREES

Richard E. Litz

Tropical Research and Education Center
Institute of Food and Agricultural Sciences
University of Florida
Homestead, Florida 33031

INTRODUCTION

The production of tropical fruits exceeds that of temperate fruits (8). Tropical fruits are not only important components of the diet in many countries, but their exportation provides a major source of income. This is particularly true of those fruit crops that can be grown on a large-scale plantation basis, e.g., bananas, papayas, etc. Additional export income is generated by the export of crude secondary products that can be recovered from some tropical fruits, e.g., papain from papaya. Thus, in areas of the tropics with few recoverable natural resources, tropical fruit production not only enhances the quality of life, but also provides a valuable and often irreplaceable source of income.

Many tropical fruits have been domesticated for several thousand years. The mango, for example, is known to have been prized in India for at least 4,000 years (5), and has become closely associated with the culture and religion of India. Both ancient and modern cultivars of perennial tropical fruit trees have been derived from seedling trees that resulted from uncontrolled pollinations. The horticultural qualities of superior cultivars are normally conferred by complexes of genes. These unique gene combinations are lost during sexual reproduction. Vegetative propagation has therefore been used for several thousand years to preserve the unique horticultural characteristics of highly-prized tree selections. Air layering, rooting of cuttings, and graftage methods were developed for species that could not reproduce asexually by suckering, e.g., bananas and plantains, or by the formation of adventitious embryos, e.g., some <u>Citrus</u> species and cultivars, and some mango cultivars.

Among naturally polyembryonic fruit trees, cultivars are propagated by seed. In the course of time, many clones have been modified by the isolation of bud sports or somatic mutations. The modern banana and plantain cultivars, both sterile triploids, have been entirely derived from earlier cultivars by means of somatic mutations during several thousand years of cultivation.

Perennial fruit trees can have juvenile periods that last 6-15 years. This, together with the genetic complexity of these trees, has meant that conventional plant breeding approaches have had relatively little impact on cultivar improvement. Despite the importance of tropical fruits as export commodities of many underdeveloped countries, there have been very few genetic studies that involve these plants.

Under tropical conditions, disease, insect, and environmental stresses are constant factors that limit production of crop plants. The monoculture of a few cultivars of tropical fruit trees in large commercial plantings has frequently had devastating results. Panama disease eliminated the large-scale cultivation of 'Gros Michel' bananas in Central America earlier in this century. Similar disease outbreaks have occurred and have threatened other crops in a large production area, e.g., papaya ringspot virus in the Caribbean region. Because of the narrow genetic base of many popular tropical fruit tree cultivars, there is often no genetic protection against the outbreak of new diseases. Many ancient fruit tree clones decline due to infection by 1 or more microorganisms. Vegetative propagation ensures the transmission of infections and the spread of disease to new production areas. Furthermore, the international distribution of vegetatively propagated plant cultivars can be impeded by the infection of important fruit tree cultivars with systemic plant pathogens.

Advances in plant cell and tissue culture, and in somatic cell genetic approaches to crop improvement, have been demonstrated to have considerable potential for the improvement of many agronomic crops of the temperate, developed countries. These crops have also benefitted from nearly a century of plant breeding and genetic studies. The application of somatic cell genetics to plant improvement is dependent on the development of efficient methods for regenerating plants from cell culture. However, it has been difficult to demonstrate in vitro regeneration pathways for woody perennial plants (1) because of the apparent loss of regenerative potential in mature tissues of woody plants. Although differentiation has been demonstrated from callus derived from cultured embryos and seedlings of several tree species, callus derived from explants of mature origin has lost most of its regenerative potential (2). Thus, it has not always been feasible to induce in vitro regeneration from superior tree selections or cultivars. In order to restore the in vitro regenerative potential of tissues of mature trees, the trees must be rejuvenated. Roots possess a degree of juvenility that is expressed in many species by the ability to form suckers. The root-shoot zone is also juvenile, and repeated hard pruning can stimulate juvenile growth from the base of the trunk. Alternatively, tissues or organs within the mature tree that still possess a degree of juvenility must be identified. Floral parts possess a high degree of regenerative potential in vitro. It has been suggested that this is due to the occurrence and closeness of cells undergoing meiosis which may have a rejuvenating effect on surrounding cells (30). Regenerative callus has been recovered from inflorescence tissue of 2 palm species, Howeia forsteriana Becc. and Chamaedorea costaricana Oerst. (37).

Adventitious embryos are produced in vivo from the inner-integument and nucellus in ovules of several plant species (32,35). According to Melchior (27), abnormal, extensive growth of the nucellus occurs in species in at least 172 plant families. Adventitious nucellar embryony has been reported in species representing 16 plant families (Tab. 1). Since the original report by Rangaswamy (34) that embryogenic callus could be induced from the nucellus excised from fertilized polyembryonic Citrus ovules, relatively few plant species, representing even fewer plant families, have been regenerated in vitro from the nucellus or nucellar callus. However, the in vitro responses of Citrus nucellar callus have been thoroughly studied, and have been the subject of several reviews (3,23,40). The observation by Rangan et al. (33) that somatic embryogenesis could

Tab. 1. Polyembryony in tropical fruits.

Plant families with polyembryonic species	Polyembryonic tropical fruit species	Common name
Anacardiaceae	Mangifera indica	mango
Bombacaceae		
Buxaceae		
Cactaceae		
Capporaceae		
Cucurbitaceae		
Euphorbiaceae		
Guttiferae	Garcinia mangostana	mangosteen
Haemodoraceae		
Liliaceae		
Malphigiaceae		
Meliaceae	Lansium domesticum	langsat
Myrtaceae	Eugenia spp.	rose apple
		Malay apple
	Myrciaria cauliflora	jaboticaba
Rosaceae		
Rutaceae	Citrus spp.	orange
		grapefruit
		mandarin
		lemon
Urticaceae		

be induced in vitro from the nucellus of 3 monoembryonic *Citrus* cultivars was particularly significant because this demonstrated that the in vitro responses of the excised nucellus of monoembryonic and polyembryonic species were similar.

Despite the obvious implications of in vitro studies involving *Citrus* for other woody, perennial plants, there have been only a few reports of somatic embryogenesis from the excised nucellus or from cultured ovules of other woody plants. Zatyko et al. (43) described the induction of somatic embryogenesis from the nucellus of cultured ovules of *Ribes rubrum*, a monoembryonic species. Somatic embryogenesis has also been induced from the nucellus in cultured ovules of monoembryonic *Vitis vinifera* (29) and of monoembryonic *Malus domesticum* (7).

Adventive embryony commonly occurs in trees of the tropical rain forests. Many tropical fruit tree species in addition to *Citrus* are naturally polyembryonic, including mangosteen (*Garcinia mangostana*), mango (*Mangifera indica*), several species of *Eugenia* or *Syzygium*, langsat (*Lansium domesticum*) (11), and jaboticaba (*Myrciaria cauliflora*) (41). The experimental induction of somatic embryogenesis from cultured ovules and nucellus of tropical woody fruit trees excluding *Citrus* is the major subject of this chapter. Somatic embryogenesis from nucellar explants of naturally polyembryonic and monoembryonic mango, from ovule explants of entirely polyembryonic tropical fruit species, and from entirely monoembryonic tropical fruit trees will be discussed. In addition, somatic embryogenesis from tissue cultures of herbaceous, arborescent plants of papaya, banana, and plantain will be reviewed.

REGENERATION OF TROPICAL FRUIT TREES

Woody Fruit Trees

Naturally polyembryonic fruit trees. The mango is one of the most extensively grown fruit trees. Its annual production is exceeded only by that of grapes, *Citrus*, *Musa* (bananas and plantains), and apples (8). Depending on their geographical origin, mango cultivars can be either monoembryonic or polyembryonic. The Indochina and Philippine cultivars are polyembryonic, and are seed-propagated. Adventitious embryos originate from the nucellus. In some cultivars, including 'Carabao', 'Pico', 'Cambodiana', 'Olour', and 'Strawberry', the zygote apparently aborts at a very early stage of seed development (39). The Indian cultivars are monoembryonic and are propagated vegetatively by grafting.

Maheshwari and Rangaswamy (26) described the induction of somatic embryos from polyembryonic *Citrus* nucellar tissue in vitro. Although they alluded to the induction of mango nucellar somatic embryos in the title of this report, and were doubtlessly aware of the embryogenic potential of mango nucellus, they were evidently unable to reproduce with mango their success with Citrus. The stimulation of callus from mango cotyledons was described by Rao et al. (36). Adventitious roots were differentiated from the callus after a few weeks. The callus did not lose its organogenetic potential after repeated subculturing. Neither shoot formation nor somatic embryogenesis was observed.

Litz et al. (21) reported that somatic embryos could be produced from the nucellus of cultured, polyembryonic mango ovules 40 - 60 days after pollination on modified Murashige and Skoog (MS) medium containing coconut water, ascorbic acid, and 60 g/l sucrose. The enlarged, lobular nucellus was dissected from the micropylar end of the ovules, and was subcultured on the same medium. Globular somatic embryos developed from the nucellar explants. The response appeared to be cultivar-dependent. Explants from highly polyembryonic cultivars responded more favorably than explants from less polyembryonic cultivars. The ability to induce large numbers of somatic embryos on this medium formulation are limited; however, by culturing excised nucellus and globular, adventitious embryos from polyembryonic mango cultivars on medium with 1.0-2.0 mg/l 2,4-dichlorophenoxyacetic acid (2,4-D), it is possible to obtain a rapidly growing, lobular callus, particularly in liquid medium, where it assumes a distinctive pseudobulbil appearance (Fig. 1). The callus is initially white, but despite frequent subculturing, becomes dark brown after a few weeks. This change in appearance of the callus appears to be normal, and is not a sign of senescence. Activated charcoal and ascorbic acid both appear to inhibit the color change, but do not alter the embryogenic nature of the callus. Somatic embryogenesis occurs in the presence of 2,4-D, although the embryoids rarely develop beyond the globular stage. Following subculture into liquid MS medium without growth regulators, somatic embryogenesis also occurs (Fig. 2). Intermediate stage somatic embryos possess large cotyledons and small hypocotyls. Mature somatic embryos have attained lengths of 5-6 cm prior to germination. At this time 15 polyembryonic mango cultivars have been regenerated via somatic embryogenesis (Tab. 2).

The regeneration of somatic embryos of monoembryonic mango cultivars from nucellar explants is also possible (13), although the rate of success is less than that obtained with polyembryonic cultivars. The optimum time for excision of the nucellus occurs before the zygotic embryo has expanded to fill the embryo sac. The nucellus de-differentiates to form callus on modified MS medium containing 0.5-2.0 mg/l 2,4-D. The callus is loose and pale green. Somatic embryogenesis occurs from callus 5-7 weeks after culturing. On medium without 2,4-D, embryogenesis can occur directly from the nucellus, but at a very low frequency (Fig. 3). Callus formation from the nucellus in the absence of growth regulators has not been

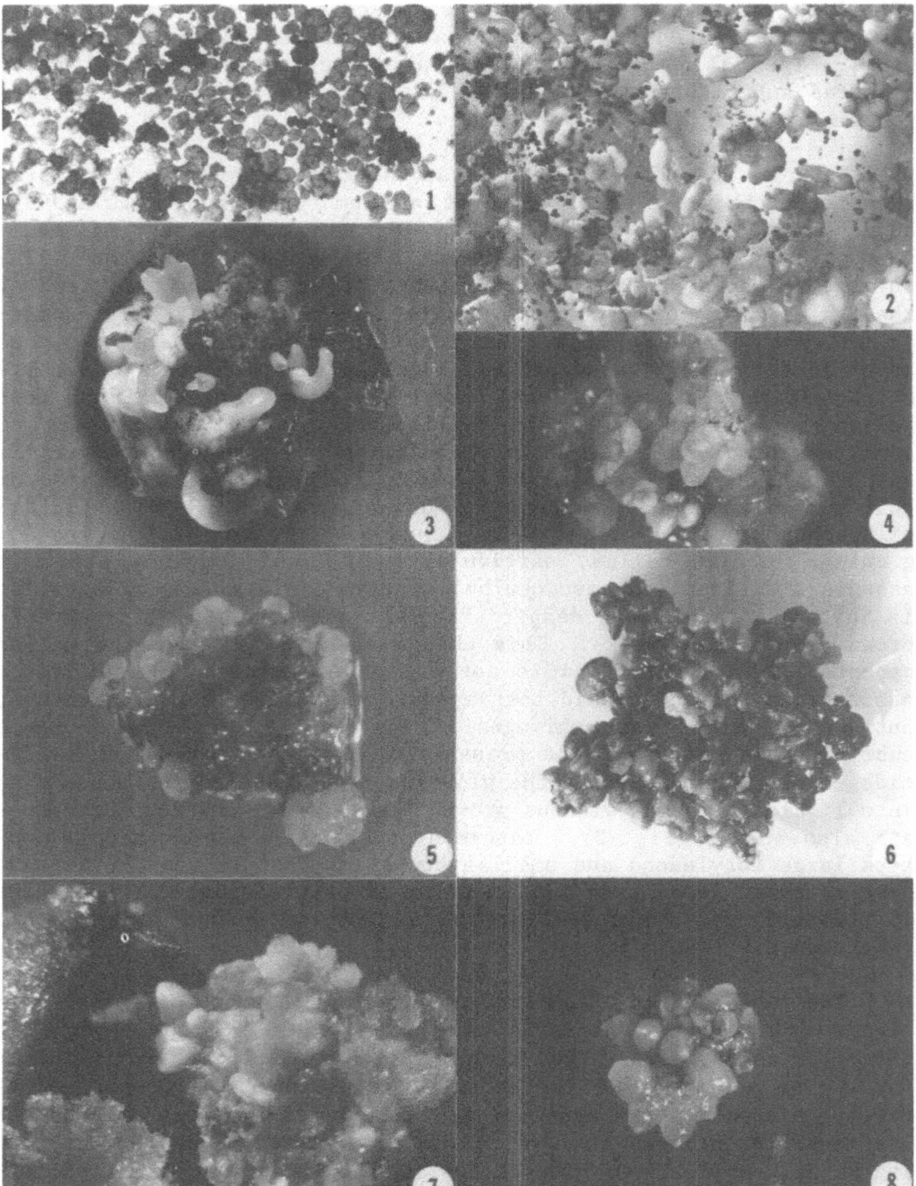
1
2
3
4
5
6
7
8

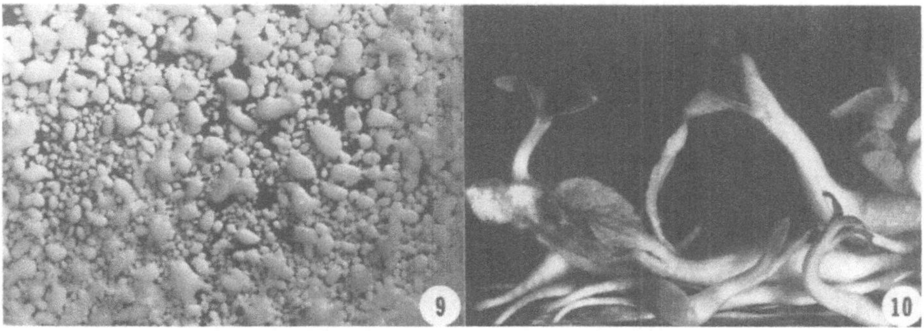

Fig. 1-10. Fig. 1. Embryogenic nucellar callus of polyembryonic 'Ono' mango in suspension culture. MS medium with 2.0 mg/l 2,4-D. Fig. 2. Somatic embryos of polyembryonic 'Sabre' mango in suspension culture. MS medium without growth regulators. Fig. 3. Direct induction of somatic embryos of 'Tommy Atkins' mango from nucellus explant. MS medium without growth regulators. Fig. 4. Somatic embryogenesis from callus derived from adventitious embryo of Malay apple. MS medium with 0.5 mg/l 2,4-D. Fig. 5. Somatic embryogenesis from ovular callus of jaboticaba. MS medium with 1.0 mg/l 2,4-D. Fig. 6. Jaboticaba somatic embryos on MS medium without growth regulators. Fig. 7. Somatic embryogenesis from nucellar callus of 'Tsukumo' loquat. MS medium with 1.0 mg/l BA and 1.0 mg/l 2,4-D. Fig. 8. Somatic embryogenesis from callus of 'Dwarf Cavendish' (Musa group AAA) on MS medium with 10.0 mg/l dicamba and 0.1% (w/v) activated charcoal. Fig. 9. Suspension culture of globular somatic embryos of C. papaya x C. cauliflora. MS medium with 1.0 mg/l 2,4-D. Fig. 10. Germinating C. papaya x C. cauliflora somatic embryos in liquid MS medium without growth regulators.

observed. These observations are in general agreement with previous studies involving somatic embryogenesis of monoembryonic Citrus species and cultivars (10,33). Only 3 monoembryonic mango cultivars have been regenerated from nucellar explants at this time (Tab. 2).

During maturation, necrotic areas appear on the cotyledons and hypocotyls of mango somatic embryos, and these gradually spread over the entire embryo. With a few exceptions, it has not been possible to prevent this from occurring by the addition of reducing agents to the medium or by alteration of the medium composition in other ways. Secondary embryos often develop from the expanded hypocotyl of germinating somatic embryos, although this can sometimes be controlled by the inclusion of low concentrations of cytokinin in the medium.

Tab. 2. Somatic embryogenesis in Mangifera indica cultivars.

Polyembryonic cultivars	Monoembryonic cultivars
Cambodiana	Irwin
Chino	Ruby
Heart	Tommy Atkins
Jame Saigon	
Kensington	
Manzano	
Micogensis	
Nam Dank Mai	
Ono	
Parris	
Peach	
Sabre	
Stringless Peach	
Tuehan	
Turpentine	

There are several fruit and spice trees within the family Myrtaceae. Many species, e.g., guava (Psidium guajava L.) and clove (Eugenia caryophyllum L.), have considerable economic significance. Polyembryony occurs in a number of species and genera within the Myrtaceae, including a number of fruit species (Tab. 1). Among the polyembryonic species that grow in south Florida, the jaboticaba (Myrciaria cauliflora D.C. Berg.) and 2 Eugenia species, i.e., the rose apple (E. jambos L.) and the Malay apple (E. malaccensis Lam.), were used as models to demonstrate the morphogenetic potential of the nucellus from naturally polyembryonic trees of this family.

Adventitious embryos in jaboticaba and Malay apple ovules are nucellar in origin (31,41), whereas in the rose apple they originate from cells of the inner integument of the ovule (31). Unfertilized ovules and ovules from very young fruit of these species are very small and difficult to dissect. Consequently, it is impractical to remove the nucellus from ovules prior to and just after fertilization. However, callus can readily be induced from the adventitious embryos of ovules removed from immature fruits (14,15).

The stage of fruitlet and ovule development is important. Ovules excised from very young fruitlets usually do not respond well in vitro. Moreover, as the seed coat develops around the embryonic mass, the induction of embryogenic callus from the adventitious embryos becomes increasingly difficult. Modified MS medium containing 60 g/l sucrose and 0.5 - 2.0 mg/l 2,4-D has been highly effective for stimulating embryogenic callus for all 3 species (Fig. 4, 5, and 6). Somatic embryogenesis has not been observed when other auxins, i.e., indoleacetic acid (IAA) and naphthaleneacetic acid (NAA), have been supplied to the medium. It is perhaps for this reason that

Kong and Rao (12) were unable to demonstrate somatic embryogenesis from seedling callus of Eugenia grandis in the presence of NAA. High concentrations of 2,4-D stimulate root development from adventitious embryos of the rose and Malay apples, and cytokinins stimulate maturation of the adventitious embryos of these species. Germinated jaboticaba embryos form branched plantlets in the presence of benzyladenine (BA), whereas proliferation of axillary buds occurs with the rose and Malay apples.

Although the jaboticaba and the rose and Malay apples are relatively minor tropical fruit trees, the Eugenia species are closely related to several important, monoembryonic tropical fruit and spice trees within the same genus. Similar approaches can possibly be used to regenerate these other species in vitro, particularly as it has been demonstrated that monoembryonic, as well as polyembryonic, species and cultivars within a genus can be regenerated from ovule or nucellar explants by somatic embryogenesis.

Monoembryonic fruit trees. The loquat (Eriobotrya japonica L.) in the Rosaceae is a widely grown subtropical fruit tree. The seeds are monoembryonic, and cultivars are propagated by grafting onto seedling rootstocks. The fruits of many cultivars possess only a single seed, whereas other cultivars can contain 6 or more seeds. The nucellus is enlarged, particularly in the single-seeded cultivars; however, naturally occurring polyembryony has never been reported.

'Tsukumo' loquat was used as a stock plant in a preliminary study to induce somatic embryogenesis from nucellar callus (Litz, unpub. data). Fertilized ovules were aseptically removed from immature fruitlets of various developmental stages. Correlation of fruitlet size with seed development was difficult to establish as this experiment was conducted relatively late in the flowering season. Only those ovules in which the zygotic embryo had not filled the embryo sac were used for in vitro studies. Following removal of the embryo, the bisected ovules were cultured on modified MS medium containing 60 g/l sucrose and different concentrations and combinations of BA, kinetin (KIN), 2,4-D, and NAA so that the nucellus was in contact with the medium. Unlike other systems that have been described in this review, optimum conditions for callus induction were provided by a mixture of auxin (1.0-2.0 mg/l 2,4-D) and cytokinin (0.25-5.0 mg/l KIN). The callus was loose in texture and white. In the presence of 0.5-10.0 mg/l BA and 1.0-2.0 mg/l 2,4-D, the callus was much more compact and slowly growing. Following subculture of callus into liquid culture media of the same formulations of BA and 2,4-D or into medium containing 0.5 mg/l 2,4-D as the only growth regulator, callus growth continued, although very slowly. This callus was compact, lobular, and brown in appearance. Somatic embryogenesis occurred from this callus after several weeks, but only in medium containing 2,4-D alone (Fig. 7). These results represent only preliminary observations, and the complete sequence of embryo development leading to germination has not been observed.

Herbaceous Fruit Trees

Bananas and plantains (Musaceae). Although they are genetically and morphologically fairly similar, the bananas (*Musa* group AAA) and plantains (*Musa* groups AAB and ABB) are usually distinguished from each other according to use. Bananas include the dessert bananas, whereas plantains are normally grown as a starchy, staple crop. Both types of plant are vegetatively propagated from offshoots or suckers. Both bananas and plantains are sterile triploids. All modern cultivars have arisen during several hundred years in cultivation by somatic mutations within existing clones. There is consequently a high degree of genetic vulnerability to disease and environmental stress (38).

Although bananas and plantains have been propagated in vitro on a commercial scale for at least 10 years (24,25), the procedure has involved the stimulation of axillary bud development. Cronauer and Krikorian (4) reported the induction of somatic embryogenesis from callus of 'Saba' and 'Pelipita' of the plantain group ABB. Embryogenic callus was induced from the leaf sheaths and shoot bases of proliferating shoot tip cultures in the presence of 0.1-1.0 mg/l BA and 1.0 mg/l 2,4,5-trichlorophenoxyacetic acid (2,4,5-T). The differentiation of somatic embryos of the plantain, *Musa* group AAB, which contains the most important plantain clones, e.g., 'Horn', and of the dessert banana, *Musa* group AAA, however, was not reported. Studies in our laboratory with cultivars in *Musa* groups AAA and AAB have resulted in the definition of a medium that will stimulate embryogenic callus from explants from plantains (AAB) and dessert bananas (AAA) (Jarret and Litz, unpub. data). Embryogenic callus was induced from the leaf bases and along the vascular tissue on MS medium containing 10.0-20.0 mg/l 2-methoxy-3,6-dichlorobenzoic acid (dicamba) with 0.1% (w/v) activated charcoal (Fig. 8). Neither Cronauer and Krikorian (4) nor our research group have been able to obtain normal germination of *Musa* somatic embryos. Nevertheless, these somatic embryos are morphologically similar to the diploid *Musa* species, although somewhat larger in size (4).

Carica spp. The papaya *Carica papaya* L. (Caricaceae) is an important plantation and dooryard fruit tree throughout the tropical regions of the world. The babaco *C. heilbornii* Badillo n.m. pentagona is also beginning to be exploited commercially. Other *Carica* species are grown primarily as dooryard trees. The papaya is grown for fresh fruit and puree, and for the production of the proteolytic enzyme, papain. *Carica* spp. plants may be dioecious or hermaphroditic; the 'Solo' papaya cultivars of Hawaii are highly inbred hermaphrodites. All papayas are propagated by seed, although rooting of cuttings is possible.

DeBruijne et al. (6) first reported the induction of embryogenic callus from seedling stem explants of papaya (Tab. 3), although they were unable to recover plantlets from these somatic embryos. Callus derived from seedling stem segments can differentiate both adventitious shoots and somatic embryos on the same medium formulation (42), although normal germination of the embryoids was not observed by these workers. Embryogenic callus has been obtained from peduncle explants of a related species, C. stipulata Badillo, which produced somatic embryos following the addition of 0.5% (w/v) activated charcoal to the growth medium (17). Normal germination and plantlets were reported. Jordan et al. (9) induced embryogenic callus from hypocotyl segments of C. candamarcensis Hook. f., and plantlets were recovered from the somatic embryos. They were unable to obtain differentiation from callus derived from mature tissues.

Litz and Conover (18) described the induction of polyembryony in cultured papaya ovules following interspecific hybridization with C. cauliflora Jacq. which is incompatible with papaya. Polyembryony was stimulated by the presence of either cytokinin or auxin, and occasionally in the absence of growth regulators, although the response was highly dependent on the maternal genotype (20). Embryogenic papaya ovular callus could be induced on MS medium with 60 g/l sucrose and either 20% coconut water (19) or, more efficiently, 2.0 mg/l 2,4-D (Fig. 9) (20). Proliferation of globular somatic embryos by budding from other somatic embryos occurs in the presence of 2,4-D. Maturation and germination of interspecific C. papaya x C. cauliflora embryoids occurs in the absence of growth regulators (Fig. 10) (20,28).

Tab. 3. Somatic embryogenesis in Carica papaya and related species.

Species	Explant[1]	Growth Regulators Auxin	Growth Regulators Cytokinin	Reference
C. papaya	stem S	NAA	2iP	6
C. papaya	petiole S	NAA	KIN	42
C. papaya x C. cauliflora	ovule S	coconut water		19
C. papaya x C. cauliflora	ovule S	2,4-D		20
C. stipulata	peduncle M	NAA	BA	17
C. candamarcensis	hypocotyl S	NAA	KIN	9

[1]Explant of mature (M) or immature (S) origin.

Different regeneration pathways have been well-defined for papaya and for closely related Carica species (16,23). Papaya is therefore unique among fruit trees, and could easily become a model for the application of somatic cell genetic approaches to improvement of this crop.

DISCUSSION AND CONCLUSION

The conditions for induction of embryogenic callus from tissue explants differ greatly, depending on the herbaceous or woody nature of tropical fruit trees. Somatic embryogenesis has been described for several Carica species and from callus of mature and immature origin, including seedling stems and petioles, peduncles, and ovules. Musa somatic embryos have been recovered from leaf bases and along the leaf sheaths. In contrast, somatic embryogenesis in tissue cultures of woody tropical fruit trees has been confined to callus derived from the nucellus or from adventitious embryos of nucellar or integumental origin. At this time, the induction of somatic embryogenesis has been reported from the excised nucellus and cultured ovules of several polyembryonic Citrus species and cultivars (40), mango (21,22), jaboticaba (15), and 2 Eugenia species (14). Somatic embryogenesis has also been described occurring directly from the isolated nucellus of monoembryonic Citrus species and cultivars (33), of monoembryonic mango cultivars (13), and of a monoembryonic loquat cultivar (Litz, unpub. data). The use of the nucellus, immature adventitious embryos, and intact ovules as explants from woody, tropical fruit trees demonstrates the morphogenetic potential of the nucellus for these and probably many other tropical fruit and forest trees.

The cultural conditions for induction of embryogenic callus from ovule explants from woody fruit trees are simple. In most cases modified MS medium containing 1.0-2.0 mg/l 2,4-D, a reducing agent, and 60 g/l sucrose has been the optimum formulation for inducing embryogenic callus. For each species, it is necessary to determine the stage of ovule development that is most suitable for explanting, e.g., it is most convenient to remove the zygotic embryo from monoembryonic ovules before it has filled the embryo sac. Culture of the intact, undissected ovule is also effective for polyembryonic species and cultivars; however, the induced callus and the regenerated somatic embryos will in most cases be a mixture of zygote- and nucellus-derived embryos.

The germination of somatic embryos of tropical fruit trees has been difficult to demonstrate with any consistency. Unlike Citrus, somatic embryos of other tropical woody fruit trees, e.g., mango, jaboticaba, Eugenia spp., do not germinate on a relatively simple, defined medium. Although Carica spp. somatic embryos have developed normally in vitro to produce plants (9,17,19,20), other researchers have stressed the abnormal development of somatic embryos in the same genus (6,42). Musa somatic embryos also have not germinated normally to produce plantlets (4). Until these developmental problems can be overcome, the application of somatic cell genetic approaches to tropical fruit tree improvement and the use of in vitro systems for germplasm preservation, disease elimination, and propagation will be unsuccessful.

The implications of this work to other tropical fruit and forest trees are intriguing. Because of the widespread occurrence of crassinucellate and polyembryonic ovules among largely tropical plant families (Tab. 1), further studies of the in vitro morphogenetic potential of ovules and excised nucellus from a range of tropical tree species are surely justified.

ACKNOWLEDGEMENTS

I am indebted to the Rockefeller Foundation which has supported much of my research, particularly that involving Carica and Musa. I am also grateful to the Florida Mango Forum for their interest, and to the support of USDA Cooperative Agreement No. 58-7B30-9-116. Thanks also to Sara Walker, Rose Hendrix, and Callie Sullivan.

REFERENCES

1. Bonga, J.M. (1982) Vegetative propagation in relation to juvenility. In Tissue Culture in Forestry, J.M. Bonga and D.J. Durzan, eds. Martinus Nijhoff/Dr. W. Junk Publishers, The Hague, pp. 387-412.
2. Brown, C.L., and H.E. Sommer (1982) Vegetative propagation of dicotyledonous trees. In Tissue Culture in Forestry, J.M. Bonga and D.J. Durzan, eds. Martinus Nijhoff/Dr. W. Junk Publishers, The Hague, pp. 109-149.
3. Button, J., and J. Kochba (1977) Tissue culture in the citrus industry. In Applied and Fundamental Aspects of Plant, Cell, Tissue and Organ Culture, J. Reinert and Y.P.S. Bajaj, eds. Springer-Verlag, Berlin, pp. 70-92.
4. Cronauer, S.S., and A.D. Krikorian (1983) Somatic embryos from cultured tissues of triploid plantains (Musa 'ABB'). Plant Cell Rpt. 2:289-291.
5. DeCandolle, A. (1889) Origin of Cultivated Plants. Kegan Paul Trench and Co., London.
6. DeBruijne, E., E. DeLanghe, and R. van Rijck (1974) Action of hormones and embryoid formation in callus cultures of Carica papaya. Int. Symp. Fytofarm. Fytiat. 26:637-645.
7. Eichholtz, D., H.A. Robitaille, and P.M. Hasegawa (1979) Adventive embryology in apple. HortScience 14:699-700.
8. Food and Agricultural Organization of the United Nations (1982) 1982 FAO Production Yearbook. FAO, United Nations, Rome.

9. Jordan, M., I. Cortes, and G. Montenegro (1982) Regeneration of plantlets by embryogenesis from callus of Carica candamarcensis. Plant Sci. Lett. 28:321-326.
10. Juarez, J., L. Navarro, and J.L. Guardiola (1976) Obtention de plantes nucellaires de divers cultivars de clementiners au moyen de la culture de nucelle "in vitro". Fruits d'Outre Mer 31:751-761.
11. Kaur, A., C.O. Ha, K. Jong, V.E. Sands, H.T. Chan, E. Soepadmo, and P.S. Ashton (1980) Apomixis may be widespread among trees of the climax rain forest. Nature 271:440-442.
12. Kong, L.S., and A.N. Rao (1982) In vitro plantlet production of some tropical tree species. In Tissue Culture of Economically Important Plants, A.N. Rao, ed. COSTED, Singapore, pp. 185-190.
13. Litz, R.E. (1984a) In vitro somatic embryogenesis from nucellar callus of monoembryonic Mangifera indica L. HortScience 19: 715-717.
14. Litz, R.E. (1984b) In vitro responses of adventitious embryos of two polyembryonic Eugenia species. HortScience 19:720-722.
15. Litz, R.E. (1984c) In vitro somatic embryogenesis from callus of jaboticaba, Myrciaria cauliflora. HortScience 19:62-64.
16. Litz, R.E. (1984d) Papaya. In Handbook of Plant Cell Culture, Vol. II. Crop Species, W.R. Sharp, D.A. Evans, P.V. Ammirato, and Y. Yamada, eds. Macmillan Publishing Co., New York, pp. 349-368.
17. Litz, R.E., and R.A. Conover (1980) Somatic embryogenesis in cell cultures of Carica stipulata. HortScience 15:733-735.
18. Litz, R.E., and R.A. Conover (1981) In vitro polyembryony in Carica papaya L. ovules. Z. Pflanzenphysiol. 104:285-288.
19. Litz, R.E., and R.A. Conover (1982) In vitro somatic embryogenesis and plant regeneration from Carica papaya L. ovular callus. Plant Sci. Lett. 26:153-158.
20. Litz, R.E., and R.A. Conover (1983) High frequency somatic embryogenesis from Carica suspension cultures. Ann. Bot. 51:683-686.
21. Litz, R.E., R.J. Knight, and S. Gazit (1982) Somatic embryos from cultured ovules of polyembryonic Mangifera indica L. Plant Cell Rpt. 1:264-266.
22. Litz, R.E., R.J. Knight, and S. Gazit (1984) In vitro somatic embryogenesis from Mangifera indica L. callus. Scientia Hort. 22:233-240.
23. Litz, R.E., G.A. Moore, and C. Srinivasan (1985) In vitro systems for propagation and improvement of tropical fruits and palms. Hort. Rev. Vol. 7: (in press).
24. Ma, S., and C. Shii (1972) In vitro formation of adventitious buds in banana shoot apex following decapitation. J. Hort. Sci. China 18:135-142.
25. Ma, S., and C. Shii (1974) Growing bananas from adventitious buds. J. Hort. Sci. China 20:6-12.

26. Maheshwari, P., and N.S. Rangaswamy (1958) Polyembryony and in vitro culture of embryos of Citrus and Mangifera. Indian J. Hort. 15:275-282.
27. Melchior, H. (1964) A. Engler's Syllabus der Pflanzenfamilien, Vol II. Angiospermen, Berlin-Nikolassee.
28. Moore, G.A., and R.E. Litz (1984) Biochemical markers for Carica papaya, C. cauliflora and plants from somatic embryos of their hybrid. J. Amer. Soc. Hort. Sci. 109:213-218.
29. Mullins, M.G., and C. Srinivasan (1976) Somatic embryos and plantlets from an ancient clone of the grapevine (cv. Cabernet-Sanvignon) by apomixis in vitro. J. Exp. Bot. 27:1022-1030.
30. Nozeran, R. (1978) Polymorphisme des individus issus de la multiplication végétative des vegetaux superieurs, avec conservation du potentiel génétique. Physiol. Veg. 16:177-194.
31. Pijl, L. van der (1934) Über die Polyembryonie bei Eugenia. Rec. Trav. Bot. Neer. 31:113-187.
32. Rangan, T.S. (1982) Ovary, ovule and nucellus culture. In Experimental Embryology of Vascular Plants, B.M. Johri, ed. Springer-Verlag, Heidelberg, pp. 105-129.
33. Rangan, T.S., T. Murashige, and W.P. Bitters (1968) In vitro initiation of nucellar embryos in monoembryonic Citrus. HortScience 3:226-227.
34. Rangaswamy, N.S. (1958) Culture of nucellar tissue of Citrus in vitro. Experimentia 14:111-112.
35. Rangaswamy, N.S. (1982) Nucellus as an experimental system in basic and applied tissue culture research. In Tissue Culture of Economically Important Plants, A.N. Rao, ed. COSTED, Singapore, pp. 269-286.
36. Rao, A.N., Y.M. Sin., N. Kathagoda, and J. Hutchinson (1982) Cotyledon tissue culture of some tropical fruits. In Tissue Culture of Economically Important Plants, A.N. Rao, ed. COSTED, Singapore, pp. 124-137.
37. Reynolds, J.F., and T. Murashige (1979) Asexual embryogenesis in callus cultures of palms. In Vitro 5:383-387.
38. Rowe, P. (1984) Breeding bananas and plantains. Plant Breed. Rev. 2:135-155.
39. Singh, L.B. (1960) The Mango, Leonard Hill Ltd., London.
40. Spiegel-Roy, P., and J. Kochba (1980) Embryogenesis in Citrus tissue cultures. In Advances in Biochemical Engineering, A. Fiechter, ed. Springer-Verlag, Heidelberg, pp. 27-48.
41. Traub, H. (1939) Polyembryony in Myrciaria cauliflora. Bot. Gaz. 101:233-234.
42. Yie, S.T., and S.I. Liaw (1977) Plant regeneration from shoot tips and callus of papaya. In Vitro 13:564-568.
43. Zatyko, J.M., I. Simon, and C.S. Szabo (1975) Induction of polyembryony in cultivated ovules of red currant. Plant Sci. Lett. 4:822-825.

MICROPROPAGATION OF TWO TROPICAL CONIFERS: *PINUS OOCARPA* SCHIEDE AND *CUPRESSUS LUSITANICA* MILLER

Edgar O. Franco* and Otto J. Schwarz

Department of Botany
The University of Tennessee
Knoxville, Tennessee 37996-1100

ABSTRACT

Plantlet regeneration of 2 tropical conifer species (*Pinus oocarpa* Schiede and *Cupressus lusitanica* Miller) was accomplished using explants obtained from embryonic and seedling tissues. Adventitious buds were induced on *P. oocarpa* cotyledonary explants obtained from 7- and 10-day-old seedlings. The explants were placed on a defined nutrient medium containing cytokinin alone or in combination with an auxin for 21 days and then were transferred to a culture medium without hormones. Further bud development and shoot elongation occurred in this medium. Hypocotylary explants obtained from 4-week-old *C. lusitanica* seedlings were cultured for 21 days on nutrient medium supplemented with various cytokinins alone or in combination with an auxin. Further development of buds and subsequent shoot elongation occurred when the explants were transferred to nutrient medium without growth regulators. Elongated shoots of both species were rooted after treatment with an auxin. Successful transfer of the rooted plantlets to soil under greenhouse conditions has been achieved for both species.

INTRODUCTION

Pinus oocarpa Schiede and *Cupressus lusitanica* Miller are conifers of considerable economic potential in Central and South America. Provenance trials have been established along with related

*Present address: Facultad de Agronomia, Universidad de San Carlos, Cindad Universitaria, Zona 12, Guatemala.

genetic tree improvement programs for these 2 species in over 30 countries world-wide (13,16,17,19,20). Greaves (15) called for an immediate exploration, evaluation, and conservation of P. oocarpa genetic resources because of prolonged and excessive exploitation of its natural stands which has brought about a serious erosion of its natural genetic base. This call for action is in concert with the widely recognized need for the genetic preservation and improvement of many tropical forest species (18,24). The application of currently held and rapidly developing plant tissue culture technology to the resolution of these problems in the conifers is of great interest (11,12). However, these and other authors (3,31,54) also warn of the limitations of currently held and future technology with respect to its application to the mass clonal propagation of economically important forest tree species. Current methodology does not allow the routine in vitro cloning of mature trees selected for their elite properties. Those systems of micropropagation that have become well established (i.e., adventitious budding of embryonic and young seedling tissues) are limited by the cost per plantlet produced in vitro, the concern that clonal uniformity will not be maintained, and the inability to screen for desirable traits only accessible in mature individuals. Nonetheless, several useful applications for the plantlets obtained by the micropropagation of immature tissues more than justify the effort. First, this methodology can make possible the clonal multiplication of costly seed obtained as a result of controlled pollination. Second, it can provide a useful model system for basic physiological and biochemical studies of the hormonal control of morphogenesis in vitro (1). In addition, the knowledge gained in developing the methodology for juvenile tissues should provide a basis for the mass micropropagation of mature elite trees.

This paper describes the methodology for plantlet regeneration in P. oocarpa and C. lusitanica. The methodologies reported are patterned heavily after the existing body of conifer micropropagation literature (see Ref. 3 and 11).

MATERIALS AND METHODS

Seed Source

Open-pollinated P. oocarpa seed was obtained from the International Forest Seed Company, Birmingham, Alabama. The seeds were collected in El Chol Baja Verapaz, Guatemala. Seed of open-pollinated C. lusitanica, collected in Tecpán, Chimaltenango, Guatemala, were obtained from the Seed Export Company, Guatemala City.

Aseptic Culture of Seedlings

Seedlings of both conifers received the same culture treatment. Seeds were surface sterilized for 30 min in a water solution containing 1.05% sodium hypochlorite (1:4 Clorox[R]/water) and 0.2% Tween[R] 20. The seeds were then washed 3 times with sterile distilled water and held until needed in sterile petri dishes lined with moist, sterilized filter paper. Seeds were planted in tall (100 x 80 mm), sterile culture dishes containing a 1:1 v/v mixture of sand and vermiculite moistened with sterile distilled water. These culture dishes containing the culture mix were sterilized by autoclaving 3 days prior to seed placement. The seeds were germinated in a growth chamber at 25° ± 1°C under a 16-hr photoperiod. A low level of illumination (2 $\mu E/m^2$ sec, at plant level) was provided by fluorescent lamps (Gro-lux[R]).

Adventitious Shoot Formation on Excised Cotyledons of Pinus Oocarpa

Cotyledons were excised from axenically grown seedlings 7 and 10 days after sowing and were placed in plastic petri dishes (100 x 15 mm) containing 25 ml of basal medium [Murashige and Skoog Modified (MSM) (8); Tab. 1] supplemented with 3% sucrose, 1.0% Bacto agar (Difco Laboratories; 0140-01), and the following growth regulators in 6 combinations: 6-benzylaminopurine (BA) (25 and 50 μM) and 1-naphthaleneacetic acid (NAA) (0.10 and 25 nM). The media was adjusted to pH 5.5 and autoclaved at 15 psi at 121°C for 15 min. Petri dishes were sealed with laboratory film (Parafilm[R] "M"; American Can Company) and placed in a growth room maintained under a 16-hr photoperiod at 25° ± 2°C. Illumination (60 $\mu E/m^2$ sec, at agar surface) was provided by fluorescent lamps (Gro-lux[R]). After 2 weeks, the explants were transferred to plastic petri dishes containing 25 ml basal medium (MSM; Tab. 1) without growth regulators, supplemented with 2% sucrose and 1.0% Bacto agar.

Rooting of Adventitious Shoots of Pinus Oocarpa

Elongated shoots (≥5 mm) derived from various bud induction experiments were pooled and randomly selected for use in the rooting trials. Shoots were transferred to culture tubes (20 x 150 mm) containing 20 ml of medium. The medium [Cupressus Basal Medium (CBM); Tab. 1] contained mineral salts (23), iron (29), organic compounds (30), 0.7% Bacto agar, sucrose at 0.5 or 1.0%, and NAA at 0.1 or 1.0 μM. Shoots were maintained in the root induction medium for 15 days, then transferred to half-strength Gresshoff and Doy Modified medium [GDM (36); Tab. 1]. Rooting data were taken after 8 weeks.

Adventitious Shoot Formation on Hypocotylary Explants of Cupressus Lusitanica

Hypocotylary explants of 4-week-old seedlings (consisting of the intact cotyledonary node, including the stem apex and 5 mm of hypocotyl) were placed in culture tubes (20 x 150 mm) containing 20 ml of CBM medium (Tab. 1). The medium was supplemented with 3% sucrose, 0.7% Bacto agar, and the following growth regulators in vari-

Compound	Concentration (mg/l) MSM	GDM	CBM
Macronutrients			
NH_4NO_3	825	-	720
KNO_3	950	1000	950
$CaCl_2 \cdot 2H_2O$	220	150	220
$MgSO_2 \cdot 7H_2O$	185	250	185
KH_2PO_4	85	-	68
$(NH_4)_2SO_4$	-	200	-
$NaH_2PO_4 \cdot H_2O$	-	90	-
$NaHPO_4$	-	30	-
KCl	-	300	-
Micronutrients			
$FeSO_4 \cdot 7H_2O$	6.0	27.8	27.8
Na_2EDTA	7.2	37.3	37.3
H_3BO_3	3.1	3.0	2.4
$ZnSO_4 \cdot 7H_2O$	5.25	3.0	4.5
KI	0.4	0.75	0.375
$NaMoO_4 \cdot 2H_2O$	0.15	0.25	-
$CuSO_4 \cdot 5H_2O$	0.013	0.25	0.01
$CoCl_2 \cdot 6H_2O$	0.013	0.25	-
$MnSO_4 \cdot H_2O$	8.45	10.0	7.0
$(NH_4)_6Mo_7O_{24} \cdot 4H_2O$	-	-	0.093
Vitamins and other organics			
Myo-inositol	250.0	10.0	100.0
Thiamine HCl	2.5	1.0	0.5
Nicotinic acid	-	0.1	5.0
Pyridoxine HCl	-	0.1	0.5
Folic acid	-	-	0.5
Biotin	-	-	0.05

Tab. 1. Composition of 3 basal media used in the micropropagation of P. oocarpa and C. lusitanica. The media are: MSM [Murashige and Skoog Modified medium (8)]; GDM [Gresshoff and Doy Modified medium (36)]; and CBM [Cupressus basal medium (assembled using minerals from Lind and Staba, Ref. 23; iron from Murashige and Skoog, Ref. 29; and organics from Nitsch and Nitsch, Ref. 30)].

ous combinations: BA (1 or 5 µM), kinetin (6-furfurylaminopurine) (5 µM), N_6-(2-isopentenyl)-aminopurine (2iP) (5 µM), and indole-3-butyric acid (IBA) (5.0 or 50 nM). The medium was adjusted to pH 5.5 and autoclaved at 15 psi at 121°C for 15 min. After 21 days, the explants were transferred to CBM medium without hormones, supplemented with 2% sucrose and 0.7% Bacto agar. Explants were kept in this medium for 7 to 10 weeks. The cultures were maintained in a growth chamber under a 16-hr photoperiod at 25° ± 2°C. Illumination (40 µE/m^2 sec, at agar surface) was provided by fluorescent lamps (Gro-lux^R).

Rooting of Adventitious Shoots of Cupressus Lusitanica

Elongated shoots (≥5.0 mm) derived from bud induction experiments were planted in pots containing a mixture of peat, vermiculite, and sand (4:2:1). The shoots were maintained in a high-humidity environment (i.e., pots were enclosed in clear plastic bags) under a 24-hr photoperiod. Illumination (30 µE/m^2 sec, at plant level) was provided by fluorescent lamps (Gro-lux^R). The shoots were watered daily with sterile water and twice a week for the first 2 weeks with a water solution of 15 µM NAA. After 7½ weeks, the shoots were carefully removed from the rooting mixture in order to score for the presence of roots, number of roots per shoot, and length of each primary root.

Numerical Data Analysis

All experiments were performed at least twice. Where appropriate, means were subjected to Duncan's multiple range test at the 0.05 significance level.

Histological Methods

Tissue for histological observation was killed and fixed in CRAF III according to Sass (33), dehydrated in an ethanol-tertiary butyl alcohol series, embedded in paraffin, and sectioned at 10 µm on a rotary microtome. Sections were stained with Gill's hematoxylin (14).

RESULTS AND DISCUSSION

Pinus Oocarpa

Cotyledonary explants of 7- and 10-day-old seedlings were tested for their ability to produce adventitious buds using various combinations of BA and NAA (Tab. 2). BA, alone or in combination with NAA, has been shown to induce adventitious buds in conifers (see review, Ref. 11). Preliminary studies (data not shown) indi-

Tab. 2. Adventitious bud induction on cotyledons derived from 7- and 10-day-old seedlings of *P. oocarpa* treated with various combinations of BA and NAA.[a]

Hormonal treatments BAP (μM)	NAA (nM)	Explants forming buds (%)	Range of buds/explant (Number)	Mean number of buds/explant[b]
Explants from 7-day-old seedlings (average length 4.25 mm)				
0.0	0.0	0	0	0[c]
25.0	0.0	52	1-7	4[A]
25.0	10.0	42	1-5	2[A]
25.0	25.0	38	1-10	4[A]
50.0	0.0	63	1-23	7[A]
50.0	10.0	57	1-18	5[A]
50.0	25.0	33	1-14	6[A]
Explants from 10-day-old seedlings (average length 14 mm)				
0.0	0.0	0	0	0[c]
25.0	0.0	52	1-10	4[A]
25.0	10.0	53	1-13	5[A]
25.0	25.0	67	1-20	7[A]
50.0	0.0	68	1-13	5[A]
50.0	10.0	38	1-19	7[A]
50.0	25.0	39	1-7	3[A]

[a]Fifty-one to 66 explants were used per treatment. Data were taken after 7 weeks.

[b]Means with the same uppercase superscript letter (A) are not significantly different at the 0.05 level.

[c]Data were not included in the statistical analysis.

cated that, at equimolar concentrations with BA, 2iP and kinetin also induced adventitious buds on the cotyledonary explants, but at a much lower frequency than BA (i.e., percentage of explants producing buds). BA, alone or in combination with NAA, induced adventitious bud formation in P. oocarpa. Little differences were observed in the percentage of explants forming buds at the different hormonal concentrations tested when cotyledons derived from 10-day-old seedlings were compared with cotyledons derived from 7-day-old seedlings. Statistical analysis showed no significant difference at the 5% significance level in the average number of buds produced per explant by cotyledons from 7- and 10-day-old seedlings at all hormone concentrations tested. However, explants derived from 10-day-old seedlings were observed to produce fewer leaf-like structures or phylloids (3) and provided buds that elongated more uniformly and rapidly than buds derived from explants of 7-day-old seedlings. The extent of bud induction was found to be dependent upon the age of the explant in Pinus silvestris L. (39). Increasing levels of BA were necessary for the induction of adventitious buds as the tissue became more differentiated (i.e., explants were derived from 2- and 5-day-old seedlings). The age-related response to adventitious bud production described by Tranvan (39) is not readily apparent in our results with P. oocarpa. Further studies patterned after Wochok and Abo El-Nil (40), comparing a much greater explant-source age difference, are in progress.

The chronology of morphological events typical of adventitious bud initiation and development on excised cotyledons of P. oocarpa is shown in Fig. 1. The observations recorded for adventitious bud organogenesis in P. oocarpa cotyledonary explants are similar to those reported for other conifers (1,2,3,6,7,10,22,32,35,36). Cotyledons (average length 14 mm) were excised from axenically cultured seedlings 10 days after sowing (Fig. 1A). Each seedling provided from 5 to 7 cotyledonary explants that were transferred to a single petri dish containing solid basal medium and sucrose plus hormones (in this instance, 25 μM BA and 25 nM NAA). This combination of hormones was not unique in its ability to induce organogenesis. As Tab. 2 indicates, all combinations of BA and NAA tested resulted in adventitious bud production. After 2 weeks in culture, the explants remained bright green and had slightly elongated and curled to form a crescent shape. In addition, those explants that subsequently produced buds also began to swell. Extensive changes in surface morphology became apparent by the third week in culture (Fig. 1B). All surfaces of the explant became rough and succulent, and many dome-shaped swellings appeared. This surface morphology is typical of cotyledonary and other seedling tissues that are in the process of organogenesis (4,5,10,36,41). Figure 1C clearly shows the presence of a conical or tube-shaped protrusion (picture center) that may be analogous to the leaf-like outgrowths or phylloids described by Bornman (3) in Picea abies cotyledonary needle explants or the cataphylls or primary needles reported by Sommer et al. (36) and others (4,21).

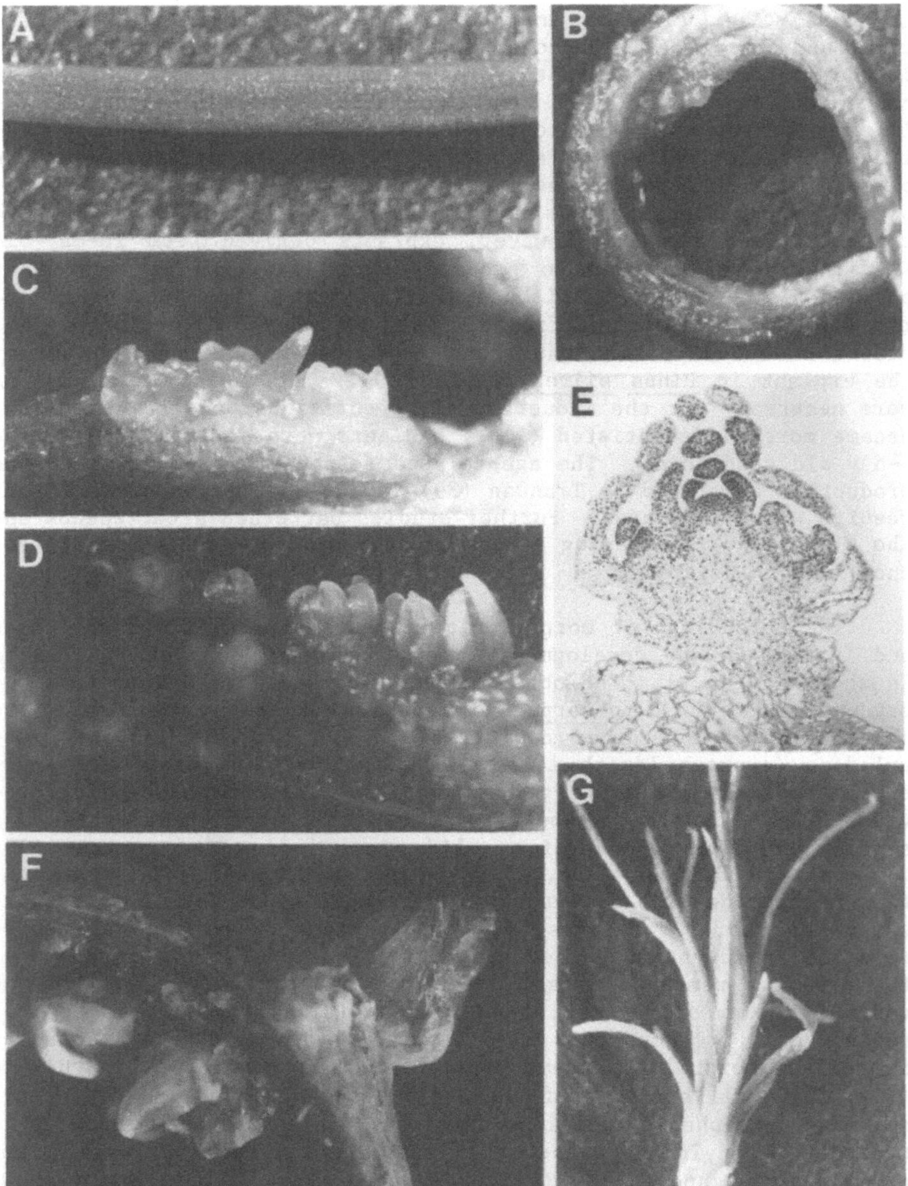
A
B
C
D
E
F
G

←

Fig. 1. Adventitious bud development on cotyledonary explants obtained from 10-day-old P. oocarpa seedlings. Explants were placed on MSM medium supplemented with 25 µM BA and 25 nM NAA. After 15 days, they were transferred to MSM medium without hormones (see Materials and Methods for details of protocol). A. Portion of cotyledon obtained from a 10-day-old seedling prior to culture, showing initial surface morphology (30X). B. Cotyledon showing changes in surface morphology that accompany the early stages of bud formation (3 weeks in culture). Note the many dome-shaped swellings along both sides of the cotyledon (25X). C. Close-up of explant surface showing typical surface morphology. Note presence of conical structure (picture center). D. Well-developed adventitious buds present after 5 weeks in culture (40X). E. Longitudinal section of an adventitious bud after 5 weeks in culture (100X). F. Buds formed on both sides of the explant beginning to elongate after 6 weeks in culture (30X). G. Elongated shoot after 10 weeks in culture (10X).

Well-developed adventitious buds are present after 5 weeks in culture (Fig. 1D and E). At this stage of development, the adventitious buds contained a well-developed shoot apex, needle primordia, and needles. Bud formation was observed over the entire surface of the explant; however, they were usually more heavily clustered at the distal end of the cotyledon. From the sixth to tenth week in culture, buds elongated rapidly but not synchronously. Elongated shoots were removed from the explants for rooting as they attained 5 mm in length. Smaller shoots continued to elongate and a few new buds appeared when the remaining explant tissue was subcultured in the same medium.

Rooting of the elongated (≥5 mm) shoots of P. oocarpa was achieved using a root induction medium containing NAA (Tab. 3). The basic experimental design for the rooting experiments was closely patterned after the procedure originally developed for Pinus taeda L. (25,26,28) and subsequently applied to Pinus monticola Dougl. (27). Two concentrations of NAA were tested at 2 sucrose levels. Controls (data not shown), cultured in media without hormones, and supplemented with sucrose at 0.5 and 1.0% failed to produce roots. Maximal rooting (46%) occurred when shoots were treated with 1.0 µM NAA at 0.5% sucrose. The mean length of roots produced in the 0.1 µM NAA/1.0% sucrose treatment was slightly longer (8.6 vs. 6.7 mm); however, the efficiency of root induction was less than half that produced under maximal rooting conditions. Callus was observed at the base of the shoots after 10 to 12 days in the root induction medium. After 15 days, the shoots were transferred to the same basal medium without growth regulators. Nascent roots were observed

Tab. 3. Root formation on P. oocarpa adventitious shoots as a function of various NAA and sucrose concentrations.[a]

Hormone	Sucrose concentration (%)	Shoots rooted (%)[b]	Mean length of roots (mm)
NAA 0.1 μM	0.5	10	6.0
NAA 0.1 μM	1.0	20	8.6
NAA 1.0 μM	0.5	46	6.7
NAA 1.0 μM	1.0	25	5.3

[a]Thirty shoots tested per treatment.

[b]Data taken after 8 weeks. Values reported are the combined results of two replications.

after 2 weeks. Roots continued to develop and elongate over an 8-week period (Fig. 2). The rooted shoots (plantlets) were subsequently transferred to a nonsterile soil mixture consisting of peat, vermiculite, and sand (4:2:1). The plantlets were maintained under high relative humidity conditions for 3 weeks and then brought to ambient moisture levels over an additional 7-day period.

Cupressus Lusitanica

Adventitious buds have been induced in vitro in several species of the Cupressaceae (9,37,38). In all of these studies BA was present in the media that produced adventitious budding. Thomas et al. (37) stated that of the 3 cytokinins tested (BA, kinetin, and zeatin), only BA was effective in inducing adventitious bud formation in 5 species of Cupressaceae.

Adventitious buds were formed on hypocotylary explants of C. lusitanica after treatment with BA, kinetin, and 2iP alone or in combination with IBA (Tab. 4). Duncan's multiple range test showed no significant difference between BA (1 μM) and kinetin (5 μM), in the mean number of adventitious buds produced per explant forming buds. However, CBM medium containing BA at 1 μM resulted in twice as many explants producing buds. BA (5 μM), alone or in combination with IBA, produced the greatest number of buds per explant. Approximately 25-30% of buds produced by these explants were fleshy and transparent in appearance. These shoots elongated infrequently (less than 1%). The 2iP (5 μM) was the least effective of the 3 hormones tested in inducing organogenesis, on the basis both of percent explants forming buds and of the mean number of buds per explant. Adventitious buds were formed mainly on the cotyledons and epicotyl leaves and were formed only infrequently on the hypocotyl.

Fig. 2. Rooting of P. oocarpa shoots. A. Elongated shoot prior to root induction and elongation treatment. B. Rooted plantlet 8 weeks later. C. Plantlet hardened off to greenhouse environment.

Tab. 4. Adventitious bud production on hypocotylary explants of C. lusitanica as a function of hormonal treatment.[a]

Hormonal treatments	Explants forming buds (%)	Range of buds/explant (Number)	Mean number of buds/explant[b]
Control	0	0	0[c]
BAP 1 μM + IBA 0 M	80	1-30	9[A]
BAP 1 μM + IBA 5 nM	80	1-25	11[A]
BAP 1 μM + IBA 50 nM	87	1-30	11[A]
BAP 5 μM + IBA 0 M	77	1-52	16[B]
BAP 5 μM + IBA 5 nM	60	1-60	24[B]
BAP 5 μM + IBA 50 nM	68	1-45	15[B]
2iP 5 μM + IBA 0 M	40	1-30	6[C]
2iP 5 μM + IBA 5 nM	37	1-10	3[C]
2iP 5 μM + IBA 50 nM	40	1-5	3[C]
Kinetin 5 μM + IBA 0 M	63	2-30	10[A]
Kinetin 5 μM + IBA 5 nM	53	1-20	8[A]
Kinetin 5 μM + IBA 50 nM	73	1-18	8[A]

[a]Thirty explants were used per treatment. Data were taken after 7 weeks in culture.

[b]Means with different uppercase superscript letters (A, B, or C) are significantly different at the 0.05 level.

[c]Data were not included in the statistical analysis.

The addition of IBA to the various cytokinin treatments had little effect on the mean number of buds produced per explant. However, as the concentration of IBA increased, bud formation was increasingly accompanied by callus formation at the base of the hypocotyl. Thomas et al. (37) reported a similar effect in hypocotylary explants of Biota orientalis. Promotion of callus formation at the basal end of the hypocotyl occurred in all the tested media that contained an auxin.

Adventitious bud development on the cotyledons of C. lusitanica hypocotylary explants is illustrated in Fig. 3. Early stages of bud formation, indicated by the presence of isolated groups of cells

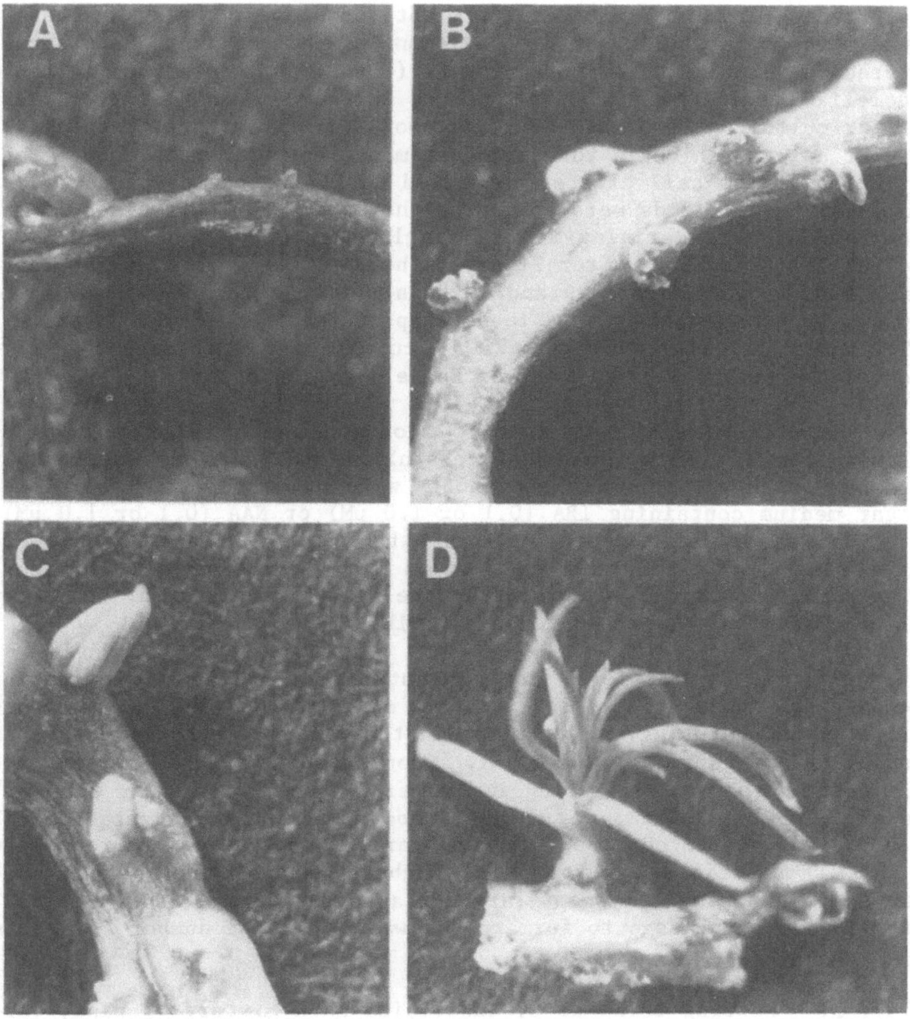

Fig. 3. Adventitious bud development on cotyledons of C. lusitanica. Hypocotylary explants were cultured on CBM medium supplemented with 1 µM BA and 50 nM IBA. After 3 weeks, the explants were transferred to half-strength CBM medium without hormones (see Materials and Methods for details of protocol). A. Small clusters of cells protruding from the cotyledon surface after 3 weeks in culture (30X). B. Buds showing well-formed structure after 4 weeks in culture (25X). C. Buds with well-developed leaves beginning to elongate after 6 weeks in culture (15X). D. Elongated shoots obtained after 8 weeks in culture. Note the multiple sets of leaves (10X).

protruding from the surface of the cotyledon, were observed after 2½ weeks in culture (Fig. 3A). Cotyledons of Thuja plicata Donn., cultured on defined media containing BA (1 to 10 μM), produced multiple domelike swellings on their adaxial surfaces within 3 to 6 weeks after culturing (10). These meristemoids differentiated into single shoot apices, which in turn produced many leaf primordia. Differentiation of the cell clusters observed in C. lusitanica proceeded in a like manner. Buds were formed on the upper and lower sides of the cotyledons, each arising from a single, spatially isolated cluster of cells (Fig. 3B). After 21 days, the explants were transferred to CBM medium (see Materials and Methods) without hormones, where the newly formed buds continued to develop and elongate (Fig. 3C and D). Typically, after 7 to 10 weeks in culture, several sets of well-developed leaves were present on the elongating shoots.

Initial attempts at rooting elongated shoots (≥5 mm) of C. lusitanica in vitro were unsuccessful. In 2 sets of experiments (data not shown) in which the shoots were transferred to solidified agar medium containing IBA (0.1 or 1.0 μM) or NAA (0.1 or 1.0 μM), callus was produced at the base of the shoots. These shoots were transferred after 7 to 15 days to an agar medium without hormones. The callus continued to enlarge at the base of the shoots, and after 7½ weeks in culture no roots were visible.

Rooting was achieved under nonsterile conditions by planting elongated shoots (≥5 mm) in a mixture of peat, vermiculite, and sand (4:2:1) and watering them 4 times in the first 2 weeks with a water solution of NAA (15 μM) (Fig. 4). Shoots from the same explant were placed in the same planting container. Data were taken after 7½ weeks (Tab. 5). The percentage of shoots forming roots was obtained by pooling shoots obtained from the various bud induction treatments (Tab. 4) described earlier. The highest percentage of shoots derived from a single hypocotylary explant that formed roots was 83%; the lowest percentage to form roots was 0%. The number of roots

Tab. 5. Rooting trial of C. lusitanica in a mixture of peat, vermiculite, and sand (4:2:1) under continuous light. Shoots were treated twice a week at evenly spaced intervals for 2 weeks with a water solution of NAA (15 μM). Data were taken after 7½ weeks.

Number of shoots tested	Shoots rooted (%)	Average number of primary roots/shoot	Mean root length (mm)
92	30	2.6	3.5

Fig. 4. Rooting of C. lusitanica shoots. A. Elongated shoot (≥ 5 mm) ready to be planted in a mixture of peat, vermiculite, and sand (4:2:1). B. Shoot growing in rooting mixture. C. Rooted shoots (plantlets) showing well-developed roots after 7½ weeks in the rooting medium.

produced per shoot varied from 1 to 6, and their length varied from 0.25 mm to 4.0 cm. A similar approach was used for rooting adventitious shoots obtained from the cotyledonary tissues of western hemlock (Tsuga heterophylla) (7).

Plantlets of C. lusitanica were brought to ambient greenhouse conditions over a 10-day period by gradually removing the plastic tent that enclosed the rooting containers. Several of these plantlets have been recently outplanted in southwest Florida. Their survivability and growth rates are being monitored.

SUMMARY

We have reported protocols for successfully inducing organogenesis and plantlet formation from explants of embryonic and young seedling tissues of P. oocarpa and C. lusitanica. The in vitro propagation of these species using juvenile material provides an opportunity to produce cloned individuals from seed produced by controlled crossing experiments. The production of plantlets from both species required a series of hormone-containing inductive treatments of relatively short duration, followed by incubation on hormone-free media. Research in progress is aimed at increasing the efficiency of plantlet production and understanding the biochemical and developmental events that accompany organogenesis in conifers.

ACKNOWLEDGEMENTS

We wish to thank the Latin American Scholarship Program for American Universities for supporting E.O. Franco during his graduate training program and Robert Beaty for providing the micrograph of an adventitious bud of P. oocarpa Schiede.

REFERENCES

1. Aitken, J., K.J. Horgan, and T.A. Thorpe (1981) Influence of explant selection on the shoot-forming capacity of juvenile tissue of Pinus radiata. Can. J. For. Res. 11:112-117.
2. Arnold, S. von, and T. Eriksson (1978) Induction of adventitious buds on embryos of Norway spruce grown in vitro. Physiol. Plant. 44:283-287.
3. Bornman, C.H. (1983) Possibilities and constraints in the regeneration of trees from cotyledonary needles of Picea abies in vitro. Physiol. Plant. 57:5-16.
4. Campbell, R.A., and D.J. Durzan (1975) Induction of multiple buds and needles in tissue cultures of Picea glauca. Can. J. Bot. 53:1652-1657.
5. Campbell, R.A., and D.J. Durzan (1976) Vegetative propagation of Picea glauca by tissue culture. Can. J. For. Res. 6:240-243.

6. Cheng, T.-Y. (1975) Adventitious bud formation in culture of Douglas fir (Pseudotsuga menziesii [Mirb.] Franco). Plant Sci. Lett. 5:97-102.
7. Cheng, T.-Y. (1976) Vegetative propagation of western hemlock (Tsuga heterophylla) through tissue culture. Plant and Cell Physiol. 17:1347-1350.
8. Cheng, T.-Y. (1977) Factors affecting adventitious bud formation on cotyledon culture of Douglas fir. Plant Sci. Lett. 9:179-187.
9. Coleman, W., and T.A. Thorpe (1976) Induction of buds in tissue cultures of four different conifers. Plant Physiol. 57:67 (Supplement).
10. Coleman, W.K., and T.A. Thorpe (1977) In vitro culture of western red cedar (Thuja plicata Donn.) I. Plantlet formation. Bot. Gaz. 138:298-304.
11. David, A. (1982) In vitro propagation of gymnosperms. In Tissue Culture in Forestry, J.M. Bonga and D.J. Durzan, eds. Martinus Nijhoff/Dr. W. Junk Publishers, The Hague, pp. 72-108.
12. Durzan, D.J. (1980) Prospects for the mass propagation of economically important conifers by cell and tissue culture. In Plant Cell Cultures: Results and Perspectives, F. Sala, B. Paris, R. Cella, and O. Ciferri, eds. Elsevier/North-Holland Biomedical Press, Amsterdam, pp. 283-288.
13. Dyson, W.G. (1973) An east African provenance trial of Cupressus lusitanica Miller. In Proceedings of a Joint Meeting on Tropical Provenance and Progeny Research and International Cooperation, J. Burly and D.G. Nikles, eds. Commonwealth Forestry Institute, Oxford, pp. 124-128.
14. Gill, G.W., J.K. Frost, and K.A. Miller (1974) A new formula for half-oxidized hematoxylin solution that neither overstains nor requires differentiation. Acta Cytol. 18:300-311.
15. Greaves, A. (1982) Pinus oocarpa. Forestry Abst. 43:503-532.
16. Greaves, A., and R.H. Kemp (1977) International provenance trials of Pinus oocarpa Schiede. In Proceedings of a Joint Workshop on Progress and Problems of Genetic Improvement of Tropical Forest Trees, Vol. 2, D.G. Nikles, J. Burley, and R.D. Barnes, eds. Commonwealth Forestry Institute, Oxford, pp. 552-561.
17. Gutierrez, M. (1977) Initiation of a genetic improvement program of Cupressus lusitanica Mill. and Pinus patula Schlecht. and Cham. in Colombia. In Proceedings of a Joint Workshop on Progress and Problems of Genetic Improvement of Tropical Forest Trees, Vol. 2, D.G. Nikles, J. Burley, and R.D. Barnes, eds. Commonwealth Forestry Institute, Oxford, pp. 824-826.
18. Hawkes, J.G. (1976) Introduction. In Tropical Trees, Variation, Breeding and Conservation, J. Burley and B.T. Styles, eds. Academic Press, London.
19. Kemp, R.H. (1973) Status of the G.F.I. international provenance trial of Pinus oocarpa Schiede. In Proceedings of a Joint Meeting on Tropical Provenance and Progeny Research and International Cooperation, J. Burley and D.G. Nikles, eds. Commonwealth Forestry Institute, Oxford, pp. 76-82.

20. Kemp, R.H. (1977) Pinus oocarpa Schiede: Research and development needs. In Proceedings of a Joint Workshop on Progress and Problems of Genetic Improvement of Tropical Forest Trees, Vol. 2, D.G. Nikles, J. Burley, and R.D. Barnes, eds. Commonwealth Forestry Institute, Oxford, pp. 655-660.
21. Koleveska-Pletikapíc, B., S. Jelaska, J. Berljak, and M. Vidakovic (1983) Bud and shoot formation in juvenile tissue culture of Pinus nigra. Silvae Genetica 32:115-119.
22. Konar, R.N., and M.N. Singh (1980) Induction of shoot buds from tissue cultures of Pinus wallichiana. Z. Pflanzenphysiol. 99:173-177.
23. Lind, M.L., and J. Staba (1961) Peppermint and spearmint tissue culture. I. Callus formation and submerged culture. Lloydia 24:139-145.
24. Longman, K.A. (1976) Conservation and utilization of gene resources by vegetative multiplication of tropical trees. In Tropical Trees Variation, Breeding and Conservation, J. Burley and B.T. Styles, eds. Academic Press, London, pp. 19-24.
25. Mehra-Palta, A., R.H. Smeltzer, and R.L. Mott (1978) Hormonal control of induced organogenesis: Experiments with excised plant parts of loblolly pine. Tappi 61:37-40.
26. Mott, R.L., and H.V. Amerson (1981) A tissue culture process for the clonal production of loblolly pine plantlets. In North Carolina Agricultural Research Service Technical Bulletin, No. 271, Raleigh, North Carolina, 14 pp.
27. Mott, R.L., and H.V. Amerson (1981) Tissue culture plantlets produced from Pinus monticola embryonic materials. Forest Sci. 27:299-304.
28. Mott, R.L., R.H. Smeltzer, A. Mehra-Palta, and B.J. Zobel (1977) Production of forest trees by tissue culture. Tappi 60:62-64.
29. Murashige, T., and F. Skoog (1962) A revised medium for rapid growth and bioassays with tobacco tissue cultures. Physiol. Plant. 15:473-497.
30. Nitsch, J.P., and C. Nitsch (1965) Néoformations de fleurs in vitro chez une éspice de jours courts: Plumbago indica. Ann. Physiol. Vég. 7:251-258.
31. Patel, K.R., and G.P. Berlyn (1982) Genetic instability of multiple buds of Pinus coulteri regenerated from tissue culture. Can. J. For. Res. 12:93-101.
32. Reilly, K., and J. Washer (1977) Vegetative propagation of radiata pine by tissue culture: Plantlet formation from embryonic tissue. N.Z. J. For. Sci. 7:199-206.
33. Sass, E.J. (1958) Botanical Microtechnique, 3rd ed. The Iowa State University Press, Ames, Iowa. 228 pp.
34. Sharp, W.R., and D.A. Evans (1981) Patterns of plant regeneration: Opportunities for cloning or production of genetic variability. IAPTC Newsletter 35:2-8.
35. Sommer, H.E., and C.L. Brown (1974) Plantlet formation in pine tissue cultures. Am. J. Bot. 61:11 (Supplement).

36. Sommer, H.E., C.L. Brown, and P.P. Kormanic (1975) Differentiation of plantlets in longleaf pine (Pinus palustris Mill.) tissue cultured in vitro. Bot. Gaz. 132:196-200.
37. Thomas, M.J., E. Duhoux, and J. Vazart (1977) In vitro organ initiation in tissue cultures of Biota orientalis and other species of the Cupressaceae. Plant Sci. Lett. 8:395-400.
38. Thomas, M.J., and H. Tranvan (1982) Influence relative de la BAP et de l'IBA sur la néoformation de bourgeons et de racines sur les plantules du Biota orientalis (Cupressacées). Physiol. Plant. 56:118-122.
39. Tranvan, H. (1979) In vitro adventitious bud formation on isolated seedlings of Pinus silvestris L. Biologia Plantarum (Praha) 21:230-233.
40. Wochok, Z.S., and M. Abo El-Nil (1977) Conifer tissue culture. Proc. Int. Plant Prop. Soc. 27:131-139.
41. Yeung, E.C., J. Aitken, S. Biondi, and T. Thorpe (1981) Shoot histogenesis in cotyledon explants of radiata pine. Bot. Gaz. 142:494-501.

PROPAGATION OF COFFEE

M.R. Sondahl, T. Nakamura, and W.R. Sharp

DNA Plant Technology Corporation
2611 Branch Pike
Cinnaminson, New Jersey 08077

ECONOMIC IMPORTANCE

Coffee is the most important agricultural commodity in the international market (US $9-12 billion annually), and its production is restricted to tropical countries. Coffee export revenues are the main source of hard currency for producing countries, thus permitting access to modern technologies and services. The world production of coffee in the 1981/82 season was 96.4 million bags of 60 kg. Some of the leading coffee-producing countries in the 1981/82 harvest were: Brazil (34.2%), Colombia (14.5%), Indonesia (5.4%), Ivory Coast (4.8%), Mexico (3.7%), Guatemala (2.8%), and El Salvador (2.5%).

There are 2 commercially important coffee species: *Coffea arabica* L. and *C. canephora* Pierre (Robusta coffee). Quality beverage is produced from *C. arabica* which is cultivated at higher altitudes. This species represents 70% of the commercial coffee of the world and about 99% of Latin American production. *Coffea canephora* is usually grown in tropical areas at lower altitudes. Eighty percent of the African production is of this type. On a very reduced scale, *C. liberica* is grown in Liberia, Surinam, and Malaysia, *C. racemosa* in Mozambique, and *C. dewevrei* in Ivory Coast and Zaire. These species produce beans of lower quality that are only acceptable in the local market.

The production cost of a coffee plantation varies from country to country. If one takes a typical Brazilian coffee farm with an average production of 1500 kg/ha (25 bags of 60 kg), the costs of fertilization and phytosanitary control are 30% of the total production costs. Improvement of coffee plants should focus on these areas of production cost as well as on the quality of the final product.

COFFEE SOMATIC EMBRYOGENESIS AS A MODEL SYSTEM

The high-frequency recovery of somatic embryos from adult tissues (leaves) of coffee plants has made somatic embryogenesis an attractive model system for other tree species. Similar results have been obtained with some fruit trees using the nucellar tissue as explant source. In both coffee and fruit trees, the degree of success of somatic embryogenesis varies according with the genotypes in culture. A common feature of all high-frequency somatic embryogenesis (HFSE) is the presence of an "embryogenic tissue." This tissue differentiates from particular cells: embryogenic mother cells. Embryogenic tissue can be characterized as a friable tissue, containing small and spherical cells (15-20 µm diameter), with dense cytoplasm, a prominent nucleolus, a basophilic nucleus, and a fast cell-cycle time. Another general feature of HFSE systems is the sequential approach used during initial phases of culture: high auxin/cytokinin ratio during the primary culture (induction medium), and low auxin/cytokinin ratio or no plant growth regulator during secondary culture (condition medium). In coffee, the following phases for HFSE can be recognized:

a) Induction medium (cell division and redetermination)
b) Condition medium (differentiation)
c) Embryo rescue (isolation)
d) Embryo germination (shoot and root growth)
e) Plantlet hardening
f) Transfer to soil

There are many general aspects that have been identified as controlling factors of somatic embryogenesis:

a) Related to donor tissues:

1. Plant species or varieties
2. Source of explant
3. Pre-treatment of explant

b) Related to growth medium:

1. Inorganic and organic constituents
2. Plant growth regulators (concentration ratio and types)
3. Total osmolality

c) Related to growth conditions:

1. Light (quality, energy, photoperiod)
2. Temperature (thermoperiod)
3. Gas exchange
4. Subculture regime
5. Tissue selections during subculture

The success of HFSE with other perennial species may be improved from these established systems. The present literature on somatic embryogenesis describes the use of a high salt medium for primary culture (75% among dicots and 82% for monocots). The addition of 2,4-dichlorophenoxyacetic acid (2,4-D)/Kinetin (KIN) (22%) or 2,4-D alone (15%) has prevailed among dicots for the induction phase. In the case of monocots, 2,4-D alone has been utilized in 68% of the successful induction media.

Vegetative propagation has been primarily accomplished by induction of development of pre-existing meristems (apical or axillary). Adventitious shoots (organogenesis) and high-frequency somatic embryos (embryogenesis) have been more recently utilized as a means of vegetative propagation. Coffee plants can be successfully propagated by means of high frequency somatic embryogenesis (Fig. 1, A-C) or by axillary bud development (Fig. 1, D-F).

TISSUE CULTURE OF COFFEE

Leaf Culture

Mature leaves from orthotropic or plagiotropic branches of greenhouse plants are surface sterilized in 1.6% sodium hypochlorite (30% commercial bleaching agent) for 30 min and rinsed 3 times in sterile double-distilled water. If the coffee plants are growing in the field, the surface sterilization is much more difficult. The following procedure has given some degree of success for field material: 2.6% sodium hypochlorite for 30 min, rinse in sterile water, incubate in sealed petri dishes overnight followed by another exposure to 2.6% sodium hypochlorite for 30 min, then rinse 3 times with sterile water. It has been found that 70% ethanol and solutions of $HgCl_2$ are toxic to coffee leaves; on the other hand, solutions of sodium or calcium hypochlorite are effective and nontoxic. Before immersing the leaves in the sterilizing solution, it is advisable to wash them by hand with 1% detergent solution, and rinse in distilled water. If ozone water (water enriched with O_3) is available, a 5-min immersion is recommended. The addition of antibiotics and systemic fungicides in the saline-sugar medium used for the 3-day preculture period has been advantageous in reducing bacterial and fungal contaminations from field materials.

Leaf explants of about 7 mm^2 are cut, excluding the midvein, margins, and apical and basal portions of the leaf blade. The elimination of the midvein also excludes the domatia, consequently lowering the number of contaminated leaf explants. Domatia are deep pores (ca. 0.18 mm diameter) located in the acute angle formed by the midvein with secondary veins on the abaxial side of the coffee leaf. Leaf sections are usually cut on top of sterile filter paper (or paper towel) that is changed frequently to avoid cross contami-

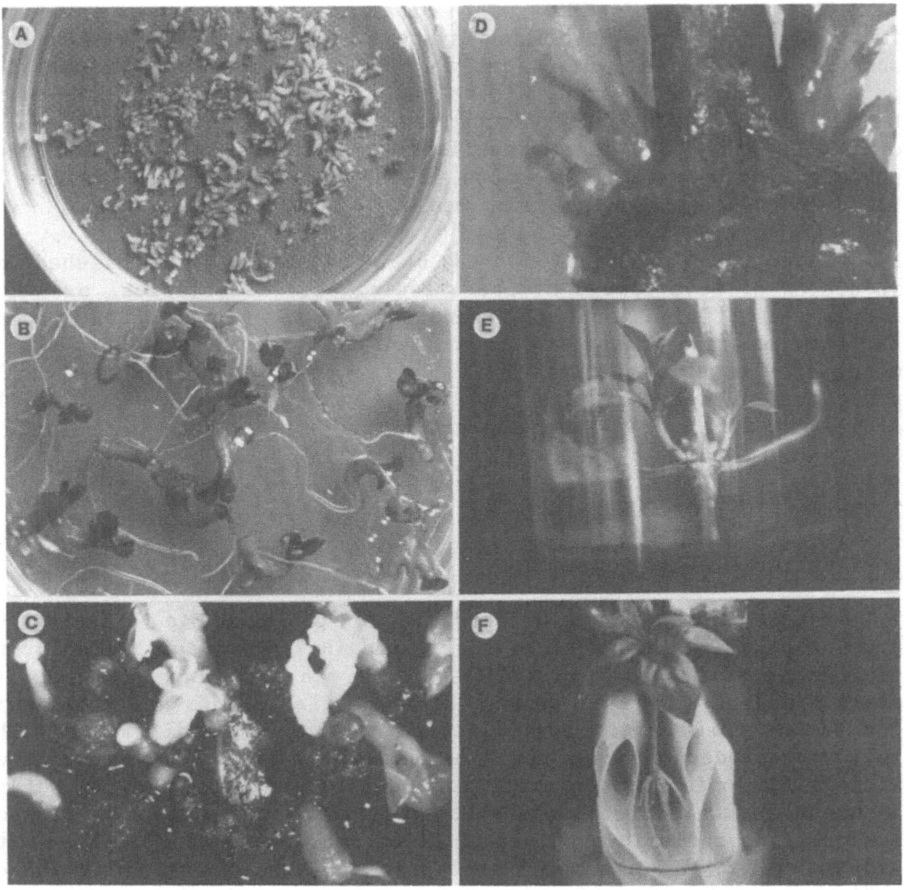

Fig. 1. Propagation of coffee by somatic embryogenesis and axillary bud culture. (A) High frequency somatic embryogenesis (or indirect embryogenesis) from a single coffee leaf explant. Embryos growing in liquid basal medium for ca. 5 weeks following isolation of embryogenic tissue from secondary culture (1.0X). (B) Germinating embryos on solid base medium (1.2X). (C) Differentiation of low frequency somatic embryos (or direct embryogenesis) during secondary culture of a single coffee leaf explant. This embryogenic pathway provides a low number of embryos. Note that not all embryos present a normal morphological development. (D) Early stage (ca. 3 weeks) of axillary bud development from culture of an orthotropic node. Note the growth of 3 orthotropic buds from each side of the node (10X). (E) Shoot growth from a cultivated orthotropic node following ca. 8 weeks in primary culture (1.0X). (F) Rooting of coffee shoot derived from an axillary bud using a filter paper culture (1.0X).

nation. All sides of a leaf explant must be cut since callus proliferation occurs only from cut edges. Histological studies have demonstrated that the callus tissue originates from mesophyll cells of leaf explant (15).

The leaf explants are precultured in 20 x 100 mm petri dishes containing a solidified saline-agar medium [half-strength Murashige and Skoog (MS) salts with 0.06 M sucrose (10)]. The preculture of the leaf explants can be in the dark or in the light (no difference has been detected between these two conditions) for about 72 hr. This preculture has been found to be very useful for the selection of viable explants and the elimination of contaminated leaf pieces. The abaxial surface of the leaf explants, which is clearly distinguished by a dull, pale-green coloration (in contrast to the shiny, dark-green coloration of the adaxial surface), is always placed upwards. French square bottles of about 50 ml capacity containing 10 ml autoclaved basal medium made up of MS inorganic salts (10), 30 μM thiamine-HCL, 210 μM L-cysteine, 550 μM meso-inositol, 117 mM sucrose, and 8 g/l Difco Bacto agar are used. In primary culture, an "induction medium" containing a combination of KIN (20 μM) and 2,4-D (5 μM) is used, and the cultures are incubated in the dark at 25 ± 1°C for 45-50 days. The composition of this "induction medium" was found to be ideal for HFSE induction of _C. arabica_ cv. Bourbon (12).

Secondary cultures are established under conditions of a 12-hr light period at 24-28°C by subculturing 45- to 50-day-old tissues onto a "conditioning medium" containing half-strength MS organic salts (except KNO_3, which is added at 2x concentrations), sucrose (58.4 mM), KIN (2.5 μM), and naphthaleneacetic acid (NAA) (0.5 μM). Following transfer to the conditioning medium, the massive parenchymatous type of callus growth ceases and the tissues slowly turn brown. Two sequences of morphogenetic differentiation have been characterized in secondary cultures of leaf explants in _Coffea_: low-frequency somatic embryogenesis (LFSE or direct embryogenesis) and high-frequency somatic embryogenesis (HFSE or indirect embryogenesis). Adopting the standard culture protocol described for _C. arabica_ cv. Bourbon, LFSE is observed after 13-15 weeks and HFSE after 16-19 weeks of secondary culture. LFSE appeared 3-6 weeks before the visible cluster of HFSE (13,14, 15).

The appearance of isolated somatic embryos developing into normal green plantlets in numbers ranging from 1 to 10 per culture is typical of LFSE. In contrast, the occurrence of HFSE follows a unique developmental sequence: a white friable tissue containing globular structures defined as _embryogenic tissue_ develops from the nonproliferating brown callus cell mass; the globular structures appear to develop synchronously for a period of 4-6 weeks. The embryogenic tissue gives rise to somatic embryos and finally to plantlets, but this latter developmental process lacks the synchrony of the earlier stage. The amount of this embryogenic tissue varies, but on the average, about 100-200 somatic embryos develop from such clusters of embryogenic tissue. To speed up development and increase the percentage of fully developed plantlets, it is advisable to isolate the embryogenic tissues and grow them under light conditions at 26°C in 5-10 ml of liquid basal medium devoid of growth regulators for 4-6 weeks (embryo isolation). After this period, the torpedo-shaped somatic embryos and young plantlets can be plated onto saline-agar medium containing 0.015-0.03 M sucrose in the presence of light (embryo germination). Individual plantlets bearing a developed tap root are removed from the agar medium, gently washed, and immediately transferred to small pots inside a humid chamber. After a hardening period of 1 month, they can be exposed to normal atmospheric humidity and transferred to a greenhouse. Another way of hardening the plantlets is to use approximately 100 ml of saline-agar medium without sucrose in 250 Erlenmeyer flasks closed with cotton plugs and paper. The Erlenmeyer flasks are exposed to sunlight in a shaded portion of the greenhouse (usually protected with plastic screens that filter about 60-70% of the sun's rays). After about 2 months, the plantlets have good leaf and root development and can be transferred to small pots containing a light soil or promix.

This procedure has been applied to 4 different species of coffee and to many C. arabica cultivars. Currently, somatic embryos from leaf tissues of some morphological coffee mutants have been obtained in order to evaluate the genotype stability of this methodology. Differentiation has occurred in the following single gene mutants: Erecta (Er), Purpurascens (pr), Angustifolia (ag), Nanna (na), San Ramon (Sr), and Volutifolia (vf). Embryogenic tissues and/or somatic embryos at different stages of development have been obtained for each of the mutants. The genetic analysis of plants derived from the leaf explant culture of these mutants will provide new insights on the feasibility of somatic embryo production for vegetative propagation of plants. Moreover, it will possibly permit a further distinction between organogenesis and embryogenesis in regard to genetic stability and cytological origin.

The culture of cotyledonary leaves of coffee has not been described in the literature. The feasibility of this explant source was tested with 3-month-old seedlings of C. arabica cv. Mundo Novo using the same methodology described for mature leaves, except the concentration of 2,4-D during primary culture (conditioning medium) was varied. Little contamination was observed among all 550 cotyledonary segments cultivated after 2 months in culture. The lowest concentration of 2,4-D (1 μM) also induced the differentiation of adventitious roots. A concentration of 5 μM 2,4-D, which is the same concentration used for mature leaves, was adopted for future cultures of cotyledonary leaf explants. Embryogenic tissue was obtained profusely after 4 months of secondary culture on induction medium.

Apical Meristem Culture

The development of a protocol for meristem culture is a prerequisite for the future cryopreservation of coffee germplasm. Kartha et al. (4a) isolated 0.3 mm apical meristems from sterile seedlings of C. arabica cv. Caturra Rojo and Catuai. Culture media consisting of MS salts and B5 vitamins, supplemented with benzyladenine (BA) or zeatin (ZEA) (5-10 M) and NAA (1 M), induced multiple shoots, whereas lower concentrations of cytokinins produced single shoots. Rooting occurred only in media with half-strength MS salts devoid of sucrose in the presence of 1 M indolebutyric acid (IBA).

In our laboratory, apical meristems of 3-month-old C. arabica cv. Mundo Novo have been excised and plated on saline-sucrose agar plates. After 5 days of preculture of 192 meristems, the meristems were transferred to primary culture after discarding the contaminants (32%) and oxidation (4%). The culture media consisted of MS salts (0.1x microsolution), pyridoxine (15 μM), cysteine (500 μM), sucrose (0.087 M), IAA (0.5 μM), and various concentrations of BA. After 10 weeks of culture, the meristems cultured on media containing 5 μM BA were the most vigorous. The rooting medium adopted consisted of half-strength MS salts (0.1x microelements), sucrose (0.058 M), vitamins, IBA (5-10 μM), and NAA (5-10 μM). In all cases, only long and single roots were obtained. In order to form secondary roots and root hairs, the plantlets were subcultured on sterile sand:vermiculite (2:3) substrate. Apical meristems of C. arabica cv. Catuai were cultivated with the same protocol described for cv. Mundo Novo with similar results.

Axillary Bud Culture

Characteristically, coffee plants have multiple arrested orthotropic buds and 2 plagiotropic buds at each stem node. The plagiotropic buds differentiate only after the 10th-11th node of a developing seedling, whereas the orthotropic buds are present in the first node (cotyledonary node). The removal of the apical meristem results in the development of 2 orthotropic shoots, one at each leaf axil, at the most apical node. At each leaf axil of plagiotropic branches, there are 4 serially arrested buds. These buds will differentiate into flower buds under proper environmental conditions. Sometimes, the uppermost bud of that series at one or both sides of a node will develop as a vegetative bud instead of a floral bud, leading to a secondary plagiotropic branch.

The presence of these arrested buds has been explored as a means of vegetative propagation of coffee plants. Nodal explants of aseptically grown C. arabica plants have been cultured on MS medium supplemented with BA (44 μM) and indoleacetic acid (IAA) (0.6 μM), under a 16-hr photoperiod (2000 lux) at 25 ± 0.5°C (2). Shoot development occurs on the average of 2.2 per node after 2-5 weeks.

The use of NAA (1.1 μM) in complete darkness encourages good rooting of these young coffee shoots. It is recommended that at least 3-month-old plants grown in vitro be used as a source of nodal sections, and that the leaves be retained during culturing. Similar techniques have been used by Dublin (3) with Arabusta plants. Shoot development was observed in medium supplemented with malt extract (400 mg/l) and BA (4.4 μM).

Orthotropic Buds

Nodal cultures were established from orthotropic branches of coffee. The donor plants were derived from the greenhouse (ca. 10 months old), nursery (1-2 years old), or field plots (15-20 years old). Green orthotropic shoots containing 4-6 nodes were excised, washed with 1% detergent and sterile water, and surface sterilized with 2.6% sodium hypochloride for 30 min under continuous agitation (150 rpm). Shoots excised from the nursery or field-grown plants were allowed to incubate in a humid chamber overnight, and then were submitted to a second sterilization as described above. The apical meristems were eliminated from all orthotropic shoots. The uppermost node beneath the apical meristem was coded No. 1. Whenever possible, the nodes were cultivated with the attached leaf pair trimmed to one-half to one-third of the original size.

The primary medium consisted of a modified B5 medium (3a) containing pyridoxine (15 μM), nicotinic acid (15 μM), thiamine (30 μM), inositol (550 μM), cysteine (500 μM), sucrose (87 mM), BA (25-50 μM), IAA (10 μM), activated charcoal (2.5 g/l), PVP-40 (1 g/l), and agar (7 g/l). Round screw cap bottles of 150-ml capacity charged with 25 ml of medium were used. The bottles were inoculated with individual nodes and sealed with a 16.5 μM PVC film. This film has the following permeability characteristics: CO_2 (10 cm^3/cm^2 per 24 hr); oxygen (2 cm^3/cm^2 per 24 hr); nitrogen (0.4 cm^3/cm^2 per 24 hr); and water vapor (10 mg/cm^2 per 24 hr).

The nodal cultures were maintained in a growth room under 500 lux illumination with a 14-hr photoperiod. Contamination frequencies were very high during primary culture. Sixty percent of the uppermost nodes from small plants and 90-100% of the lower nodes (4th-6th) were contaminated. Average contamination rates from 824 nodal explants from 4 different cultivars of field-grown coffee plants were 43% with 42% oxidation.

The development of the arrested orthotropic buds from nodal cultures becomes visible during the third week of culture (Fig. 1D). The number of developed buds per node varies from 1 to 6 and is partially controlled by the level of cytokinin. Higher cytokinin levels induce the development of higher numbers of buds, but they may cause adverse effects on the subsequent growth of the derived shoots and lower the rooting frequencies. Pertaining to the cytokinin source, BA has proven more effective than KIN at 25 μM (2.1 versus 1.2 developed buds/node) and 50 μM (2.3 versus 1.4 buds/node), respectively.

The effects of gibberellic acid (GA) on the elongation of shoot internodes were tested. Unsatisfactory results were found with the addition of GA in the solid medium sterilized either by autoclave or membrane filters. Effective treatments consisted of 25-50 ppm of GA applied directly to the developing shoots at the time of excision for rooting.

When the developed shoots are 3 nodes long (ca. 60 days of primary culture; see Fig. 1E), they can be excised and rooted. The best rooting frequencies (60%) were found with shoots treated with IBA (10 μM). A double-layer medium can be used for rooting, containing basic medium supplemented with IBA (10 μM) in the top layer and the basic medium with added charcoal (2.5 g/l) in the bottom. Another rooting technique consists of treating the undeveloped shoots with an auxin solution for 10 days and then transferring the shoots to a basic solid medium or on a paper bridge (Fig. 1F). Shoots treated with a talc preparation of the auxin also gave high frequencies of rooting (60%) on the basic solid medium.

After excision of the first developed shoots, the nodal explants are transferred to fresh primary medium in order to recover the remaining arrested orthotropic buds. A nodal explant may survive 2-3 passages and so permit the recovery of most of the arrested buds. In several instances, 9-10 developing buds were recovered after the third bimonthly transfer (6 months of continuous culture). The developed buds derived from the original nodal explants were called primary buds or shoots. The buds developed from the nodes of primary shoots were designated secondary buds or shoots.

In one experiment of nodal culture from 1-year-old coffee plants (*C. arabica* cv. Mundo Novo), the frequencies of primary and secondary bud development per node were recorded during 6 months of continuous culture with bimonthly transfers. An average frequency of 1.33 developed buds/node was observed after the first 60 days of culture. At the end of the secondary culture period (120 days), an accumulated frequency of 4.5 buds/node was recorded. A final accumulated frequency of 7.5 buds/node was obtained at the end of the third culture period (180 days).

These data demonstrate the potential to explore the presence of arrested orthotropic buds for vegetative propagation of superior coffee plants by nodal culture technique (11,16). A 45% recovery (85 buds/19 nodes) of all arrested buds was recorded at the end of the second culture period (120 days). Taking into account the experimental data, a theoretical estimation of the multiplication rate can be made as follows:

a. First growth period (6 months) - 26 shoots with 3 nodes derived from nodal cultures of one orthotropic branch containing 6 nodes.

b. Second growth period (3 months) - 104 shoots with 3 nodes.

c. Third growth period (3 months) - 415 shoots with 3 nodes.

This protocol permits the recovery of 415 shoots/6-node branch after 12 months of successive culture. Since a mature coffee plant has several orthotropic branches that can be used , a high number of plants can be obtained by this technique within the time frame of 1 year. In conclusion, the final number of propagated shoots will vary according to the number of orthotropic branches utilized from each donor plant, the losses due to contamination and oxidation, and the efficiency of rooting.

THEORETICAL ASPECTS OF SOMATIC EMBRYOGENESIS

Working Hypothesis

In vitro embryogenesis has been frequently associated with the presence of an organized tissue which contains small spherical cells (10-20 μm diameter) dividing actively (short cell-cycle time). These cells have basophilic nuclei and prominent nucleoli typical of high RNA activity. This tissue enlarges by successive cell division, but later in development the daughter cells become cemented to each other, assuming polarity in the course of differentiation. Globular structures become visible leading to the final forms of heart- and torpedo-shape of embryo formation. This tissue is recognized in all cases of HFSE and is referred to as "embryogenic tissue" (ET). The degree of the initial phase of mitotic activity before formation of globular structures as well as the time lapse to the appearance of ET may vary among species and explant sources. In the case of *Coffea* leaf cultures, this ET is very distinct and has been fully characterized. It is important to emphasize that ET should be traced back to Embryogenic Mother Cells (EMC), i.e., cells at a particular regulatory stage of the cell cycle that assume an embryogenic pathway triggered by specific culture conditions.

Two general patterns of development of in vitro embryogenesis are discernible: (a) direct embryogenesis--embryos originate directly from tissues in the absence of callus proliferation [i.e., epidermal cells of hypocotyl in wild carrot and *Ranunculus sceleratus* (5,8)]; (b) indirect embryogenesis--callus proliferation and embryogenic tissue are associated with embryo development (i.e., secondary phloem of domestic carrot, leaf tissue of coffee, pollen of rice and some other Gramineae, etc.). Direct embryogenesis proceeds from preembryogenic determined cells (PEDC), while indirect embryogenesis requires the redetermination of differentiated cells, callus proliferation, and induction of embryogenic determined cells (IEDC).

Apparently, PEDCs await either synthesis of an inducer substance or removal of an inhibitory substance, requisite to resumption of mitotic activity and embryogenic development. Conversely, cells undergoing IEDC differentiation require a mitogenic substance to reenter the mitotic cell-cycle and/or exposure to specific concentrations of growth regulators. Arrestment may occur at any step in this process.

Each phase of the cell-cycle (G_1, S, G_2, and M) contains a specific synthetic machinery which is under regulatory control. The time duration of S and M is fairly constant in a given tissue system, while that of G_1 and G_2 is variable. The principal control points of the mitotic cell-cycle are situated in G_1 and G_2 (17). The decision to proliferate and the fate of daughter cells following mitosis have been suggested to occur at these critical points.

We postulate that a specific exogenous growth regulator concentration(s) or concentration ratios in the culture medium have a dual role in the onset of embryogenesis. First, growth regulators are responsible for the initiation of cell division, i.e., reentry of cells into the mitotic cell-cycle from G_0 (a special nonoperational state of the cell-cycle) or either the G_1 or G_2 principal control points of Van't Hof. Second, growth regulators have either a direct or indirect role in the control of cytoplasmic factor(s) synthesis during G_1 and G_2. At the time of cytokinesis, a quantal mitotic division would occur resulting in 2 phenotypically unlike daughters committed to different developmental patterns.

It should be understood that preceding cell divisions may influence subsequent differentiation: events in cycling cells may determine or limit the subsequent pathways of cytodifferentiation upon which the cells can embark. There may be a long time interval separating a regulatory cell division and the appearance of visible signs of cytodifferentiation. The presumption is that during a quantal cell division one of the daughter cells remains meristematic, whereas the other is determined to become a differentiated mother cell.

Here it is important to introduce the concept of multiple phenotypic populations in plant tissue cultures (18). Explant materials used for the establishment of primary cultures usually consist of an array of phenotypic cellular populations which can be characterized on the basis of cellular morphology, biochemical characteristics, and mitotic cell-cycle time. These cells have different developmental destinies imposed upon them because of their different DNA template sequence availabilities. Furthermore, some of these cell populations are actively dividing while others are in mitotic arrest (G_0, G_1, or G_2).

It would be naive to expect that a specific ratio of growth regulators would have a uniform effect on assorted phenotypic populations of cells. Rather, one would expect a particular growth regulator ratio to affect mitosis and/or redetermination of a particular phenotypic cellular population. Therefore, it is suggested that the embryogenic mother cells are descendants of a population of cells receptive to a critical embryogenic-inducing growth regulator concentration or a concentration ratio of growth regulator substances (auxin/cytokinin). Such cells are redetermined as to their developmental commitments during a quantal mitotic division in vitro which results in 2 daughters, one of which is probably the embryogenic mother cell. It appears that growth regulators (in particular auxins and cytokinins) have 2 functions in the conditioning culture medium (primary culture medium): (a) determination of the embryogenic mother cells (EMC) and (b) synchronization of EMCs. Thereafter, this cellular population most likely resides in a state of mitotic arrest until subculture to an embryogenic induction medium, where the auxin (frequently 2,4-D) is removed from the culture medium. This probably allows for the release of the EMC from mitotic arrest and subsequent embryo development.

The developmental sequences leading to ET are not fully explained at present. It is possible that growth regulators induce differentiation of a distinct population of cells according to one of the following phenomena: (a) proliferation of a unique phenotypic population from the original explant (cambium, secondary phloem, etc.); (b) lengthening of the mitotic cell-cycle time of a particular cell population by interference with one or more control points and redetermination; (c) determination and arrest of a distinct population of cells in the cell-cycle with a blockage of G_0, G_1, or G_2.

There is little information in the literature relating growth regulators to activities of the cell-cycle. Kinetin may cause a blockage at the G_1/S interface (9). Kinetin may also affect the oxidation of carbohydrates by inhibiting glycolysis and consequently increasing carbohydrate oxidation through the pentose shunt (9). Carrot tissues with active cell division promoted by kinetin have a lower oxygen consumption rate than tissues growing without kinetin (9). The transcription and translation processes of the different cellular populations comprising the callus tissues are obviously differentially affected in regard to their cell-cycle phase intervals on different culture media. Growth regulators added to the culture medium probably control morphogenesis indirectly or nonspecifically through the control of cell division and cell aging. It is interesting to note that aging of callus, omission of sucrose from medium, or gamma irradiation are reported to stimulate embryogenesis in unfertilized ovule tissues of <u>Citrus sinensis</u> 'Shamouti' (6). There is a report of a positive correlation between increasing 2,4-D concentration and increasing the mean generation time in carrot cell suspension (1). It also was found that the mitotic duration increased and the mitotic index decreased with increasing 2,4-D concentrations (0.5 to 70 µM; Ref. 1). These results were partially interpreted as a blockage or lengthening of G_1 and G_2, and the lengthening of prophase and metaphase relative to anaphase and tele-phase (1). It is interesting that in carrot, somatic embryogenesis occurs only after removal of 2,4-D, and that the threshold for embryo inhibition is 0.1 μM of 2,4-D.

Peaks in peroxidase activity are associated with the appearance of somatic embryos in orange ovular tissues. An isoenzyme analysis of these tissues reveals a cathodic band to be distinctly associated with the embryogenic process (7). Shortening of G_1 and an increase in peroxidase activity were described in association with development of hooded barley primordia (4).

Coffee System

Following transfer to the inducing medium, the massive parenchymous type of callus growth ceases and the tissues slowly turn brown. Two sequences of morphogenetic differentiation have been characterized in secondary cultures of leaf explants in *Coffea*: low-frequency somatic embryogenesis (LFSE or direct embryogenesis) and high-frequency somatic embryogenesis (HFSE). Adopting the standard cultural protocol described for *C. arabica* cv. Bourbon, LFSE is observed after 13-15 weeks and HFSE after 16-19 weeks of secondary culture. LFSE has appeared 3-6 weeks before the visible cluster of HFSE. LFSE is defined by the appearance of a few isolated somatic embryos that are not associated with a large mass of embryogenic tissue.

The fact that HFSE is so effectively triggered (up to 60%) by the addition of 2,4-D in combination with kinetin during primary culture is striking. Other sources of auxins (IBA and NAA) in combination with kinetin are not very effective in the induction of HFSE (10-20%). However, NAA in combination with kinetin induces LFSE in up to 60% of the cultured bottles. Moreover, the removal of 2,4-D and the lowering of NAA and kinetin concentrations in the induction medium of the secondary cultures were found essential to the development of somatic embryos of HFSE in *Coffea* spp. The effective and preferential role of 2,4-D in triggering the developmental sequences leading to HFSE cannot be fully explained at present. Calluses induced by 2,4-D and NAA appear to be phenotypically similar upon visual observation. However, it is possible that 2,4-D induces the differentiation of a distinct population of cells. More experimental work needs to be done in order to explain the effective physiological difference between 2,4-D and NAA in the induction of somatic embryogenesis in coffee leaf tissues.

Two distinct cell populations occur in the coffee callus mass during culture: one consisting of elongated cells and the other consisting of small spherical cells. Leaf explants develop a cream-colored soft callus tissue consisting of elongated parenchma-like cells (ca. 140 x 25 μm) during primary culture. After 50 days in primary culture, the callus tissues became brown and callus proliferation is no longer observed.

Later, following 16-19 weeks in secondary culture on an induction medium in which the growth regulator 2,4-D is removed, differentiation of a friable embryogenic tissue from the callus tissue is apparent. This friable embryogenic tissue becomes isolated from the original callus mass as proliferation proceeds with cells being sloughed off onto the surrounding culture medium. The embryogenic tissue consists of small spherical cells which divide very frequently during the early development.

During this proliferation period, all of the spherical daughter cells resulting from these divisions become separate from one another. At a subsequent time, the daughters originating from the spherical cells no longer separate, but remain as a coherent tissue. This developmental change results in the development of the globular embryos, the first organized structures in embryo development of Coffea cells in vitro. More recently, it has been found that mainly LFSE can be induced from coffee leaf explants cultured on basic medium supplemented with cytokinin (no auxin). The differential sensitivity to plant growth regulators, the time lapses between LFSE and HFSE, and the pattern of LFSE differentiation suggest that the mother cells of LFSE come from a distinct cell population which is different from the one responsible for HFSE. A developmental model for somatic embryogenesis for coffee leaf explants is presented in Fig. 2.

Summary of Hypothesis

Culture conditions, especially the plant growth regulators, have an important role in the induction of mitotic activity of different cell populations of cultured tissues. Depending on the particular regulatory state of a population of cells, the culture conditions will promote different cell-cycle times and blockages. Somatic embryos can be traced back to embryogenic mother cells (EMC) which are associated with 2 developmental sequences: preembryogenic determined cells (PEDC) and induced embryogenic determined cells (IEDC). Environmental and cultural stresses can preferentially affect different cell populations, permitting regulatory shifts and synchronization in cycling cells (determination of EMCs). By enhancing the EMC population of a particular system, more detailed analysis could be made (cytological and biochemical levels) as well, as will increase the yield of somatic embryos. It is expected that the labelling of DNA would allow for close monitoring of gene(s) action(s) during EMC determination and subsequent development. PEDC developmental sequence should differ from IEDC during initial stages, and so comparison between these 2 pathways will be profitable for understanding the control of somatic embryogenesis.

The model for somatic embryo differentiation (Fig. 3) suggests that embryogenic mother cells (EMC) are always associated with either PEDC or IEDC. Moreover, the model also would explain the

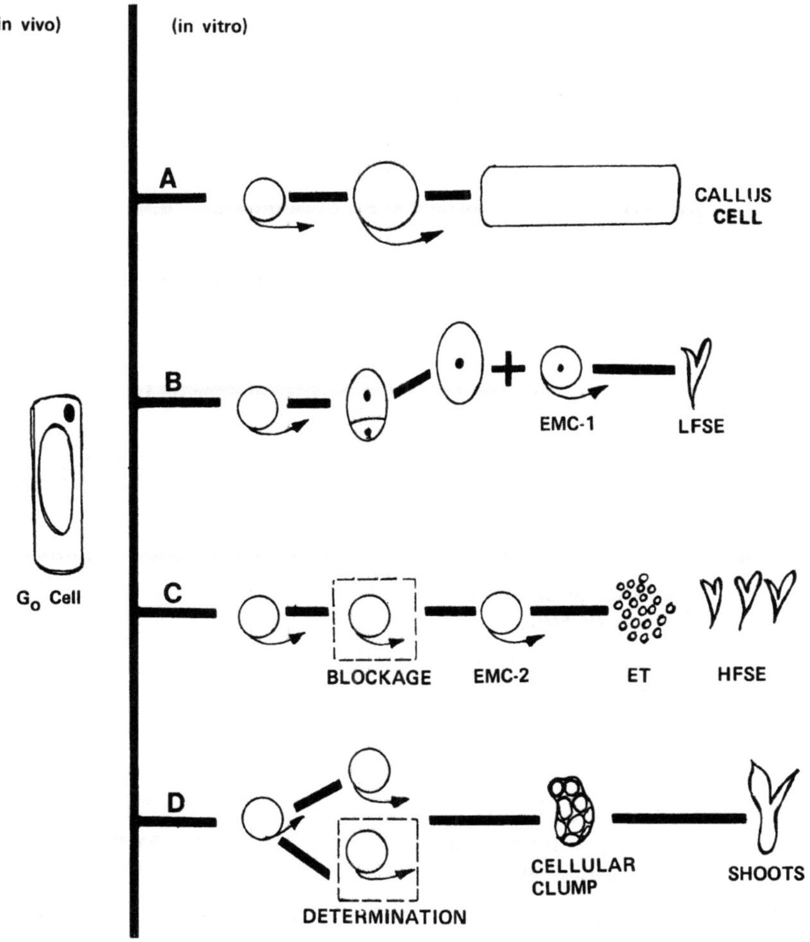

Fig. 2. Developmental model for somatic embryogenesis of coffee from leaf explants. (A) G_0 spongy mesophyll cells reenter the cell cycle in the presence of 2,4-D and kinetin. (B) At least 2 populations of cells are present in the callus mass: callus cells (long, highly vacuolated) and embryogenic cells (spherical, dense cytoplasm). (C) A sub-population of embryogenic cells follows the pattern of direct embryogenesis (low frequency somatic embryogenesis) still under primary culture; embryo germination occurs upon transfer to secondary medium. (D) Removal of 2,4-D during secondary cultures permits the differentiation of the second sub-population of cells - embryogenic tissue. (E) Presence of completely differentiated somatic embryos derived from sub-population I (low frequency) and sub-population II (high frequency). Note that due to the differentiation process, there is ca. a 1-2 month time difference between these 2 groups of somatic embryos.

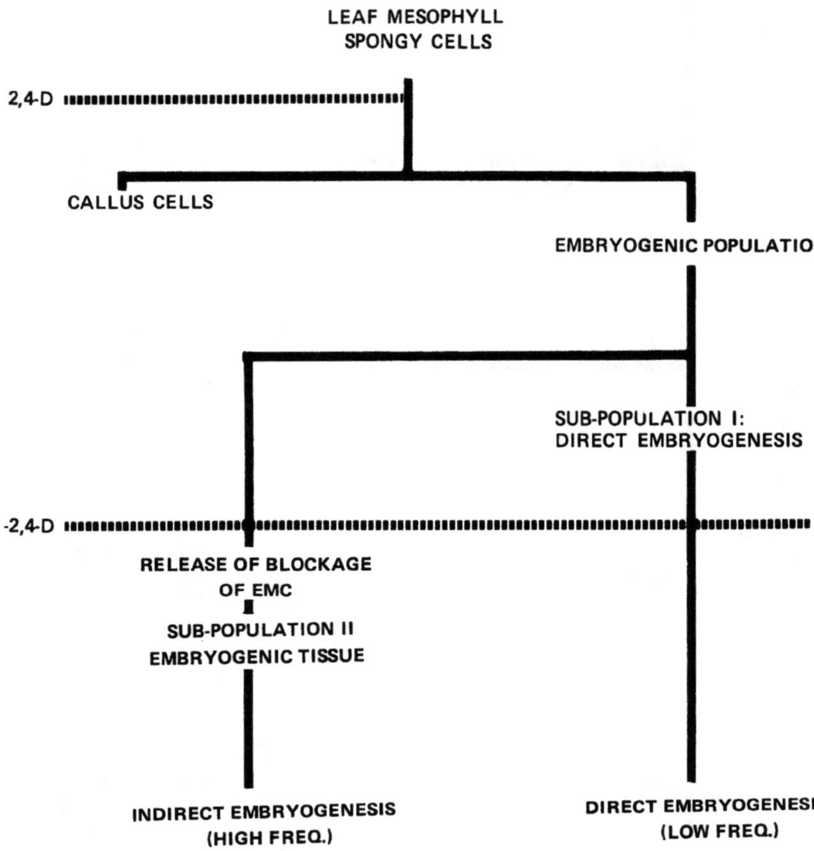

Fig. 3. Differentiation model for in vitro plant regeneration. Routes A, B, C, and D refer to 4 distinct pathways of differentiation following in vitro cell divisions. Route A is the successive cell divisions leading to a callus cell. Route B is unequal cell division, redetermination leading to an embryogenic mother cell (EMC-1) and direct embryogenesis. Route C is redetermination through a cell blockage and release of an embryogenic mother cell (EMC-2); embryogenic tissue derived from successive cell divisions of EMC-2 and, finally, embryo formation (high frequency). Route D is the organogenetic pathway leading to shoot formation that some cell populations, during in vitro cell divisions, become determined to follow. Note that this model proposes that both embryogenesis and organogenesis could be traced back to one single cell which is predetermined in vivo, or becomes induced by in vitro culture conditions.

origin of an organogenetic pathway, which is a deviation of the normal embryogenetic pathway. Organogenesis takes place at an early phase of embryogenic tissue formation whereby the daughter cells remain attached to each other in a cell clump without bipolarity.

REFERENCES

1. Baylis, M.W. (1977) The effects of 2,4-D on growth and mitosis in suspension cultures of Daucus carota. Plant. Sci. Lett. 8:99.
2. Custer, J.B.M., G. Vanee, and L.C. Buijs (1980) Clonal propagation on Coffea arabica L. by nodal culture. In Proceedings IX International Colloqium of Coffee, Assoc. Scient. Intern. Coffee, London.
3. Dublin, P. (1980) Multiplication vegetative in vitro de l'Arabusta. Cafe Cacao The. 24:281-290.

3a. Gamborg, O.L., R.A. Miller, and K. Ojima (1968) Nutrient requirements of suspension cultures of soybean root cells. Exp. Cell Res. 50:151-158.

4. Gupta, V., and G.L. Stebbins (1969) Peroxidase activity in hooded and awned barley at successive stages of development. Biochem. Genet. 3:15.

4a. Kartha, K.K., L.A. Mronginski, K. Pahl, and N.L. Leung (1981) Germplasm preservation of coffee (Coffea arabica) by in vitro culture of shoot apical meristems. Plant Sci. Lett. 22:301-307.

5. Kato, H., and M. Takeuchi (1966) Embryogenesis from the epidermal cells of carrot hypocotyl. Sci. Papers College Gen. Educ., University of Tokyo, 16:245.
6. Kochba, J., and J. Button (1974) The stimulation of embryogenesis and embryoid development in habituated ovular callus from the Shamouti orange (Citrus sisensis) as affected by tissue age and sucrose concentration. Z. Pflanzenphysiol. 73:415.
7. Kochba, J., S. Lavee, and P. Spiegel-Roy (1977) Differences in peroxidase activity and isoenzymes in embryogenic and non-embryogenic Shamouti orange ovular callus lines. Plant Cell Physiol. 18:463.
8. Konar, R.N., and K. Nataraja (1965) Experimental studies in Ranunculus sceleratus L.: Development of embryos from the stem epidermis. Phytomorphology 15:132.
9. Macleod, R.D. (1968) Changes in the mitotic cycle in lateral root meristems of Vicia faba following kinetin in treatment. Chromosoma 24:177.
10. Murashige, T., and F. Skoog (1962) A revised medium for rapid growth and bioassays with tobacco tissue cultures. Physiol. Plant. 15:473.
11. Nakamura, T., and M.R. Sondahl (1981) Multiplicacao in vitro de genes ortotropicas em Coffea spp. Proceedings O Congresso Brasileiro Pesquisa Cafeiras, Sao Lourenco, Brazil, pp. 162-163.

12. Sondahl, M.R., and W.R. Sharp (1977) High frequency induction of somatic embryos in cultured leaf explants of *Coffea arabica* L. *Z. Pflanzenphysiol.* 81:395-408.
13. Sondahl, M.R., and W.R. Sharp (1979) Research in *Coffea* spp. and applications of tissue culture methods. In *Plant Cell and Tissue Culture: Principles and Applications.*, W.R. Sharp, P.O. Larsen, E.F. Paddock, and V. Raghavan, eds. Ohio State University Press, Columbus, Ohio, pp. 527-584.
14. Sondahl, M.R., J.L. Salisbury, and W.R. Sharp (1979a) SEM characterization of embryogenic tissue and globular embryos during high frequency somatic embryogenesis in coffee callus cells. *Z. Pflanzenphysiol.* 94:185-188.
15. Sondahl, M.R., D.A. Spahlinger, and W.R. Sharp (1979b) A histological study of high frequency and low frequency induction of somatic embryos in cultured leaf explants of *Coffea arabica* L. *Z. Pflanzenphysiol.* 94:101-108.
16. Sondahl, M.R., and T. Nakamura (1980) Propagacao vegetativa in vitro de *Coffea* spp. In *Proceedings 8 Congresso Brasileiro Pesquisa Cafeeiras.*, Campos Jordao, p. 129.
17. Van't Hof, J., and C.J. Kovacs (1972) Mitotic cycle regulation in the meristem of cultured roots: The principal control point hypothesis. In *The Dynamics of the Meristem Cell Populations.*, M.W. Miller and C.C. Ceuhnert, eds. Plenum Press, New York, pp. 13-33.
18. Webster, P.L., and D. Davidson (1968) Evidence from thymidine-^{3}H labeled meristems of *Vicia faba* of two cell populations. *J. Cell Biol.* 39:332.

TISSUE CULTURE AND IMPROVEMENT OF WOODY PERENNIALS: AN OVERVIEW

D.J. Durzan

Department of Pomology
University of California
Davis, California 95616

INTRODUCTION

In the United States, we are cutting more forest trees than we grow (4) and in several cases removing more orchards than we plant (65). In both instances breeders are attempting to capture the existing genetic gains for domesticating trees and developing new varieties and cultivars (46,66,75,81,88,94 and Zimmerman, this volume). Classical breeding methods, while successful, are slow and costly (12,58,67,93).

We now have access to the heredity of trees in ways that were never before possible. Through cell and tissue culture we have the opportunity to mass propagate elite trees with superior traits. Through somaclonal variation and recombinant DNA technologies, we can direct improvements in the traits we hope to have in our future forests and orchards (23). This overview will explore advances and opportunities related to the theme of applications of developmental systems on tissue culture systems and tree species.

PRODUCTION AND UTILIZATION CYCLES

When we examine the genetic inputs into production cycles and the attributes of trees, we find a number of constraints to domestication, breeding, and improvement. These include biological problems associated with the long, complex life cycles, size, and the uncertainties in the identification and control over elite traits for orchards, forest ecosystems, and new crops.

In the United States, 1.25 billion forest trees are planted each year (4). This translates to 1.5 million plants a day. Large numbers are indeed a challenge for domestication, our production facilities, and for cell and tissue culture scientists. An improved system of clonal propagation of trees, especially those difficult to root, is needed. Hence, mass propagation systems employing cell and tissue culture to a) capture genetic gains particularly in proven mature individual trees (Timmis, pers. comm.), and b) to introduce new genetic variation, have emerged in tree production systems (3,31,34,37).

In production horticulture, the numbers are usually less but often the value of individual specimens is higher. For example, over 25 million Christmas trees were harvested nationwide in 1983 with a value of over $375,000,000. None as yet are produced by micropropagation. By contrast, exotic species, e.g., Eucalyptus, are now micropropagated on a massive scale for biomass, e.g., 50,000 from a single nodal segment (37). However, species newly introduced to the United States will represent only a small fraction of the existing forest resources.

For the tissue culturist many traits are common to the establishment of clonal procedures for orchards and forests (Tab. 1). Trueness to type, fruit, and nut (seed) production, and tree-machine compatibility for mechanical harvesting are now sought eagerly. In

Tab. 1. Comparison of some exemplary traits in trees grown for commercial purposes. Huxley (45), Smith (81), and Widolm (91) have defined the characteristics of agricultural crops for food/feed/fiber and biomass crops.

FORESTRY	POMOLOGY
1. Rapid growth rate for wood production	1. Fruit and nut production (early to late)
2. Resistance to disease, insects, and pests, and tolerance to control measures	2. Resistance to disease, insects, and pests, and tolerance to control measures
3. Desirable wood quality (fiber length, specific gravity, etc.)	3. Desirable quality for commodity (color, flavor, nutritive value, aroma, etc.)
4. Responsiveness to silvicultural practices and tolerant of extreme environmental variables	4. Minimal orchard practices (pruning, spraying, thinning) and tolerant to extreme environmental variables
5. Good form (heavily stemmed, fine branching for lumber: strong apical dominance for Christmas trees)	5. Good form (compatible with machine harvesting and orchard practices)

pomology, tree size is problematic; whereas, in forestry, wood production is paramount. In both, the need for new disease and pest-resistant cultivars, varieties, and germplasm is especially great. These often multigenic traits can be propagated via tissue culture, an advantage over breeding methods which recombine groups of genes involved in multigenic traits.

Cloning cycles aim to improve the quality of germplasm--the genetic base (Fig. 1). The bulk materials and commodities derived from the genetic base are fed into a cycle of utility, which serves the marketplace. Under the guise of biotechnology distinctively new crops and production technologies are appearing at the frontiers of science and technology (42,87,92) (see the section Development Mechanisms in this volume).

Biotechnology promises material substitutions, transformations, interconversions, and economies of time and scale (e.g., Ref. 60) (see the section on the Applications of Developmental Systems in this volume). The integration of processes is especially important for innovation in agriculture. Hence, issues arise over patents, proprietary information, problems of germplasm preservation and enhancement, engineering limits of the technologies, conflicts among entrepreneurs, and ways to manage the ventures that have been established (e.g., Ref. 70,83).

THE NEED FOR TREE IMPROVEMENT

Demands to grow more trees of all types will continue. Already we have witnessed in Third-World countries a fuel wood crisis where, while food is abundant, it cannot always be cooked because of the lack of fuel (28). Families spend days searching for fuel wood. When family members die, they cannot be cremated by burning because of the lack of wood. In some cases the bodies are thrown into rivers and this creates other problems. In these areas, water is often limiting. Soils are saline and alkaline. Trees are needed to grow in marginal soils and adverse environments to restore the quality of life of the rural poor (28) and the well-being of civilizations in general (6,8, and Lugo, this volume).

The superiority of native trees in the western United States, when introduced and grown overseas, has been recognized by population geneticists. Coniferous forests now serve as an export source of genes even though domestication is not fully achieved (57). Trees with elite traits are being cloned by new ventures aimed at the export market. There is also a return to planting trees for conservation and for commercial use to address issues in energy, chemical feedstocks, and fuels (40,60).

Here in North America, our once remarkable eastern forests have been high-graded for their timber (62). In recent times, insects and diseases have taken their toll. Many of the superior genetic

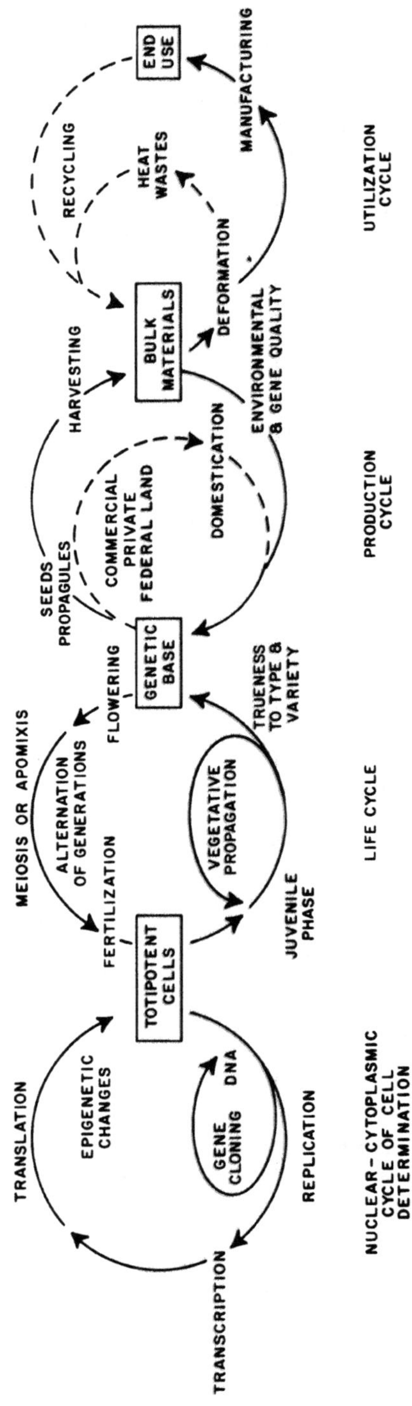
TRANSCRIPTION
TRANSLATION
EPIGENETIC CHANGES
GENE CLONING
DNA
REPLICATION
TOTIPOTENT CELLS
NUCLEAR-CYTOPLASMIC CYCLE OF CELL DETERMINATION
MEIOSIS OR APOMIXIS
ALTERNATION OF GENERATIONS
FERTILIZATION
VEGETATIVE PROPAGATION
JUVENILE PHASE
FLOWERING
TRUENESS TO TYPE & VARIETY
LIFE CYCLE
GENETIC BASE
SEEDS PROPAGULES
COMMERCIAL PRIVATE FEDERAL LAND
DOMESTICATION
HARVESTING
ENVIRONMENTAL & GENE QUALITY
PRODUCTION CYCLE
BULK MATERIALS
DEFORMATION
RECYCLING
HEAT WASTES
MANUFACTURING
END USE
UTILIZATION CYCLE

<———

Fig. 1. Tree improvement of woody perennials is based on a series of production cycles. Production is depicted arbitrarily at 4 levels (cell, life cycle, forest or orchard production, product utilization). Production is cyclical and ultimately depends on our existing but renewable germplasm resources. The output for each cycle serves as an input to another in a sequence leading to an end use. Through biotechnology we have an opportunity to shorten and engineer specificity into each cycle. Biotechnologies offer efficiency in producing end products (boxes) for each cycle.

At the cell level, genetic gains for tree improvement are captured by cell and tissue culture. The expression of totipotency and true-to-type processes involve natural polymer-producing systems (DNA, RNA, protein, cellulose). Recombinant-DNA methods permit the cloning of elite genes and their introduction into the genetic base through totipotent cells. Gene expression is mediated by interactions between the nucleus and cytoplasm, and in some instances can lead to genetic transformations or phase change in woody perennials (13). For the tissue culturist, the output of this cycle is a totipotent cell.

The life cycle is launched after fertilization or cell and tissue culture by totipotent cells. If profitable, vegetative propagation becomes a useful subcycle associated with the life cycle. Its advantage is that the additive and nonadditive genetic expression can be captured.

Genetic gains may be captured from proven mature specimens for domestication. In the production cycle, trees and their bulk materials are produced for the marketplace. Rapid wood production in juvenile trees is needed in forestry but is usually avoided where trees are grown for seed, fruit, and nut production (pomology). The perennial habit and improved cultural systems suited to specific microclimates can be exploited to minimize alternate bearing and unpredictability of the crop.

For the utilization cycle, bulk materials, derived from a diversity of germplasm, allow for substitutions of materials, agricultural chemicals, plant-machine compatibility, energy savings, and chemical field studies. We can search for ways to extend the shelf life of seeds and commodities through low temperatures, gaseous atmospheres, packaging, and transportation. Quality attributes should be identified and introduced back into the breeding strategies to improve the genetic base.

attributes of a few species (e.g., white pine, American elm) have been lost. Some others with narrower distribution (e.g., red pine, Torrey pine) have very little genetic variability for breeders to work with (33,55,80). Others with high value (e.g., Abies concolor, Picea, Pseudotsuga spp.) and with more variability are routinely cut as Christmas trees. Economically important trees are now being explored for their potential in clonal propagation using cell and tissue culture technologies.

In horticulture, many of the woody perennial crops are geographically situated in suitable soils and specific microclimates close to large urban centers. In California, one out of three jobs are related to agriculture, which represents 85% of the State's economy. The annual pomological crop production is approximately 50 million tons (almonds, walnuts, pistachios, citrus, stone fruits, etc.).

Crucial issues currently relate to genetic resource conservation, labor, water, energy, fertile land, pesticide use, and how to stay in business in the current economic climate (79, and Lugo, this volume). These will determine how, where, and when cell and tissue culture will make inroads in the nursery business, and in overcoming the current rate of loss of our germplasm resources. Propagators will continue to eagerly seek a) new breeding systems for the creation of improved cultivars, b) ways to predict and control production, and c) improved postharvest handling of commodities (15). In all of the above, the expectation is that tissue culture will accelerate the pace of tree improvement and production.

The grower, however, is caught in a seemingly impossible bind. No longer can he enjoy the ability to spread his rising costs by adding numbers through productivity stemming from the acquisition of new acres. Very little new and previously untilled virgin ground is available anymore. Water limitation, poor land type, remoteness from labor supply, climatic incompatibility, etc., are confining effective fruit and nut farming, particularly fresh varieties, to the present production areas. This means that in the future every aspect of farming will be subject to change. This will include how we view and handle soils, water, energy, money, varietal selections, pesticides, weeds, labor, automation, wastes, processing methods, marketing, etc. (49).

Forest trees do not demand the cultivation that orchards currently receive. We simply cannot spray chemicals and fertilize on similar schedules. In tree improvements, more dependence is placed on the resilience, self-reliance, tolerance, and resistance already in trees than in other breeding strategies.

Agroforestry, or the branch of agriculture dealing with the cultivation of agricultural commodities and forest trees (fiber), therefore comes as a challenge with many diverse opportunities especially for those currently engaged in cell and tissue culture. Here the current emphasis is on the production of multipurpose tree species, species improvement, and management of the agroforestry system (45).

BIOCHEMICAL AND PHYSIOLOGICAL MARKERS FOR ELITE TRAITS

The hope is that trees grown for commerce or use by the rural poor, as opposed to those for recreation and wilderness areas, are elite (Tab. 1). Unfortunately, an understanding of elite traits, in spite of the warnings (6,25), lags well behind that for most agronomic and vegetable crops. To fully appreciate the natural variation in traits we must understand the behavior of parental trees. For some woody perennials, this takes decades of observation.

We are now on the verge of mass-producing trees by cell and tissue culture, and, through genetic engineering, altering the genetic makeup of trees. However, we have little or no understanding of the molecular events and mechanisms of gene expression for many of the elite traits we seek to capture, clone, and exploit. Fortunately, greater insight into photosynthesis is available. We may eventually capitalize on this information to produce trees resistant to herbicides and suitable for biomass production (5,16,17, 61,95).

In Asia and South America, biomass and agricultural practices are being combined through multipurpose trees, e.g., Prosopis spp. For some areas, including California, Prunus spp. (e.g., almond) provide high-quality food. Pruning and thinning allow for low-ash fuel wood. Furthermore, since almonds are insect-pollinated, honey provides additional food. Almond wood shavings are suitable for mushroom production and the almond nut is prized for its high-value oils. Progress in clonal propagation of the almond has progressed significantly (51).

A major advance over the past decade, in spite of great pessimism on the part of skeptics, has been the mass-production of clonally propagated trees with a range of elite traits. These trees are now being established in forests (2,34,37,54, and Timmis, pers. comm.), in orchards (68,69,94, and this volume), and in agroforestry systems (1).

Our understanding of the molecular basis for elite traits will be accompanied by an increasing sophistication in research and how we innovate to produce new, arcane high technologies. The expression of elite traits in the cell and life cycles (Fig. 2) is beset, however, with problems in juvenility, maturity, totipotency, somaclonal aberrations, and trueness to type (58,21,77).

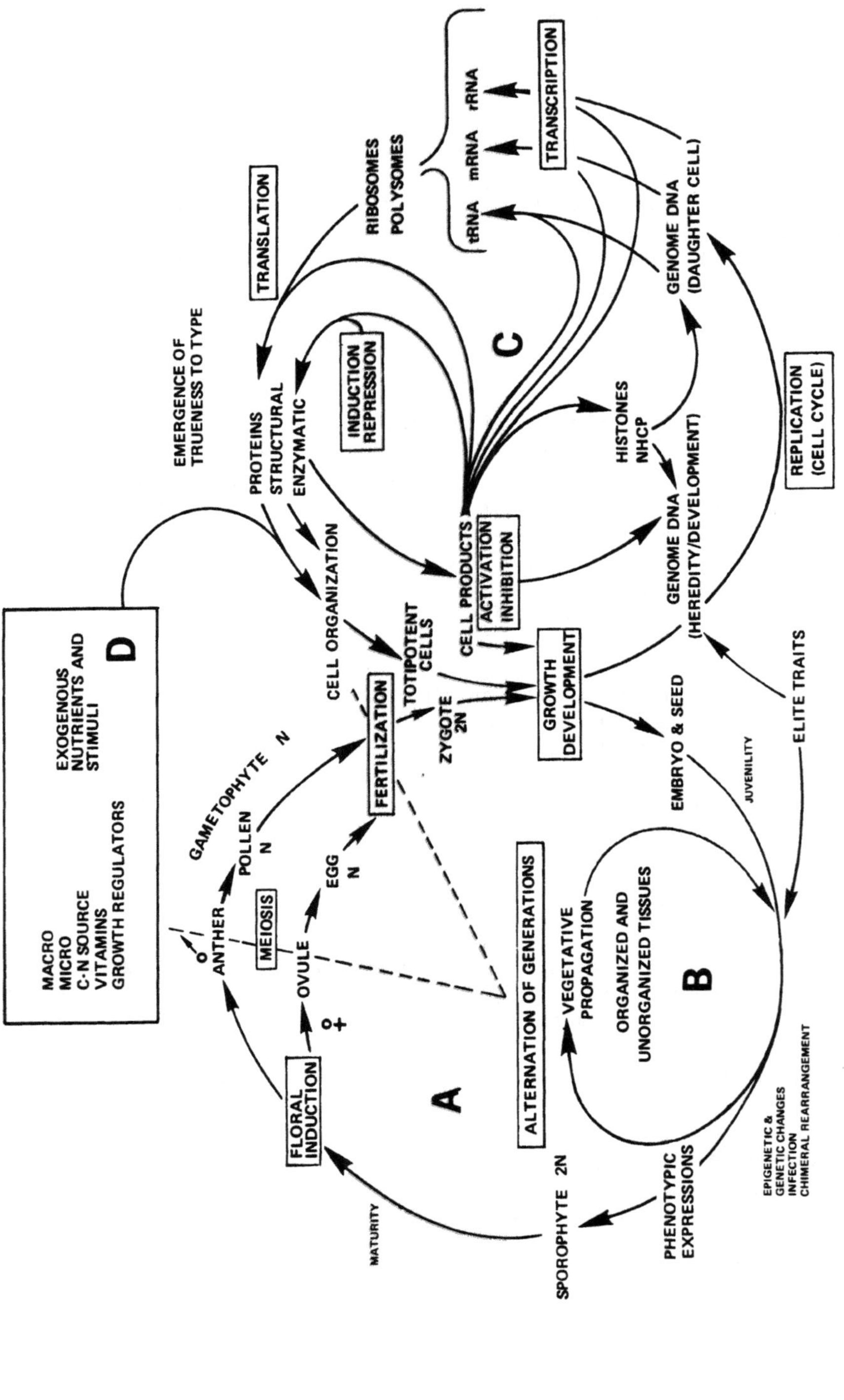
MACRO
MICRO
C-N SOURCE
VITAMINS
GROWTH REGULATORS
EXOGENOUS
NUTRIENTS AND
STIMULI
D
EMERGENCE OF
TRUENESS TO TYPE
PROTEINS
STRUCTURAL
ENZYMATIC
TRANSLATION
RIBOSOMES
POLYSOMES
tRNA
mRNA
rRNA
TRANSCRIPTION
INDUCTION
REPRESSION
C
CELL ORGANIZATION
TOTIPOTENT
CELLS
CELL PRODUCTS
ACTIVATION
INHIBITION
HISTONES
NHCP
GENOME DNA
(HEREDITY/DEVELOPMENT)
GENOME DNA
(DAUGHTER CELL)
REPLICATION
(CELL CYCLE)
GAMETOPHYTE N
POLLEN
N
ANTHER
MEIOSIS
OVULE
EGG
N
FERTILIZATION
ZYGOTE
2N
GROWTH
DEVELOPMENT
EMBRYO & SEED
JUVENILITY
ELITE TRAITS
FLORAL
INDUCTION
A
ALTERNATION OF GENERATIONS
VEGETATIVE
PROPAGATION
ORGANIZED AND
UNORGANIZED TISSUES
B
EPIGENETIC &
GENETIC CHANGES
INFECTION
CHIMERAL REARRANGEMENT
PHENOTYPIC
EXPRESSIONS
SPOROPHYTE 2N
MATURITY

Fig. 2. Parameters associated with the cell cycle with its nuclear-cytoplasmic interactions when linked to the life cycle create major barriers for successful cloning of trees having specific elite traits. A. Life cycle: tree size, complexity, physiological preconditioning, juvenility, maturity, invigoration, alternation of generations, and variations (genetic, epigenetic, chimeral; elements that move in and out of the genome, e.g., virus, mycoplasmas) are factors complicating gene expression. B. Vegetative propagation: this is currently achieved with juvenile tissues in a responsive physiological state. Where rooting of cuttings is difficult, micropropagation and somatic embryogenesis allow the breeder to capture additive and nonadditive gene expressions. C. Cell cycle: understanding the molecular basis of elite traits and the development of sequential treatments using components of the culture medium are needed to select for totipotent cells. D. The expression of totipotency with protoplasts, cell suspensions, and recovery of trees from genetically altered cells remain as major challenges for the tissue culturist. The most suitable combinations of treatments may eventually involve physiological states characterized by contour analyses covering the composition of cells as described by Christianson (see this volume).

CONTRIBUTIONS OF CELL AND TISSUE CULTURE

The past decade was oriented largely to difficult-to-root species and to cloning juvenile trees by micropropagation (24,25, and Amerson et al., and Schwartz, this volume). By comparison with population genetics, the advantage offered by cell and tissue culture is that the genetic traits having additive and nonadditive characteristics can be cloned through the expression of totipotency (Timmis, pers. comm.).

Heterozygosity already established in the genome is usually preserved by cell and tissue culture procedures. Furthermore, through explant choice there are now many different ways a haploid and diploid sporophyte can be obtained through apomixis with the aid of cell and tissue culture (e.g., Ref. 22).

By modifying nutrient media and through exposure to synthetic auxins and cytokinins, shoots and roots can be routinely induced under aseptic conditions on explants from juvenile or invigorated trees. In this form of micropropagation, which avoids the production of callus, occasional new and unexpected variability (useful and aberrant) has appeared (21,73). Generally, however, considerable uniformity has been observed (7,14,21,27,36,37,38,64,68,76).

In addition to scenarios presented by Bondi and Thorpe (9), Durzan (23), and Mott and Zimmerman (68), we have 2 general examples of what tissue culture can do for forestry in North America. First, the USDA (86) has estimated that for a number of species the gains in Tab. 2 can be achieved. Second, on a commercial scale, Hall (39) of the International Paper Company has indicated that for Loblolly pine the gains on managed plantations with first-generation improved stock can yield 120-190 cu ft/acre/yr. Asexual propagation increases the gain substantially above that of conventional genetics if the appropriate cloning systems can be developed. Cloning would contribute to a 3.7-fold increase in productivity.

For Douglas-fir and Loblolly pine, 3 strategies are proposed by Weyerhaeuser to reach target yields of 25 to 30 Mg ha^{-1} yr^{-1} (31, and Timmis, pers. comm.). First to make the process economical, the

Tab. 2. Anticipated genetic gain resulting from the first generation of breeding and its effect when improved trees are managed under intensive silviculture systems (after Ref. 86).

Tree species	Characters(s) under improvement	If genetic gain is at least:	Then the effect may be an increase in measure of:
Red pine	Height growth	3%	300 million board feet more per rotation.
White pine	Resistance to blister-rust	13%	Doubles per acre yields; 4X acres planted; losses to rust in young plantings greatly decreased.
Jack pine	Form, height	11%	Up to 70,000 cords per year.
Short-leaf pine	Height	9%	1 million board feet per acre.
White spruce	Volume	23%	Nearly 2 million board feet more per year harvested and over 5,000 additional cords per year.
Black spruce	Volume	18%	900,000 cords per rotation.
Yellow birch	Apical dominance	10%	84 million board feet more per rotation.
Black cherry	Diameter growth Apical dominance	10% 25%	Shorter rotations; 7,860 board feet per acre more per rotation.

progeny of full-sib seed are multiplied from a small number of controlled crosses in seed orchards. Second, Farnum et al. (31) aim to rejuvenate older tissue from proven trees for the cloning process. Third, they are exploring cryogenic storage of juvenile tissue until the germplasm is proven by progeny tests.

While we have a glimpse of how cell and tissue culture complements the work of breeders, physiologists, nurserymen, and forest managers, the most dramatic developments have been scientific and at the level of molecular genetics (11,56,82). New developments have depended on the way genetic material can be isolated, mapped, and expressed across phylogenetic barriers. Much remains to be learned about the nature and fidelity of gene expression in cells from perennial woody species. Tree improvement needs a vertical integration to introduce innovations into the marketplace.

For other woody plants of agricultural importance, the impact of cell tissue culture has not been defined specifically. We have, however, several general assessments of biotechnology (30,32,42, 85,92). Johnson and Wittwer (48) estimate that a 45% increase in agricultural output incorporating a 30% increase in crop yields would be attainable in the next four decades.

Nurseries throughout California are beginning to employ micropropagation methods particularly for proven cultivars and rootstocks, e.g., 84-year-old Paradox walnut, Juglans, and several Prunus spp. (Burchell and Driver Nurseries, Modesto, CA). It will not be long before this technology has a significantly wider impact on agriculture (47). Once trees can be mass-produced by cell and tissue culture we need to modify cultural and nursery practices to accommodate the propagules. Here, cell and tissue culture systems must capitalize on the perennial habit and life cycle where seeds, fruits, and nuts are sought.

In pomology, wood can be bred out of trees by selecting dwarfs (41). This facilitates production and a reduction of costs by minimizing pruning, thinning, and the application of fertilizers. Especially attractive is the use of machines that move over the tops of dwarf trees for harvesting and spraying of insects. The vigor of the seed, fruit, and nut production is captured in a few years. Trees that have become spent in bearing fruit can be uprooted and replaced by a new vigorous orchard cloned through cell and tissue culture.

This principle has been refined for the strawberry (90) which in many ways is a good model for the perennial habit. Through selection and cultural practices (12) the capacity of the perennial habit to produce fruit is brought to bear on plants treated as annuals. In this way, V. Voth and R.S. Bringhurst have increased strawberry production in California over a 30-year period from an average of 10 T per acre to over 30 T per acre. In some cases productivity has exceeded 50 T per acre. Cell and tissue culture is now making its way into strawberry-improvement programs. Strawberry plants produced by meristem culture outperformed standard plants by increasing annual fruit production by 26% (89). This shows that the benefits of cell and tissue culture extend beyond the vegetative phase and into reproduction. Unfortunately, however, strawberry plants propagated by tissue culture have shown much more unpredictable variability than is desirable for commercial purposes.

From this symposium we see that tree improvement through cell and tissue culture is indeed imminent for many woody species. Several books are available that reveal the progress especially over the past 5 years (10,29,35). Currently, the invigoration or rejuvenation of mature trees and control over morphogenesis are problems (77). While the American elm has been regenerated from cell suspension cultures (26), success with other tree species has been elusive. Much of the recent research deals with problems of somatic embryogenesis in cell suspensions (Fig. 3 and 4). Recently somatic embryogenesis from protoplasts has been reported for Sandalwood (74, and Atkins-Ozias et al. poster, this volume).

For many woody perennials, however, difficulties in clonal propagation persist. The development of trees from cells is not yet routine and many developmental errors occur (21). We are not yet sure of the "trueness to type" of plants generated by in vitro methods (50,84). Greater emphasis is now being placed on the need to certify material generated by tissue culture for commerce.

BIOTECHNOLOGY

Biotechnology has been defined in several ways. The U.S. Department of Agriculture interprets biotechnology as the use of living organisms or their components in industrial processes (Annual Priorities Report 1983). By contrast, industry has a more specific perspective. Biotechnology is viewed as "the integrated use of biochemistry, microbiology, and chemical engineering to achieve technological (industrial) application of the capabilities of microbes and cultured tissue cells" (71).

At the genetic level, protoplast, cell and tissue culture, and recombinant DNA methods (restriction endonuclease analysis, gene isolation and cloning, vector construction) are emerging as new powerful tools (56,88). Novel genetic variation can now be introduced into somatic cells through a range of vectors.

Success in manipulating the expression of DNA in genetically modified cells will depend on a) the quality of vectors for the transfer and integration of foreign DNA into the genome of proto-

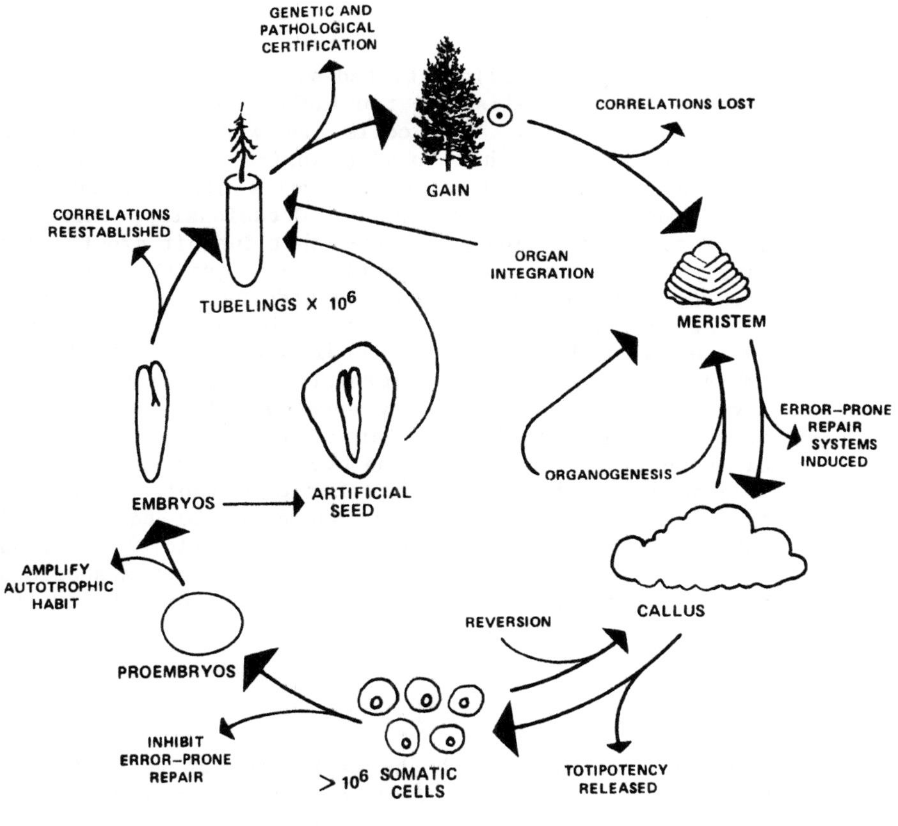

Fig. 3. Example of cellular cloning cycle for conifers that is applicable to other woody perennials. At the top meristems (nodal segments, apical and axillary buds) from mature trees or explants from seeds embodying a specific set of genetic gains are used as tissue donors in propagation systems (Timmis, pers. comm.). For mass-propagation by somatic embryogenesis, callus is grown under aseptic conditions on a chemically defined medium in agar. The formation of callus removes many of the physiological and structural correlations found in the donor and may activate error-prone repair systems that are important for true-to-type expressions of totipotency. Under the influence of auxins and cytokinins, cells and tissues may regenerate primary meristems. By contrast, in micropropagation, shoot meristems are induced directly on cotyledons and hypocotyls from seeds, or on axillary buds of invigorated shoots. The root is usually induced afterwards. Micropropagation yields up to 50,000 copies of the donor tree (38). In the cellular approach callus is reduced to

Fig. 3, continued

cells which are dedifferentiated and rapidly multiplied in diffuse light. Cells are routinely graded for size between 60 to 300 μ. When cells are washed and placed in the same medium either lacking auxins and cytokinins or containing other treatments, the cells may be transformed into proembryo-like structures. Occasionally, this step is improved by adding substances that inhibit growth especially where an overproduction of growth regulators occurs. Someday protoplasts may also be used in place of cells. Somatic embryos develop primary meristems under the somewhat heterotrophic conditions by reestablishing embryonic correlations. Alternatively, sphaeroblasts having a secondary meristem may develop under the influence of auxins and cytokinins. Here shoots or roots emerge by organogenesis. The aim of the cycle is to mass- propagate individuals with elite traits for a mechanized container planting system. One option would be to coat somatic embryos with materials to produce an artificial seed, but this technology remains undeveloped. All clones should be tested for trueness- to-type before some form of certification.

plasts from somatic cells, and b) the expression of genes in this foreign DNA. So far the vectors that show greatest promise are: tumor-inducing (Ti) plasmids, microinjection, transposons, virus-based vectors, and liposomes. Direct transformation, somatic cell fertilization, and the use of chimeric genes (44) are other possibilities.

Someday these vectors may be useful simply to introduce genetic markers into woody plants. The marker or trait, encoded in the synthetic or natural foreign DNA, should be incorporated and integrated in totipotent protoplasts. The traits sought should be expressed in daughter cells. Definition of traits at the molecular level will not be easy. It will require costly and rigorous cytogenetic and molecular genetic studies. For such manipulations to become useful the trait will have to be inserted at will, in quantity and with trueness to type during growth and development.

Unfortunately, the expression of totipotency in genetically engineered cells remains a problem even in plants easy to regenerate from cells, e.g., tobacco and carrot (Fig. 2). This has forced some laboratories to avoid cell and tissue culture, and to expose the wounds of plants to vectors. In this form of cocultivation, plants are regenerated at the wound surface from cells transformed by inte-

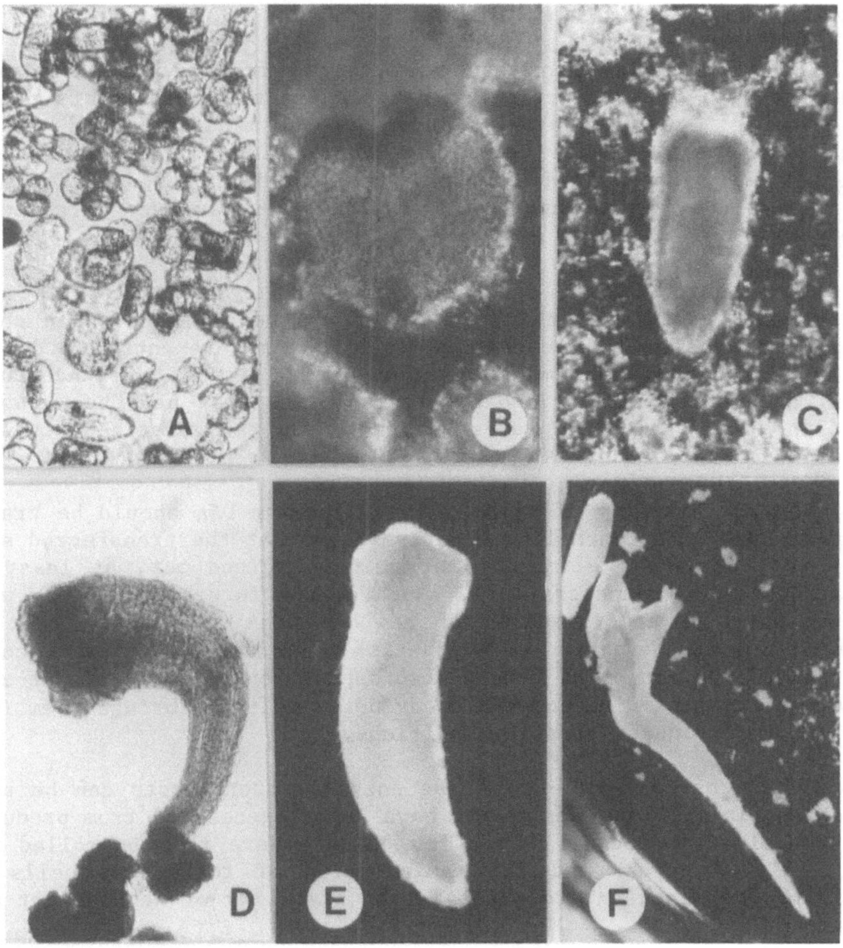

Fig. 4. Somatic embryogenesis in Prunus cerasus cv. Vladimir. Explants were taken from the petiole of a 28-year-old tree established by rooted cutting. A. Cell suspension of petiole callus < 300 μ dia. B. Heart-shaped stage, x90. C. Elongated embryos showing development of cotyledon initials. D. 10-day-old wild carrot somatic embryo for comparative purposes. E. Elongated of somatic embryoid, x25. F. Embryo development after 90 days in suspension culture, x10 (K.C. Hansen and D.J. Durzan, unpub. data).

gration vectors with genes for cytokinin expression (44). Alternatively, the use of cell wall fragments may facilitate the localized production of organs (leaves, roots, flowers) on explants during cocultivation (Dure, this volume).

Once somatic embryos can be obtained from genetically engineered cells, the production of genetically modified mature plants is possible. New technologies involving the manufacture of artificial seeds using somatic embryos (Lutz, this volume) may someday allow field plantings of thousands of genetically modified plants (59). The use of coatings for somatic embryos has been extended to woody species (52,59).

The specificity of detection of specific macromolecules, diseases, and pests, and the recovery of unique products while challenging is easier than ever before (53). Improvements are due to ELISA (enzyme-linked immunosorbent assays), tissue microencapsulation, and monoclonal antibodies--technologies that have been developed in other fields. Interest has begun in the genetic control of senescence and postharvest losses.

For the breeding of woody plants, foreign DNA should be transmitted through the germ line. Early fears that the transferred segment was completely eliminated at meiosis turned out, at least in tobacco, to be groundless (72). Much of the future technology will profit from an understanding of how proteins and enzymes act on chromatin and DNA (native or synthetic) to express genetically engineered information. The future also depends on cell and tissue culture, which simplifies complex woody perennials and brings them into the laboratory under aseptic conditions.

By use of cell-wall-degrading enzymes, protoplasts can be produced routinely; however, plants have been regenerated from products of protoplast fusion in only a few cases. Here the controlled use of specific enzymes and the manipulation of totipotent cells to avoid genome shock and to maintain the stability of development are paramount (63).

More dramatic are the shorter-term opportunities in biotechnology described in a special issue of Science (Vol. 219, February 11, 1983). Some are related to the immobilization of heat-stable enzymes and cells. The limitation for most proteins is no longer the production technology but the research needed to demonstrate efficacy of the protein and its products. New combinations, albeit artificial, stabilize reaction systems for capturing the low molecular weight and highly specific chemical transformations. These developments are relevant to the utilization cycle and involve the fiber-producing systems of woody perennials (Fig. 1). The goals are to provide alternative raw materials, increase production of scarce materials, decrease energy requirements, develop novel products or processes, and to use less toxic by-products in agriculture (43).

CONSTRAINTS TO TREE IMPROVEMENT

As yet, we have no sound, well-established, fundamentally scientific generalizations for determining in advance the best genetic source or cultural treatment of a new unit of forest or of a new orchard. The forest and the orchard have been natural units of productivity. For both we do not fully understand how to make yields self-sustaining, productive, and predictable as with other agricultural crops (42,92).

One major difficulty that we face in obtaining improved trees by traditional crosses or by recombinant DNA technology lies in the large number of characters desired. These embrace vigor of the plant parts, potential to complete the life cycle, resistance to insects and disease, hardiness, productiveness, size, and quality (shape, firmness, flavor, earliness, etc.). One major handicap is seen in our ignorance of the biology and chemistry of woody perennial trees, especially as this relates to the improvement of specific traits. We have already noted the recalitrance of many commercially important woody species when explants are grown as cell or tissue cultures. As yet we simply do not have control over morphogenesis (e.g., somatic embryogenesis) that is true-to-type and free of somaclonal aberrations.

According to Rumburg (78, see also Ref. 18) biotechnologies in agriculture depreciate and become obsolete. They must be repaired, maintained, and updated. They do not all depreciate at the same rate, hence, updating technology in some cases may be equivalent to standing still. The support base required for maintenance of appropriate technologies is proportional to the level of productivity. Furthermore, their takeover time of new technologies varies considerably (20).

Our agriculture is now moving from primitive to more advanced production and utilization cycles (19). This includes more complex agroforestry systems with shortened rotations. Hence, we must broaden our arsenal of biotechnologies. Our sciences and technologies are moving from those based in classical physics and chemistry to those more representative of biological systems, i.e., the biotechnologies are mimicking biological systems. These are referred to as being "biomimetic."

Integration and refinement of production cycles (Fig. 1) is expected to provide high-tech competitive advantages for the United States and ease problems in soil-water conservation, plant productivity, human nutrition, commodity delivery, and, perhaps, even animal productivity (e.g., aquaculture and agroforestry). We must ask how biotechnology can integrate into production schemes and how it will enable us to minimize the dependence upon nonrenewable resources. Systems will have to be in place to permit the growth of supply centers permitting transportation and the development of new markets and alternate utilization cycles. Perhaps the greatest opportunity will be for simple and appropriate technological fixes in the underdeveloped countries.

SUMMARY

The future for tree improvement is at a major turning point and has the potential of being one of the most impressive scientific efforts we have ever seen. Cell and tissue culture will continue to be a fundamental tool of considerable potential for the study of woody perennial species.

REFERENCES

1. Aitken-Christie, J., and J.A. Gleed (1984) Uses for micropropagation of juvenile Radiata pine in New Zealand. In Proceedings of the International Symposium of Recent Advances in Forest Biotechnology, J. Hanover, D. Karnosky, and D. Keathley, eds. Michigan Biotechnology Institute, Traverse City, Michigan, June 10-13, 1984, pp. 47-57.
2. Aitken-Christie, J., D.R. Smith, and K.J. Horgan (1981) Micropropagation of Radiata pine. In Coll. Intl. sur la Culture "in vitro" des Essences Forestières, AFOCEL, 77370 Nangis, France, pp. 191-196.
3. Aitken-Christie, J., and T.A. Thorpe (1984) Clonal propagation: Gymnosperms. In Cell Culture and Somatic Cell Genetics of Plants, Vol. I, Academic Press, New York, pp. 82-95.
4. Anonymous (1977) Forest Facts and Figures, American Forest Institute, Washington, D.C.
5. Arntzen, C. (1984) Genetic engineering of plants for herbicide resistance: Progress and potential. In Proceedings of the International Symposium of Recent Advances in Forest Biotechnology, J. Hanover, D. Karnosky, and D. Keathley, eds. Michigan Biotechnology Institute, Traverse City, Michigan, June 10-13, 1984, pp.163-171.
6. Bailey, I.W., and H.A. Spoehr (1929) The Role of Research in the Development of Forestry in North America, Macmillan Co., New Jersey.
7. Berlyn, G.P., and R.C. Beck (1980) Tissue culture as a technique for studying meristematic activity. In Control of Shoot Growth in Trees, C.H.A. Little, ed. Proceedings of the IUFRO Workshop, Oxlan and Shoot Physiology, Canadian Forestry Service, Fredericton, Canada, pp. 305-324.
8. Blake, R.O. (1984) Moist forests of the tropics. A plea for protection and development. World Resources Institute Journal '84, Washington, D.C., pp. 34-49.

9. Bondi, S., and T.A. Thorpe (1981) Clonal propagation of forest tree species. In Proceedings of the COSTED Symposium on Tissue Culture of Economically Important Plants, A.N. Rao, ed. Singapore, China, pp. 197-204.
10. Bonga, J.M., and D.J. Durzan, eds. (1982) Tissue Culture in Forestry, Martinus Nijhoff/Dr. W. Junk, Amsterdam, The Netherlands, 420 pp.
11. Brill, W.J. (1981) Agricultural microbiology. Sci. Amer. 245:199-215.
12. Bringhurst, R.S. (1983) Breeding strategy. In Methods in Fruit Breeding, J.N. Moore and J. Janick, eds. Purdue University Press, Indiana, pp. 147-153.
13. Brink, R.A. (1962) Phase change in higher plants and somatic cell heredity. Q. Rev. Biol. 37:1-22.
14. Brown, C.L., and H.E. Sommer (1982) Vegetative propagation of dicotyledenous trees. In Tissue Culture in Forestry, J. Bonga and D.J. Durzan, eds. Martinus Nijhoff/Dr. W. Junk, Amsterdam, The Netherlands, pp. 109-149.
15. Brown, D.S., R.S. Bringhurst, J.C. Crane, H.T. Hartmann, A.D. Webb, J.A. Beutel, J.H. LaRue, W.C. Micke, G.F. Mitchell, J. Osgood, R.G. Platt, D.E. Ramos, and D. Rough (1983) Fruit and nut crops. In A Guidebook to California Agriculture, A.F. Scheuring, ed. University of California Press, Berkeley, pp. 135-162.
16. Bungay, H. (1981) Energy: The Biomass Options, John Wiley and Sons, New York.
17. Calvin, M. (1980) Petroleum plantations for fuel and materials. In Proceedings of the Symposium on Paper Science and Technology - The Cutting Edge, 50th Anniversary, The Institute of Paper Chemistry, Appleton, Wisconsin, pp. 18-27.
18. Daedalis (1980) Modern technology. In Proceeding of the American Academy of Arts and Sciences, Vol. 109, p. 190.
19. Dahlsten, D. (1976) The third forest. Environment 18:35-42.
20. Davies, D.S., and I. Lawrenson (1978) Strategies for technologies with long lead times. In Resources or Organic Matter for the Future, L.E. St. Pierre, ed. Multiscience Publishing Company, Montreal, Canada, pp. 177-184.
21. Durzan, D.J. (1984) Potential for genetic manipulation of forest tree: totipotency, somaclonal aberration and trueness to type. In Proceedings of the International Symposium of Recent Advances in Forest Biotechnology, J. Hanover, D. Karnosky, and D. Keathley, eds. Michigan Biotechnology Institute, Traverse City, Michigan, June 10-13, 1984, pp.117-139.
22. Durzan, D.J. (1984) Explant source: Juvenile and adult phases. In Application of Plant Tissue Culture Methods for Crop Improvement, D.A. Evans, W.R. Sharp, P.V. Ammirato, and Y. Yamada, eds. Macmillan Publishing, New Jersey.
23. Durzan, D.J. (1980) Progress and promise in forest genetics. In Proceedings of the Symposium on Paper Science and Technology - The Cutting Edge, 50th Anniversary, The Institute of Paper Chemistry, Appleton, Wisconsin, pp. 31-60.

24. Durzan, D.J. (1977) Tissue culture - genetic and physiological aspects. In Proceedings of the 4th North American Tree Biology Conference, Syracuse, New York, August 9-11, 1976, pp. 137-164.
25. Durzan, D.J., and R.A. Campbell (1974) Prospects for the mass-production of improved stock of forest trees by cell and tissue culture. Can. J. For. Res. 4:151-174.
26. Durzan, D.J., and S.M. Lopushanski (1975) American elm from cell suspension culture. Can J. For. Res. 5:273-277.
27. Earle, E.D., and Y. Demarly (1982) Variability in Plants Regenerated from Tissue Culture, E.D. Earle and Y. Demarly, eds. Praeger Press, New York, 392 pp.
28. Eckholm, E. (1975) The other energy crisis: Firewood. Worldwatch Paper 1, Worldwatch Institute, Washington, D.C.
29. Evans, D.A., W.R. Sharp, and C.E. Flick (1981) Growth and behavior of cell cultures: Embryogenesis and organogenesis. In Plant Tissue Culture, T.A. Thorpe, ed. Academic Press, New York, pp. 45-113.
30. Evenson, R.C., P.E. Waggoner, and V.W. Ruttan (1979) Economic benefits from research: An example from agriculture. Science 205:1101-1107.
31. Farnum, P., R. Timmis, and J.L. Kulp (1983) Biotechnology of forest yield. Science 219:694-702.
32. Farrell, K.R., F.H. Sanderson, T.T. Vo, and M.F. Brewer (1984) Meeting future needs for United States food, fiber and forest products. In Reference Document: Needs Assessment for the Food and Agricultural Sciences, Joint Council on Food and Agriculture Sciences, USDA, Washington, D.C., pp. 9-100.
33. Fowler, D.P., and D.T. Lester (1970) Genetics of red pine. USDA For. Serv. Res. Pap. WO-8, 13 pp.
34. Franclet, A. (1981) Rajeunissement et micropropagation des ligneux. Coll. Intl. sur la Culture "in vitro" des Essences Forestières, AFOCEL, 77370 Nangis, France, pp. 55-63.
35. George, E.F., and P.D. Sherrington (1984) Plant Propagation by Tissue Culture, Exegetics Ltd., Eastern Press, Reading, Berkshire, 709 pp.
36. Gupta, P.K., A.F. Mascarenhas, and V. Jagannathan (1981) Tissue culture of forest trees - clonal propagation of mature trees of Eucalyptus citriodora Hook by tissue culture. Plant Sci. Lett. 20:195-201.
37. Gupta, P.K., U.J. Mehta, and A.F. Mascarenhas (1984) Tissue culture of the Eucalyptus for biomass production. In Proceedings of the VII International Biotechnology Symposium, New Delhi, India, February 19-25, 1984, pp. 114-115.
38. Gupta, P.K., U.J. Mehta, and A.F. Mascarenhas (1983) A tissue culture method for rapid clonal propagation of mature trees of Eucalyptus torelliana and Eucalyptus camaldulensis. Plant Cell Reports.
39. Hall, F.K. (1980) Biology and genetics. In Proceedings of the Symposium on Paper Science and Technology - The Cutting Edge, 50th Anniversary, The Institute of Paper Chemistry, Appleton, Wisconsin, pp. 15-17.

40. Hanover, J. (1984) Screening and breeding trees for high value chemicals. In Proceedings of the International Symposium of Recent Advances in Forest Biotechnology, J. Hanover, D. Karnosky, and D. Keathley, eds. Michigan Biotechnology Institute, Traverse City, Michigan, June 10-13, 1984, pp. 92-103.
41. Hansche, P.E. (1983) Response to selection. In Methods for Fruit Breeding, J. Janick and R. Moore, eds. Purdue University Press, Indiana, pp. 154-171.
42. Hardy, R.W. (1983) The outlook for agricultural research and technology. In Agriculture in the Twenty-First Century, J.W. Rosenblum, ed. Wiley-Interscience, New York, pp. 97-103.
43. Harsanyi, H. (1981) Biotechnology in the Year 2000 (Chapter 11). In Biotechnology: Present Status and Future Prospects, Proceedings Conference, R.S. First, Inc., White Plains, New York, June 1-2, 1981, 15 pp.
44. Herrera-Estrella, L., A. Depicker, M. Van Montagu, and J. Shell (1983) Expression of chimeric genes transferred into plant cells using a Ti-plasmid-derived vector. Nature 303:209-213.
45. Huxley, P.A., ed. (1983) Plant Research and Agroforestry, Pillans and Wilson Ltd., Edinburgh.
46. Janick, J. and J.N. Moore, eds. (1975) Advances in Fruit Breeding, Purdue University Press, Indiana.
47. JCFAS (Joint Council on Food and Agriculture Sciences) (1984) Reference Document: Needs Assessment for the Food and Agricultural Sciences, USDA, Washington, D.C., 328 pp.
48. Johnson, G.L., and S. Wittwer (1983) Role of technology in determining future supplies of food, fiber, and forest products in the United States. Resources for the Future - Michigan State University Report, October, 1983.
49. Johnson, W.E., and H.O. Carter (1983) Policy issues. In A Guidebook to California Agriculture, A.F. Scheuring, ed. University of California Press, Berkeley, pp. 381-387.
50. Kester, D.E. (1983) The clone in horticulture. Hort. Sci. 18:831-837.
51. Kester, D.E., L. Liu, C.A.L. Fenton, and D.J. Durzan (1985) Almonds. In Micropropagation of Fruit and Forest Trees, Y.P.S. Bajaj, ed. Springer-Verlag, Berlin (in press).
52. Kitto, S.L., and J. Janick (1983) ABA and chilling increase survival of asexual carrot embryos encapsulated with artificial seed coats. Hort. Sci. 18:104 (abstract).
53. Klausner, A., and T. Wilson (1983) Gene detection technology opens doors for many industries. Biotechnology 1:471-478.
54. Leach, G.N. (1979) Growth in soil of plantlets produced by tissue culture, Loblolly pine. Tappi 62:59-61.
55. Ledig, F.T., and M.T. Conkle (1983) Gene diversity and genetic structure in a narrow endemic, Torrey pine (Pinus torreyana Perry ex. Carr.). Evolution 37:79-85.
56. Lewin, B. (1983) Genes, John Wiley and Son, New York.
57. Libby, W.J. (1984) Forest genetic resources. In Proceedings of the Symposium on Genetic Resource Conservation for California, P.E. McGuire and C.O. Qualset, eds. University of California

and State Department of Food and Agriculture, Napa, California, April 5-7, 1984, pp. 17-18.

58. Libby, W.J., R.F. Stettler, and F.W. Seitz (1969) Forest genetics and forest-tree breeding. Ann. Rev. Genetics 3:469-494.
59. Linden, T. (1984) Planting seeds for the future. Encapsulation. In Western Grower and Shipper, January 12-14, 1984.
60. Lipinsky, E.S. (1981) Chemicals from biomass: Petrochemical substitution options. Science 212:1465-1471.
61. Lipinsky, E.S. (1978) Fuels from biomass: Integration with food and materials systems. Science 199:644-651.
62. Lower, A.R.M. (1938) The North American Assault on the Canadian Forest, Ryerson Press, Toronto, Canada, 377 pp.
63. Marx, J.L. (1984) Research News. Instability in plants and the ghost of Lamarck. Science 224:1415-1416.
64. McKeand, S.E., and R.J. Weir (1984) Tissue culture and forest productivity. J. Forestry, April 1984, pp. 212-218.
65. Micke, W. (1984) Trends in California fruit, nut and grape industries. In Workshop on Marketing California Food Commodities in the Future: Opportunities and Challenges, Pomology Extension, University of California, Davis, pp. 28-31.
66. Moore, J.N., and J. Janick, eds. (1983) Methods in Fruit Breeding, Purdue University Press, Indiana.
67. Morgenstern, E.K., M.J. Holst, A.H. Teich, and C.W. Yeatman (1975) Plus-tree selection: Review and outlook. Environment Canada, Canadian Forest Service, Publ. No. 1347, Ottawa, Canada, 72 pp.
68. Mott, R.L., and R.H. Zimmerman (1981) Trees: Round-table discussion. Env. Exptl. Bot. 21:415-420.
69. Murashige, T. (1978) The impact of plant tissue culture on agriculture. In Proceedings of the 4th International Congress on Plant Cell and Tissue Culture, T. Thorpe, ed. University of Calgary, Canada, August 20-25, 1978, pp. 15-26.
70. Office of Technology Assessment (1981) Impacts of Applied Genetics: Microorganisms, Plants and Animals, U.S. Government Printing Office, Washington, D.C.
71. O'Sullivan, D.A. (1981) Technology, Europeans collaborate on biotechnology. Chem. Eng. News, p. 32.
72. Otten, L., H. DeGreve, J.P. Hernalsteens, M. Van Montagu, O. Schieder, J. Straub, and J. Schell (1981) Mendelian transmission of genes introduced into plants by the Ti plasmids of Agrobacterium tumefaciens. Mol. Gen. Genet. 183:209.
73. Patel, K.R., and G.P. Berlyn (1982) Genetic instability of multiple kinds of Pinus coulteri regenerated from tissue culture. Can. J. For. Res. 12:93-101.
74. Rao, P. (1984) Plant regeneration through somatic embryogenesis in Sandalwood protoplasts. In Proceedings 9th Plant Tissue Culture Conference, Sugarcane Breeding Institute, Coimbatore, Tamil Nadu, India, February 4-6, 1984 (in press).
75. Rediske, J.H. (1974) The objectives and potential for tree improvement. In Yale University School Forestry and Environmental Studies Bulletin, No. 85, pp. 3-18.

76. Renfroe, M.H., and G.P. Berlyn (1984) Stability of nuclear DNA content during adventitious shoot formation in Pinus taeda L. tissue culture. Am. J. Bot. 17:268-272.
77. Romberger, J.A. (1976) An appraisal of prospects for research on juvenility in woody perennials. Acta Hort. 56:301-317.
78. Rumburg, C.B. (1984) Scientific requirements for maintaining technology and developing human capital. In Reference Document: Needs Assessment for the Food and Agricultural Sciences, Joint Council on Food and Agriculture Sciences, USDA, Washington, D.C., pp. 291-301.
79. Scheuring, A.F., ed. (1983) A Guidebook to California Agriculture, University of California Press, Berkeley, 413 pp.
80. Silen, R.R. (1978) Genetics of Douglas-fir. USDA For. Serv. Res. Pap. WO-35, 34 pp.
81. Smith, W.H. (1983) Energy from biomass: A new commodity. In Agriculture in the Twenty-First Century, J.W. Rosenblum, ed. Wiley-Interscience, New York, pp. 61-69.
82. Steward, F.C. (1968) Growth and Organization in Plants, Addison-Wesley, Reading, Massachusetts.
83. Sun, M. (1984) Weighing the social costs of innovation. News and Comment. Science 223:1368-1369.
84. Tran Than Van, K., and H. Trink (1978) Plant propagation: Non-identical and identical copies. In Propagation of Higher Plants Through Tissue Culture, K. Hughes, R. Henke, and M. Constantin, eds. Proceedings of the Symposium at the University of Tennessee, Technical Information Center, U.S. Department of Energy, April 16-19, 1978, pp. 134-158.
85. Tomes, D.T. (1984) An assessment of the impact of biotechnology on plant breeding. In International Association Plant Tissue Culture Newsletter, No. 42, pp. 2-9.
86. USDA Forest Service (1971) Genetics are sex. In Forest Service Pamphlet, Milwaukee, Wisconsin.
87. Varner, J.E. (1983) The molecular and genetic technology of plants. In Frontiers in Science and Technology: A report to the National Academy of Sciences, W.H. Freeman Company, New York, pp. 45-62.
88. Verne, R.V., ed. (1982) Genetic engineering in plants. California Ag. 36:1-36.
89. Voth, V., and R.S. Bringhurst (1982) Winter planted meristem vs. nonmeristem strawberry plants, Santa Ana, 1981-83. Pomology Strawberry Advisory Board Report (California), No. 6, 3 pp.
90. Welch, N.C., R.S. Bringhurst, A.S. Greathead, V. Voth, W.S. Seyman, N.F. McCalley, and H.W. Otto (1982) Strawberry production in California. Division of Agricultural Sciences, University of California Leaflet No. 2959, 14 pp.
91. Widolm, J.M. (1978) The selection of agriculturally desirable traits with cultured plant cells. In Propagation of Higher Plants Through Tissue Culture, K. Hughes, R. Henke, and M. Constantin, eds. Proceedings of the Symposium at the University of Tennessee, Technical Information Center, U.S. Department of Energy, April 16-19, 1978, pp. 189-199.

92. Wittwer, S.H. (1983) Epilogue. The new agriculture: A view of the twenty-first century. In Agriculture in the Twenty-First Century, J.W. Rosenblum, ed. Wiley-Interscience, New York, pp. 337-367.
93. Wright, J. (1976) Introduction to Forest Genetics, Academic Press, New York.
94. Zimmerman, R.H., ed. (1980) Proceedings conference on nursery production of fruit plants through tissue culture applications and feasibility. USDA Science and Education Administration, ARR-NE-11, Beltsville, Maryland, December 1980, 119 pp.
95. Zobel, B. (1980) Genetic improvement of forest trees for biomass production. Prog. Biomass Conv. 2:37-58.

IN VITRO COLONIZATION AND RESISTANCE OF LOBLOLLY PINE EMBRYOS INFECTED WITH THE FUSIFORM RUST FUNGUS

D.J. Gray*[1] and H.V. Amerson[2]

[1] Department of Plant and Soil Science
University of Tennessee
Knoxville, Tennessee 37901-1071

[2] Botany Department
North Carolina State University
Raleigh, North Carolina 27650

INTRODUCTION

Cronartium quercuum (Berk.) Miyabe ex Shirai f. sp. *fusiforme* is a macrocyclic rust, requiring both pine and oak or chestnut species to complete its complex lifecycle (Fig. 1). Previously known as *Cronartium fusiforme* Hedgc. and Hunt ex Cumm. (3), this fungus is the incitant of fusiform rust disease, so named because of the distinctive spindle-shaped (fusiform) galls that often form on the stems and branches of infected pines (6). Infrequently encountered until the advent of pine plantation management in the 1930s, fusiform rust incidence has been increasing yearly (5). Some pine plantations have become 100% infected within 3 to 4 years after establishment (6). Losses occur due to seedling death, excessive branching, and as a consequence of trunk gall formation. Trunk galls decrease wood quality and increase the risk of wind and fire damage (7). Fusiform rust is now considered the most destructive disease of southern conifer forests (8). Although a range of resistance has been noted among various pine species, loblolly pine (*Pinus taeda* L.), the species most frequently planted in the South, is highly susceptible to this disease (9).

---

* Present address: Agricultural Research and Education Center, University of Florida, P.O. Box 388, Leesburg, Florida 32749-0388.

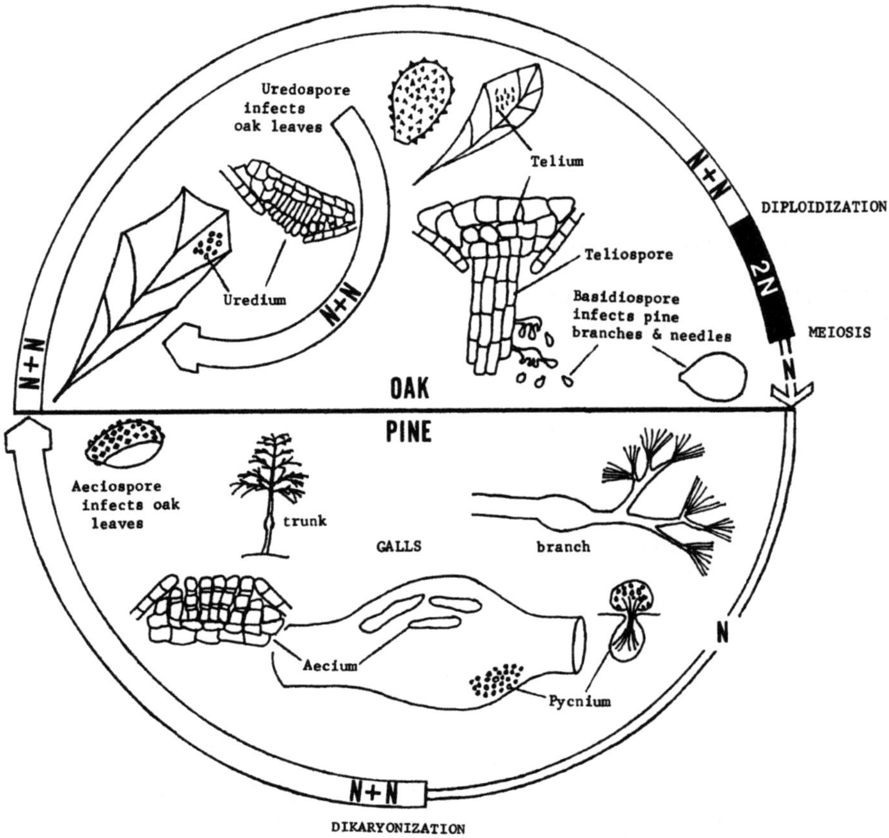

Fig. 1. Life cycle diagram of the fusiform rust fungus.

Use of resistant planting stock is presently the most feasible method of fusiform rust management (10) and several resistant seed lines have been obtained by selections from symptomless trees in nature or from plant breeding programs (16). However, conventional breeding strategies that are successful for developing disease resistance in annual species are not as useful in P. taeda due to the long pine reproductive cycle. Also, rust resistance must be effective from the seedling stage through marketable age (15-40 years). These factors have made rust assessment and breeding for resistance a long and complex process. Therefore, high priority has been placed on developing assay systems for the early prediction of long-term resistance.

A greenhouse-based method for rust assessment in slash pine seedlings evaluated only 6 months after artificial inoculation (2,19) can now account for 60-70% of the variation observed in field resistance (28), thus firmly demonstrating the usefulness of such an approach in predicting resistance levels among seed families. However, both greenhouse and longer-term field assessments depend upon the evaluation of host responses (chiefly galling) that occur late in the disease cycle long after other responses in gall-free trees may have operated to confer resistance. Recognition of such early operative responses could lead to a more rapid assay system as well as more accurate assessment via a better understanding of the disease.

A technique for axenic production and application of fusiform rust basidiospores (1) demonstrated that the in vitro infection of zygotic embryo hypocotyls of loblolly pine could be routinely accomplished. This suggested that a rapid in vitro method of rust assessment that depended upon the resolution of early occurring host cell and tissue level responses was possible. This section summarizes published studies (12-15) which demonstrated early host responses and served as the basis for eventual development of an in vitro assay (10). Previously unpublished information on host responses to fusiform rust inoculation are described in detail and some potential benefits of in vitro screening for fusiform rust resistance are discussed.

INFECTION AND COLONIZATION OF HYPOCOTYLS

The first step in establishing a reliable in vitro screening system was to determine how the fungus infected and colonized host tissues. Loblolly pine embryo hypocotyls were infected in vitro with fusiform rust by 2 separate methods of basidiospore inoculation. Basidiospores were aseptically deposited (nonquantitatively) by suspending rust telial columns over surface-sterilized embryos in a moist chamber (1). Basidiospores produced by the telial columns discharged onto underlying embryos within a few hours under these conditions. Alternatively, quantitative inoculation was effected by placing water drops containing known spore concentrations directly onto embryo hypocotyls (13). The nonquantitative inoculation method most closely approximates the events occurring in nature and is particularly useful for determining early infection events on the hypocotyl surface. The quantitative method is useful for studying host responses that may occur only at specific inoculum concentrations.

The colonization events of loblolly pine hypocotyls by the fusiform rust fungus were previously described in detail and experimental methods were provided (12,14,15). These events were similar to those occurring in slash pine (Pinus elliotii Engelm. var. elliotii) (21,23) and are briefly summarized here as well as presented diagrammatically in Fig. 2. Basidiospores germinated within 2 hours after inoculation and produced germ tubes that grew in random directions over the hypocotyl surface before developing variously shaped,

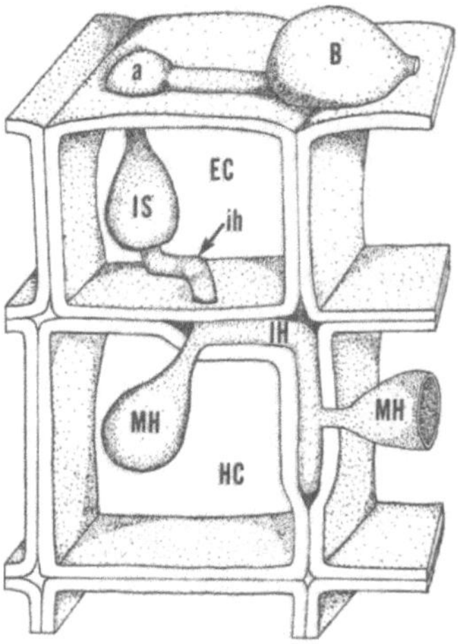

Fig. 2. Typical early colonization of loblolly pine hypocotyl tissue by the fusiform rust fungus (diagrammatic). A germinating basidiospore (B) produces a germ tube with a terminal appressorium (a). The intracellular infection structure apparatus (IS) consists of the extracellular appressorium with an expanded body, and a narrowed infection hypha (ih) formed in an epidermal cell (EC). The ih exits EC to develop an intercellular hypha (IH) which grows through intercellular spaces and between adjacent cortical cells (HC). M-haustoria (MH) are formed in HC from IH. From Gray et al. (15).

terminal appressoria. A penetration peg was produced by each appressorium and entrance into the plant occurred directly through an epidermal cell. The penetration peg grew through the outer epidermal cell wall and into the cell, invaginating the host plasmalemma. A distinctive intracellular infection structure (IS), bounded by host plasmalemma, subsequently formed within the epidermal cell (15).

The IS served a pivotal role in the colonization of cells and tissues adjacent to the infected epidermal cell. A fine infection hypha produced by the IS grew across the epidermal cell and through an inner cell wall to infect adjacent regions by either of 2 ways. Rarely, intracellular hyphae were formed directly from the IS in adjacent epidermal and cortical cells. More commonly, an intercellular hyphal network was formed by the IS. Intercellular hyphae grew both through pre-existing spaces and also intrusively through the middle lammellae of adjacent host cells. The intercellular hyphal network extended through all hypocotyl tissues, including both vascular and pith regions, as early as 7 days after inoculation. Distinctive monokaryotic (M-) haustoria were formed from intercellular hyphae in adjacent host cells (14). M-haustoria invaginated the host plasmalemma and probably functioned as feeding organs, although this role has not been proven.

HOST CELL AND TISSUE RESPONSES TO COLONIZATION

Given an understanding of typical colonization events that occurred in vitro, it was possible to examine the responses of loblolly pine hypocotyls to fusiform rust inoculation in detail. An analysis of host responses was accomplished by quantitatively inoculating hypocotyls of embryos from seed lines of known, long-term field performance. Three seed lines with divergent resistance ratings (highly susceptible, intermediate resistant, and highly resistant) were used and were designated S, I, and R lines, respectively (see Ref. 13 for details). Inoculated and uninoculated specimens were examined microscopically in order to determine if host cells and tissues elicited visually identifiable responses. Several developmental traits were discovered in inoculated embryos, but rarely or never in uninoculated specimens. Therefore, these traits were considered to be potential host responses to infection. Statistical correlations of these responses among the 3 test seed lines were studied since response correlation with a given line could provide a strong rationale for selecting resistant individuals.

In this presentation, 4 host responses to infection will be considered, including the development of periderm, hypertrophy with hyperplasia, rapid cell-tissue necrosis, and wall appositions. Except for wall appositions, similar responses have been described for older, rust-infected slash pines (22).

Periderm

Periderm was evident as an outer multilayer of collapsed thick-walled cells (Fig. 3). These resembled necrotic cells histologically. However, ultrastructural analysis (Fig. 4) revealed that starch grains and electron-opaque, tannin-like material in the cells were similar to those of conifer peridermal phellum. In addition, densely cytoplasmic cells beneath the collapsed layer were characteristic of phellogen (11). The phellogen cell in Fig. 4 possessed limited vacuolar space and an abundance of organelles, including starch-laden chloroplasts. Inner wall layerings of the phellogen cells suggested early suberization. The nucleus of the phellogen cell was not present in this section.

Although periderm was formed by hypocotyls within as early as 7 days, and only after inoculation, it occurred very rarely in embryos of all 3 lines (6 out of 145 tested) and so could not be statistically correlated with observed field resistance. However, embryos

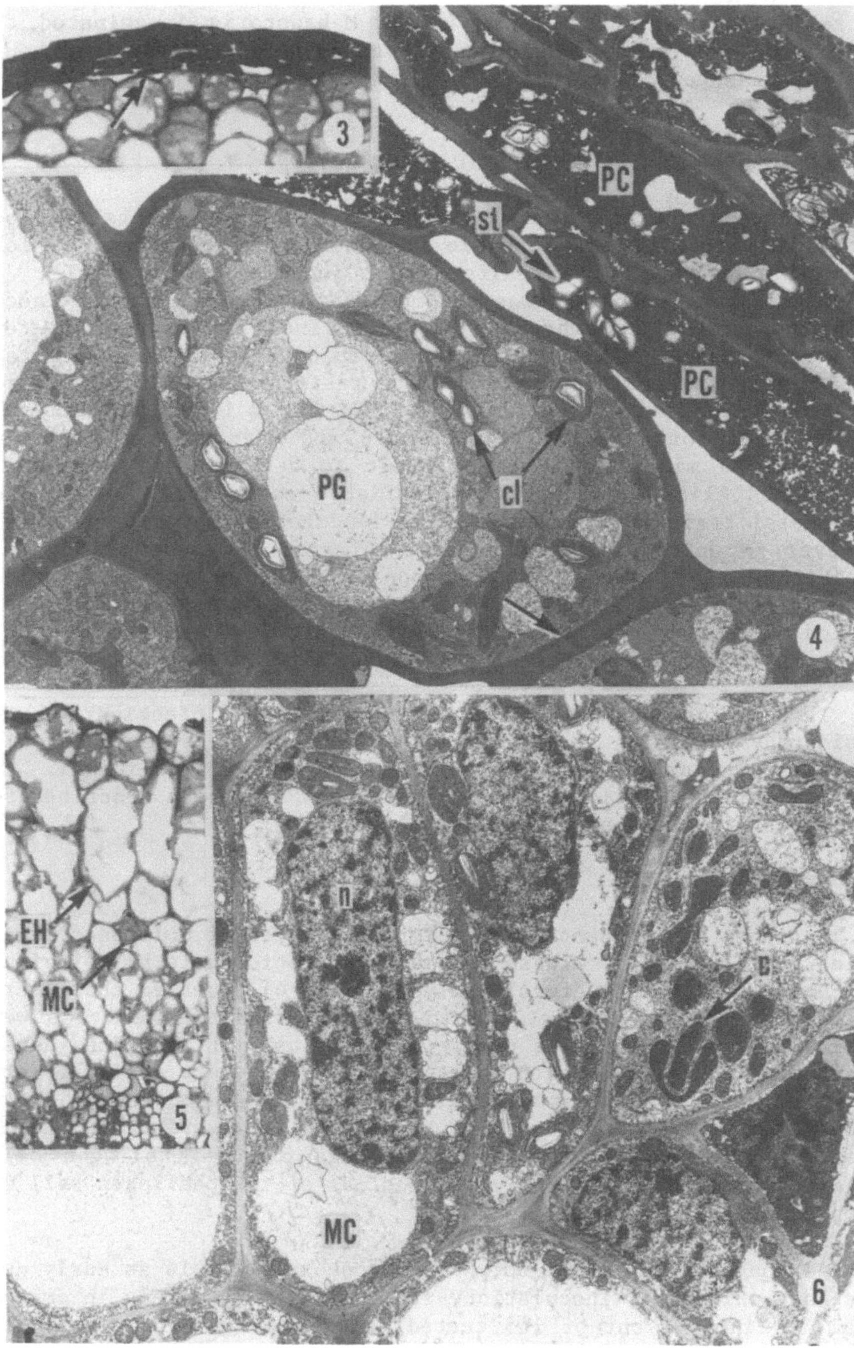
3
PC
st
PC
PG
cl
4
EH
MC
5
n
p
MC
6

←

Fig. 3-6. In vitro responses of loblolly pine embryos to fusiform rust inoculation. Fig. 3. Light micrograph of hypocotyl cross-section illustrating crushed, necrotic cells (arrow) indicative of periderm, X 800. Fig. 4. Transmission electron micrograph of hypocotyl periderm demonstrating phellum cells (PC) containing starch grains (st) and filled with electron opaque, tannin-like material. The underlying phellogen cell (PC) contains numerous starch-laden chloroplasts (cl) and wall layerings (arrow), the latter suggesting early suberization, X 12,000. Fig. 5. Light micrograph of hypocotyl cortical cross- section demonstrating enlarged hypertrophic cell (EH) and meristematic cell (MC) characteristic of hypertrophy and hyperplasia, X 700. Fig. 6. Transmission electron micrograph of meristematic cells from hypocotyl cortical region. Note the enlarged nucleus (n), lack of vacuolar space, and abundance of proplastids (p), X 11,700.

with periderm were never infected. Therefore, it is probable that this represents a viable resistance mechanism. Periderm presumably represents an impervious barrier to fungal growth.

Hypertrophy With Hyperplasia

Hypertrophy and hyperplasia of cells in cortical and vascular tissue produced enlarged and distorted regions of hypocotyl. Cells in these regions were variously shaped. Some were enlarged and highly vacuolate (Fig. 5) whereas others were highly meristematic, containing dense cytoplasm with an abundance of organelles (Fig. 5 and 6). Proplastids were particularly common in these cells. As with periderm, this response could not be statistically correlated with field resistance because it occurred so seldomly and only in inoculated, uninfected specimens. Hypertrophy and hyperplasia in uninfected hypocotyls appear to be similar to the swellings observed in gall formation in seedlings and mature trees (22) and, thus, probably represent a host response to infection.

Wall Appositions

Wall appositions were previously described in fusiform rust-infected hypocotyls (13), and in other plant-rust interactions (4,20). Their presence was determined with transmission electron microscopy (Fig. 7) and fluorescence microscopy (Fig. 8). The appositions were formed between the plasmalemma and the wall of individual host cells as a response to attempted penetration by fungal elements (i.e., IS, intracellular hyphae, or M-haustoria). Apposi-

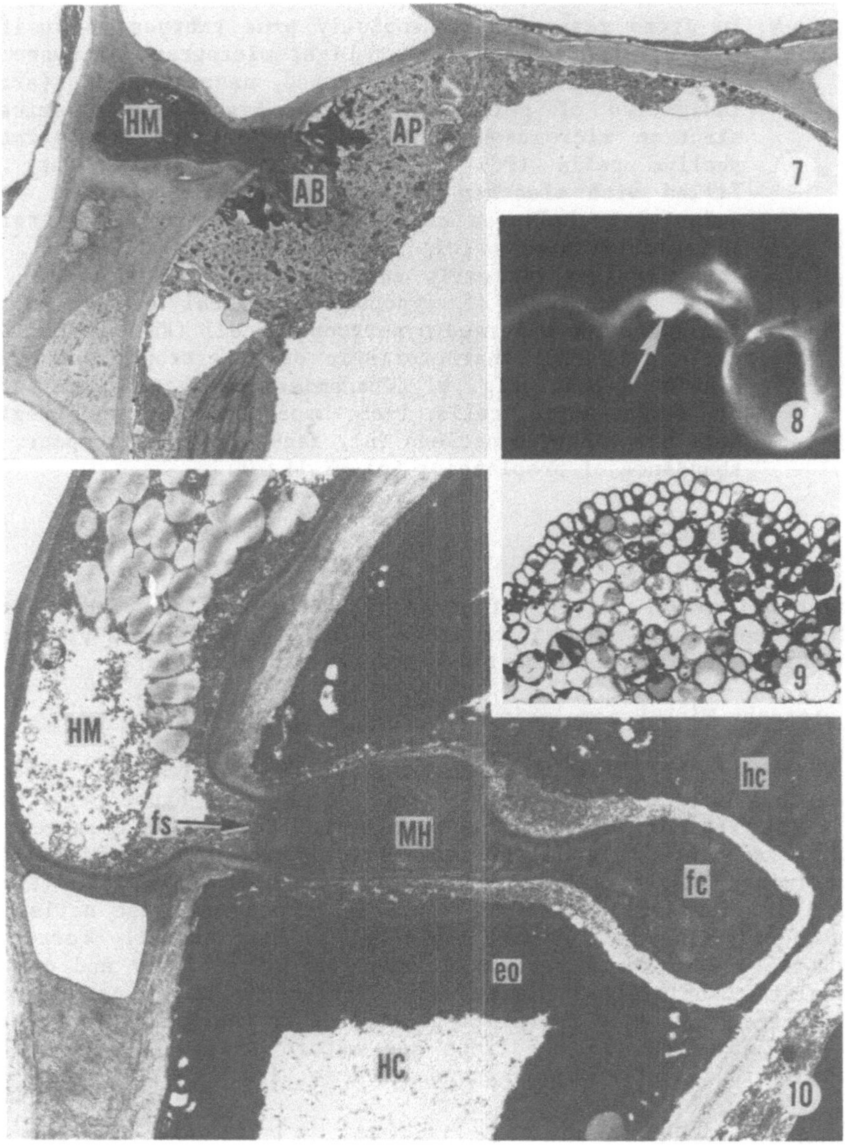

Fig. 7-10. In vitro responses of loblolly pine embryos to fusiform rust inoculation. Fig. 7. Transmission electron micrograph of wall apposition (AP) in hypocotyl cortical cell. Haustorial mother cell (HM), aborted fungal element (AB), X 8,400. From Gray and Amerson (13). Fig. 8. Fluorescent light micrograph of wall apposition (arrow) in hypocotyl epidermal cell, X 1,900. From Gray and Amerson (13). Fig. 9. Light micrograph of hypo-

tion material may be formed in abundance as a successful resistance response during early penetration resulting in complete walling off and necrosis of the intruding fungal element (Fig. 7), or it may be produced in less abundance. In the latter instance a collar of apposition material encases only the neck of M-haustoria (14) or IS, and this is probably ineffective as a barrier to fungal growth.

The early development of wall appositions in response to fungal penetration is clearly an effective method whereby individual host cells can survive fungal penetration. This response occurred commonly in all 3 test seed lines and so was not statistically correlated with observed resistance. However, it is possible that a genotype selected for an abundance of epidermal cell appositions (i.e., Fig. 8) could be essentially immune to fusiform rust invasion since the infection structures of each spore must pass through a single epidermal cell in order to be effective.

Rapid Necrosis

Rapid necrosis of hypocotyl cells and tissues occurred commonly in rust-infected specimens and was easily distinguished from a more slowly developing necrosis by the presence of a darkly stained material that was produced in the host cytoplasm and vacuole (12,13). Rapid necrosis was easily seen with light microscopy due to this darkly stained material (Fig. 9). Ultrastructurally, the material resembled tannin because of its electron opaque nature (Fig. 10). This assessment is supported by previous studies of slash pine that demonstrated the accumulation of tannins in fusiform rust-infected tissues (18,27).

The rapid necrotic response occurred both in cells that contained M-haustoria and in unpenetrated cells near infected tissue. In necrotic-penetrated host cells, only the M-haustorial body became similarly necrotic (Fig. 10). The M-haustorial neck (separated from the body by a septum) as well as extracellular fungal structures were unaffected. It appears that necrosis is initially triggered by cell penetration and this, in turn, stimulates nearby cells to re-

---

Fig. 7-10, continued

cotyl cross-section demonstrating rapid necrosis. Note densely stained, necrotic cortical cells, X 300. Fig. 10. Transmission electron micrograph of M-haustorium (MH) in a necrotic hypocotyl cortical cell (HC). Electron opaque material (eo) accumulates in HC vacuole as host (hc), and haustorial body (fc) cytoplasm becomes necrotic. Haustorial mother cell (HM) and haustorial neck, separated from the fc by a septum (fs), are unaffected, X 23,400.

spond similarly, creating a necrotic zone in advance of the invading fungal colony. The rust is inhibited by lack of suitable living host tissue for colonization. This is analogous to the hypersensitive reaction reported as a host response to other plant diseases (e.g., Ref. 26).

Rapid necrosis was the only host response that correlated with observed field resistance among the 3 test seed lines. This suggested that differences in rapid necrosis occurring early in the infection process might form the basis for long-term field resistance ratings. Fig. 11 illustrates the average amount of epidermal necrosis in hypocotyls from the 3 seed lines when data from 3 sample times and 3 inoculum levels were considered together. Although all developed some degree of rapid necrosis, the R line produced significantly more. The I and S lines appeared equivalent, but, as previously described in more detail (13), analysis of the same data over time showed a great increase in necrosis at later sample dates only in the I line. This was similar to the response of a mesothetic wheat genotype to rust infection (17) and presumably accounted for its intermediate resistance rating. Thus, all 3 lines could be separated by their ability to produce rapid necrosis over time. Subsequent to the investigations reported here, measurements of epidermal necrosis were shown to correlate highly with field resistance in 24 full-sib loblolly pine seed lines (10).

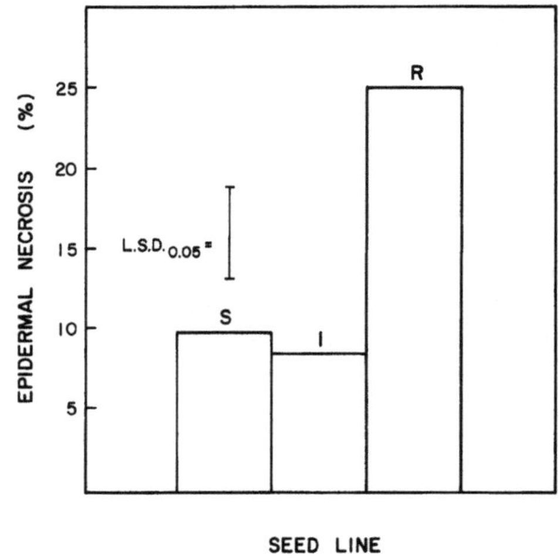

Fig. 11. Percentage of epidermal necrosis in susceptible (S), intermediate (I), and resistant (R) loblolly pine seed lines inoculated in vitro with basidiospores of the fusiform rust fungus. Each bar of the graph represents pooled data from 3 sample times after inoculation (1.5, 7, and 18 days) at 3 inoculum concentrations (540, 1,340, or 3,140 spores per embryo). From Gray and Amerson (13).

POTENTIAL APPLICATIONS OF IN VITRO RESISTANCE SCREENING FOR TREE IMPROVEMENT

The recognition of host responses that adversely influence fusiform rust disease in embryo hypocotyls strongly suggests that rapid in vitro resistance screening is possible and efforts to establish such a system are underway (10). A response such as rapid necrosis that correlates well with observed field resistance could be used to evaluate seed lines for resistance in much less time (7 days vs. 6 months) and space than is currently possible. Another application of this approach lies in the characterization of early occurring cellular events that form the basis for the most commonly observed field resistances.

Figure 12 outlines a potential of in vitro studies which couples cloning with in vitro disease assessment. Responses that are not, for one reason or another, correlated with field resistance may still be of use by incorporating an in vitro screening step into the tissue culture propagation system that currently exists for loblolly pine (25) and other conifers (24). Integration of these 2 procedures would consist of screening inoculated hypocotyls and concurrently obtaining plantlets from the cultured cotyledons of individual embryos (Fig. 12). In this way, many clones of embryos that possess rarely occurring responses to infection, such as early periderm development and hypertrophy with hyperplasia, could be produced for further examination. Desirable clones would nevertheless need to be tested over the long term in a conventional research orchard before selected elite genotypes could be used for seed production (Fig. 12). Other responses that occur commonly but are not correlated with field resistance by methods discussed here, such as wall apposition formation, may be a result of epigenetic factors that are lost in conventional breeding programs. Potentially, these may also be captured with an in vitro screening-tissue culture approach if they can be retained via vegetative propagation.

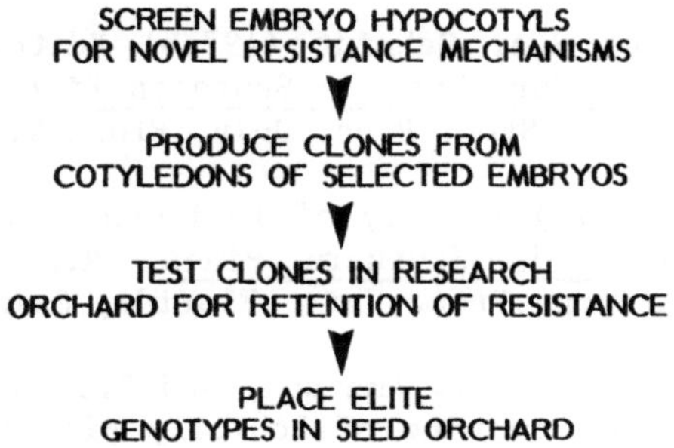

Fig. 12. Potential use of in vitro resistance screening for tree improvement.

In summary: a) Fusiform rust infection of loblolly pine embryo hypocotyls occurs in vitro in essentially the same manner as it does naturally in seedlings and mature trees. b) An analysis of host responses to inoculation suggests that in vitro resistance screening can decrease the evaluation time and potentially increase the accuracy for selecting resistant trees and seed lines (10). c) Genotypes with subtle, previously unrecognized, resistance mechanisms can potentially be obtained by incorporating in vitro screening with tissue culture propagation to capture these rarely occurring traits.

ACKNOWLEDGEMENT

We thank Amy L. Gray for help in preparing this manuscript.

REFERENCES

1. Amerson, H.V., and R.L. Mott (1978) Technique for axenic production and application of Cronartium fusiforme basidiospores. Phytopathology 68:673-675.
2. Anonymous (1982) USDA Forest Pest Management Rep. #82-1-18, 55 pp.
3. Burdsall, Jr., H.H., and G.A. Snow (1977) Taxonomy of Cronartium quercuum and C. fusiforme. Mycologia 69:503-508.
4. Chong, J., and D.E. Harder (1982) Ultrastructure of haustorium development in Puccinia coronata avenae: Some host responses. Phytopathology 72:1527-1533.
5. Cowling, E.B., R.J. Dinus, and R.A. Schmidt (1977) A status report and recommendations for research. In Management of Fusiform Rust in Southern Pines, R.J. Dinus and R.A. Schmidt, eds. Symp. Proc. Univ. Florida, Gainesville, pp. 1-9.
6. Czabator, F.J. (1971) Fusiform rust of southern pines - A critical review. USDA Forest Serv. Res. Paper SO-65, 39 pp.
7. Dinus, R.J., and R.A. Schmidt, eds. (1977a) Management of Fusiform Rust in Southern Pines, Symp. Proc. Univ. Florida, Gainesville, 163 pp.
8. Dinus, R.J., and R.A. Schmidt (1977b) Editor's preface. In Management of Fusiform Rust in Southern Pines, R.J. Dinus and R.A. Schmidt, eds. Symp. Proc. Univ. Florida, Gainesville, pp. iii.
9. Dwinell, L.D. (1977) Biology of fusiform rust. In Management of Fusiform Rust in Southern Pines, R.J. Dinus and R.A. Schmidt, eds. Symp. Proc. Univ. Florida, Gainesville, pp. 18-24.
10. Frampton, Jr., L.J., H.V. Amerson, and R.J. Weir (1983) Potential of in vitro screening of loblolly pine for fusiform rust resistance. Proc. 13th Southern Forest Tree Improvement Conf., Athens, Georgia, pp. 325-331.
11. Godkin, S.E., G.A. Grozdits, and C.T. Keith (1983) The periderms of three North American conifers, Part 2: Fine structure. Wood Sci. Technol. 17:13-30.

12. Gray, D.J. (1982) Ultrastructure and histology of the fusiform rust fungus. Ph.D. thesis, North Carolina State University, Raleigh, 122 pp.
13. Gray, D.J., and H.V. Amerson (1983) In vitro resistance of embryos of Pinus taeda to Cronartium quercuum f. sp. fusiforme: Ultrastructure and histology. Phytopathology 73:1492-1499.
14. Gray, D.J., H.V. Amerson, and C.G. Van Dyke (1982) An ultrastructural comparison of monokaryotic and dikaryotic haustoria formed by the fusiform rust fungus Cronartium quercuum f. sp. fusiforme. Can. J. Bot. 60:2914-2922.
15. Gray, D.J., H.V. Amerson, and C.G. Van Dyke (1983) Ultrastructure of the infection and early colonization of Pinus taeda by Cronartium quercuum formae specialis fusiforme. Mycologia 75:117-130.
16. Hanna, H., and L. Draper, Jr. (1977) Fusiform rust management strategies in practice: Seed source manipulation. In Management of Fusiform Rust in Southern Pines, R.J. Dinus and R.A. Schmidt, eds. Symp. Proc. Univ. Florida, Gainesville, pp. 134-144.
17. Higgins, V.J. (1981) Histological comparison of compatible, mesothetic, and incompatible reactions between Puccinia graminis tritici and wheat. Can. J. Bot. 59:161-165.
18. Jewell, F.F., and D.C. Speirs (1976) Histopathology of one- and two-year-old resisted infections by Cronartium fusiforme in slash pine. Phytopathology 66:741-748.
19. Laird, P.P., and W.R. Phelps (1975) A rapid method for mass screening of loblolly and slash pine seedlings for resistance to fusiform rust. Plant Dis. Rep. 59:238-242.
20. Littlefield, L.J., and M.C. Heath (1979) Ultrastructure of Rust Fungi. Academic Press, New York, 277 pp.
21. Miller, T., and E.B. Cowling (1977) Infection and colonization of different organs of slash pine seedlings by Cronartium fusiforme. Phytopathology 67:179-186.
22. Miller, T., E.B. Cowling, H.R. Powers, Jr., and T.E. Blalock (1976) Types of resistance and compatibility in slash pine - seedlings infected by Cronartium fusiforme. Phytopathology 66:1229-1235.
23. Miller, T., R.F. Patton, and H.R. Powers, Jr. (1980) Mode of infection and early colonization of slash pine seedlings by Cronartium quercuum f. sp. fusiforme. Phytopathology 70:1206-1208.
24. Mott, R.L. (1981) Trees. In Cloning Agricultural Plants Via In Vitro Techniques, B.V. Conger, ed. CRC Press, Boca Raton, Florida, pp. 217-254.
25. Mott, R.L., and H.V. Amerson (1981) A tissue culture process for the clonal production of loblolly pine plantlets. North Carolina Ag. Res. Serv. Tech. Bull. No. 271, 14 pp.
26. Prusky, D., A. Dinoor, and B. Jacoby (1980) The sequence of death of haustoria and host cells during the hypersensitive reaction of oat to crown rust. Physiol. Plant Pathol. 17:33-40.

27. Walkinshaw, C.H. (1978) Cell necrosis and fungal content in fusiform rust-infected loblolly, longleaf, and slash pine seedlings. Phytopathology 68:1705-1710.
28. Walkinshaw, C.H., T.R. Dell, and S.D. Hubbard (1980) Predicted field performance of slash pine families from inoculated greenhouse seedlings. USDA Forest Serv. Res. Paper SO-160, 6 pp.

LOBLOLLY PINE TISSUE CULTURE: LABORATORY, GREENHOUSE, AND FIELD STUDIES*

H.V. Amerson, L.J. Frampton, Jr., S.E. McKeand, R.L. Mott, and R.J. Weir

Departments of Forestry and Botany
North Carolina State University
Raleigh, North Carolina 27695-7612

INTRODUCTION

Conifer tissue culture had its beginnings in the late 1930s (10) and shoot regeneration cultures were first noted in 1950 (4). Since that time many species (7,17), especially those using embryonic materials for starting explants, have been cultured. Among some of the most studied species, Pinus radiata (2), Pseudotsuga menziesii (6), Pinus pinaster (8), Picea abies (22), and Pinus taeda (18), much laboratory data are accumulating. To date, little field data have been reported on the performance of tissue-cultured conifers (14), but several conifers are now established in field plantings (21) and data should be forthcoming.

Potential benefits of tissue culture technology to forest species should be viewed with a perspective of existing tree improvement practices. Currently, improved seed for commercial planting stock is produced by grafting scions from select trees into seed orchards where they intermate via wind pollination. Seed orchards are accompanied by progeny tests and periodically rogued based on genetic information to upgrade the quality of seed being harvested. One major advantage of the use of tissue culture propagation over sexual propagation via seed orchards is an additional increase in genetic gain (12). Current seed orchard designs only exploit additive genetic effects, while tissue culture

* The use of the following trade names throughout this paper does not imply endorsement of these products named, nor criticism of products not named: Captan, Furadan, Peters 15-30-15 and 20-19-18, and RL Super Cells.

propagation should ideally utilize all (additive and non-additive) genetic effects. Tissue culture propagation could conservatively increase genetic gains by one-third to one-half, but may even double genetic gains for some traits such as volume growth and disease resistance (14).

Another major advantage of tissue culture propagation over sexual propagation in conifers is a considerable decrease in the length of time between selection of improved individuals and the production of propagules from these selections. For most coniferous species a minimum of 8 to 10 years is required after scions from select trees are grafted into orchards until full-scale seed production begins. Hopefully, tissue-cultured propagules from select trees could be mass-produced 1 or 2 years following selection. Thus, tissue culture technology not only offers the opportunity to capture greater genetic gains, but also to utilize these genetic gains earlier than via the conventional seed orchard approach. Although the potential economic advantages that tissue culture technology could offer from these and other benefits (9,17,20) are tremendous, much work is needed before these techniques can feasibly be used operationally for conifer reforestation.

Currently, tissue culture propagation of loblolly pine is possible from a number of explant sources: needle fasicles (16), cytokinin-treated winter dormant buds (1), and cotyledon explants (18). The latter method using cotyledon explants will be the focus of this paper. Information will be presented on laboratory and greenhouse studies. Emphasis will be placed on regulatory controls in vitro and on field performance data obtained from over 3000 tissue-cultured plantlets currently outplanted throughout the southeastern United States.

LABORATORY STUDIES

A sequence for the tissue culture propagation of loblolly pine (3,18) was reported in 1981, and utilized the concepts of pulse timing and sequential steps that must be accomplished to progress from shoot initiation through rooting to produce tissue culture plantlets. Growth regulator pulsing was viewed as the key to success in the sequential process, and hinged on the idea that growth regulators necessary to initiate organogenic events in culture may subsequently become inhibitory to further development. Both timely application and removal were necessary to reach the desired ends. Since 1981 the propagation sequence has been and is still being amended with media, environmental, and pulse duration changes, however, the concept of pulse timing remains central for loblolly pine tissue culture.

Along with pulsing, other regulatory features in loblolly pine tissue culture are now becoming recognizable. The purpose of this section on laboratory studies is to examine some regulatory features recognized at various steps in the process in an effort to partially understand the controls that operate in pine organogenesis. To do this, we will first summarize the process and then look at controls at each step.

Culture Process

The tissue culture process begins with loblolly pine seeds, scarified at the micropylar end and partially germinated in hydrogen peroxide (typically 3 days in 1% H_2O_2 and 1 to 2 days in 0.03% H_2O_2 at 28-30°C), until the radicale and hypocotyl extend from the seed coat. Subsequent to seed coat removal and surface sterilization, the embryonic axis is aseptically excised from the female gametophyte. Next the cotyledons are surgically removed from the embryo and planted horizontally on a shoot initiation medium which is cytokinin-rich [typically 44 µM benzylaminopurine (BAP)]. Cotyledons are maintained on this medium for 14 to 28 days dependent upon light and temperature conditions. On the initiation medium cell divisions occur in the peripheral areas of the cotyledons producing a warty, meristematic surface. Cotyledons are removed from the BAP-containing medium prior to the actual visualization of shoots. The cotyledons are placed on a hormone-free medium containing charcoal to further aid cytokinin removal. On this shoot differentiation and growth medium, shoot apices become recognizable on the cotyledons and the shoots begin to elongate. Shoot growth continues during further subcultures on hormone-free medium. The multiple shoots crowded on the cotyledons are individually excised and separated from the cotyledon for further growth.

Following the shoot growth phase outlined above, shoots ca. 1-2 cm in length are transferred to root induction medium which is auxin-rich [typically α-Naphthaleneacetic acid (NAA) at 2.5 µM]. Shoots, freshly cut at the base, are implanted upright and pulsed for 6-9 days on the auxin medium. Pre-root cell divisions form near the cambial region at the stem base, resulting in a swollen, callusy stem base. Pre-root divisions may begin to organize into root primordia on the auxin medium, but to facilitate organization and rapid growth, the rooting-phase shoots are transferred to hormone-free medium. On the hormone-free medium, roots grow rapidly and plantlets typically are ready for transfer to greenhouse soil 3 to 5 weeks after the root initiation treatment.

Examination of Regulatory Features

Shoot initiation and elongation. Initiation of pre-shoot cell divisions, and ultimately shoot initiation in loblolly pine, requires the presence of a cytokinin in the initiation medium as do many other conifers (7). We have tested BAP over the range of 0-111 µM and all concentrations tested between 4.4-111 µM have produced shoots. Attempts to initiate shoots without BAP typically produced only elongate cotyledons with no visible pre-shoot meristematic tissues, or at best an occasional apical shoot at the cotyledon tip. Other cytokinins such as kinetin will produce shoots on loblolly pine cotyledons, but other cytokinins are as yet little investigated since BAP has been effective.

Although BAP is required for shoot initiation, the BAP exposure time (pulse time) is critically important in obtaining growth-competent shoots. Table 1, which examines shoots still attached to the cotyledons, clearly shows that shoots initiated for 28 days on 44 µM BAP grew poorly compared to those given only a 14-day exposure. Similar observations correlating cytokinin exposure and shoot growth have been made for lodgepole pine (19), and David (7) noted that several studies recognized cytokinin interference in shoot growth in conifers.

Further evidence of shoot growth suppression by BAP is noted in Tab. 2 which shows the response of individually excised, actively growing shoots which were placed back on BAP-supplemented medium. Inhibition occurred in all 3 BAP treatments. The highest concentration, 1.3 µM BAP, supplied continuously gave the most suppression. BAP, 1.3 µM, supplied as a 9-day pulse and 0.44 µM BAP supplied continuously permitted intermediate growth. Thus growth suppression appeared proportional to overall BAP exposure. Although BAP did clearly suppress shoot growth, the concentrations shown in Tab. 2

Tab. 1. Comparison of shoot initiation and growth as a function of initiation time. Shoots were initiated on BLG medium [modified Brown & Lawrence medium (5) with 10 mM glutamine, 1 mM KNO_3, and 10 mM KCl substituted for original nitrogen components and potassium balance] containing 44 µM BAP. BAP initiation and subsequent differentiation on hormone-free medium + charcoal occurred at 23°C, with 24 hr illumination (ca. 2200 lux cool white fluorescent and 800 lux incandescent). Elongation occurred at 22 ± 2°C with 24 hr illumination (ca. 2700 lux cool white fluorescent and 200 lux incandescent). Counts made at 4 month total.

Initiation time in days	Mean # shoots per embryo	% Shoots < 2 mm	% Shoots 2-5 mm	% Shoots ≥ 5 mm
14	46	18	34	48
	NS	*	NS	*
28	41	48	45	7

* = significant difference within column at P = .05 level.
NS = no significant difference within column.

Tab. 2. The influence of exogenous BAP supplied to adventitious shoots growing on GD 1/2 medium [modified Gresshoff and Doy 1 medium (18)]. BAP was supplied continuously in the medium for 8 weeks except in pulse treatment #3. This treatment received a 9-day pulse on GD 1/2 + 1.3 µM BAP followed by transfer to GD 1/2. Shoots grown at 22 ± 2°C with 16 hr mixed incandescent (ca. 200 lux) and cool white fluorescent light at (ca. 3200 lux) and 8 hr dark.

Treatment	# Shoots tested	Mean growth in mm at 8 weeks
1. GD 1/2 control	50	6.4 a
2. GD 1/2 + .44 µM BAP	49	4.8 b
3. GD 1/2 + 1.3 µM BAP (pulse)	50	4.7 b
4. GD 1/2 + 1.3 µM BAP	50	2.0 c

Growth values followed by the same letter are not significantly different as determined by Waller-Duncan Test at P = .05.

were sufficient to stimulate axillary bud break, so the re-addition of BAP back into the propagation process can have a useful purpose for micropropagation.

Besides being responsive to cytokinins, bud (shoot) initiation and subsequent development are both light-dependent events (Fig. 1). Examination of Fig. 1 reveals that control treatment #3 produced both pre-bud divisions and buds (shoots) in the light. In contrast, dark initiation conditions in treatment #1 prohibited pre-bud cell

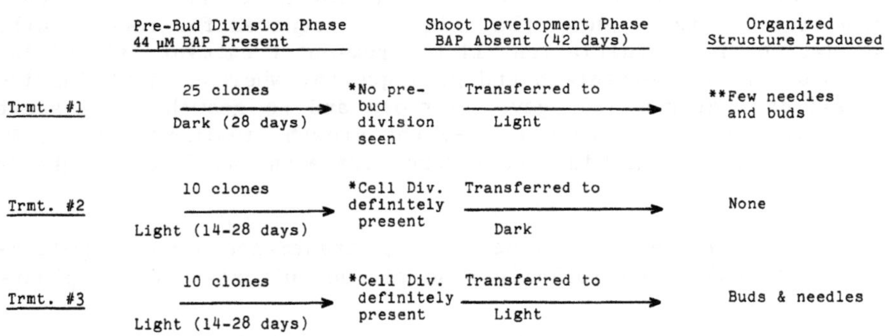

* Presence or absence of divisions determined microscopically in fresh sections.

** BAP carry-over from the dark initiation treatment accounts for the very few buds seen later.

Fig. 1. Effect of light and dark on adventitious shoot development from cotyledons. Light = 24 hr illumination at ca. 1250 lux cool white fluorescent and 750 lux incandescent. Temperature = 21°C.

divisions, and in treatment #2 existing pre-bud divisions formed in the light failed to organize into buds upon transfer to darkness. Thus in bud (shoot) initiation processes, neither cell divisions nor organization activities occur in constant darkness.

Root initiation and growth. After shoots reach a size of ca. 1 cm or more, they are suitable for the rooting phase of propagation. Controlling features of root initiation and growth have been studied using both adventitious shoots obtained from cotyledons and 1- to 2-week-old seedlings referred to as hypocotyl explants. These are seedlings which were de-rooted and then re-rooted in vitro via tissue culture procedures. Hypocotyl explants and adventitious shoots have performed similarly except that the hypocotyl explants root ca. 15-20% better, with faster root growth rates. Since their overall responses were similar, both plant types were used in rooting studies and data will be presented from each.

The process of root initiation is an auxin-mediated event in contrast to the cytokinin dependency noted for shoot initiation. Spontaneous rooting of loblolly pine tissue-cultured plantlets on auxin-free medium is rare and considerably less than 1%. Rooting occurs best in response to auxin pulses (Tab. 3). Continuous exposure to low level auxin (0.5 μM NAA) for 42 days yielded only 17% rooting, whereas pulse exposure to the same 0.5 μM NAA concentration for 12 days yielded 50% rooting. A shorter term (6-9 day), higher concentration (2.5 μM NAA) pulse further improved rooting and is now in routine use.

The overall rooting process is a light-requiring process. Attempts to root shoots in complete darkness have yielded no roots. However, simply recognizing the absence of rooting in darkness does not indicate if both root initiation and growth are light-dependent. Examination of Fig. 2 shows that light is required for root growth. Note that the light initiation, light grow-out treatment (■) yielded a good rooting percentage and good growth, whereas light initiation and dark growth (△) gave no roots and no growth. The above data coupled with dark initiation-light growth treatments (○ and □) which both produced growing roots demonstrate the need for light in

Tab. 3. Comparison of various rooting treatments (auxin applications) using loblolly pine tissue culture-produced plantlets.

Auxin & concentration in μM	# Plantlets Tested	Auxin time in Days	Growth Medium	% Rooting
NAA 0.50	90	42	-	17
NAA 0.50	30	12	No auxin 5 wk	50
NAA 2.50	128	6-9	No auxin	71

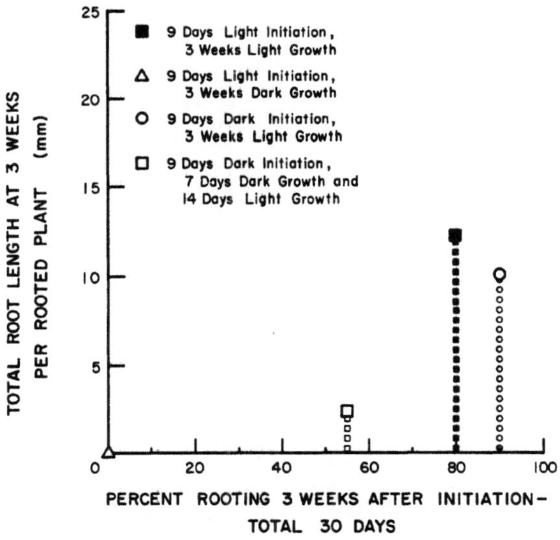

Fig. 2. Comparison of root initiation and growth in light vs. dark environments. Data collected from hypocotyl explants which were de-rooted and then cultured on GD 1/2 medium + .44 μM BAP and 2.5 μM NAA for 9 days, followed by 3 weeks growth on GD 1/2 without growth regulators. Light = 24 hr. Illumination at ca. 1830 lux warm white fluorescent and 320 lux incandescent. Temperature = 25 ± 2°C. N = 9-11 plants.

root growth. Based solely on Fig. 2, light requirements for root initiation are unclear since roots seen in the dark initiation treatments (○ and □) could possibly have originated either in the dark or as a result of auxin carry-over to the light treatment. However, microscopy of hypocotyl bases taken from treatment □ at the end of the dark period revealed that pre-root cell divisions had formed in darkness, but distinctly organized root apices were not observed. Thus root organization and growth are light-dependent, but pre-root divisions are not.

Not only is root growth dependent on the presence of light, but light quality strongly regulates the rate of root growth. Figure 3 shows a comparison of root growth on hypocotyl explants in fluorescent light only, mixed fluorescent/incandescent light, and incandescent light only. Growth in the environments containing incandescent light is much greater than in the fluorescent-only treatment. Incandescent light-enhanced root growth has occurred in numerous experiments, and is even more profound in the rooting of adventitious shoots where fluorescent-only treatments produced little if any growth.

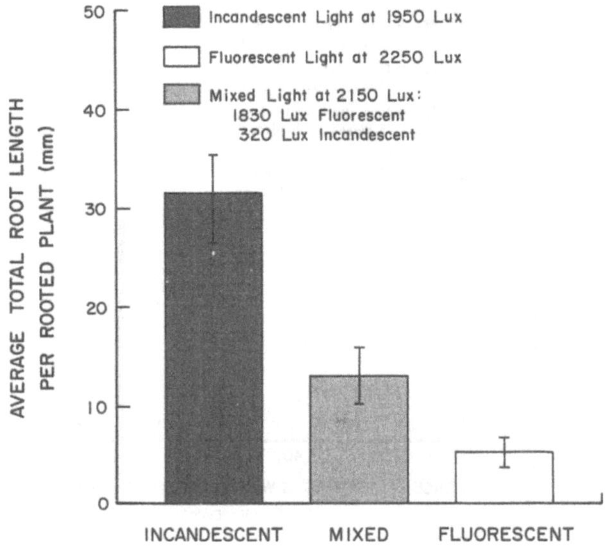

Fig. 3. Root growth in 3 different light regimes (warm white fluorescent light, warm white fluorescent/incandescent light, and incandescent light only). Data collected from hypocotyl explants which were de-rooted and re-rooted in vitro. Plants grown on GD 1/2 medium at 25 ± 2°C, and data collected after 3 weeks growth. Error bars represent 95% confidence intervals for root growth means. N = 52 plants per treatment.

The beneficial aspects of incandescent lighting on root growth are easily recognized, but are as yet poorly understood. Both phytochrome regulation and short wavelength inhibition of growth in fluorescent environments are suspected and are under investigation.

GREENHOUSE AND FIELD PROCEDURES

Plantlets are transferred from the laboratory to the greenhouse when their total shoot lengths (including needles) exceed 1-2 cm and their individual root lengths exceed 3-5 mm. Plantlets meeting these requirements are carefully removed from the agar-solidified laboratory medium and planted in 164 cc RL Super Cells containing a fine textured mix of peat, vermiculite, and perlite (2:2:1). The plantlets are grown in a mist bench the first 3 to 6 weeks in the greenhouse. After the first week, they are fertilized 3 to 5 times weekly with Peters 15-30-15 mixed at 30 ppm N. Plantlets in the mist bench are sprayed weekly with a fungicide, Captan, to reduce damping off and other disease problems. When necessary, the photoperiod of the plantlets is extended to 16 hrs using incandescent lights (approximately 4 Wm^{-2}) to prevent dormancy.

Although the initial growth of the plantlets in the greenhouse is very slow, after about 6 weeks, new vigorous growth appears. At this time plantlets are removed from the mist bench and fertilization is changed to Peters 20-19-18 mixed at 40 ppm N applied 3 to 5 times weekly. After removal from the mist bench, plantlets are watered as needed with pH 5.5 water. Generally, plantlets reach a suitable size for field planting (about 20-30 cm in height and 3-5 mm in caliper) after 6 months in the greenhouse.

Using similar procedures, seedlings are also grown in the greenhouse to establish in field tests for comparison purposes. Seedlings generally require only 4 months to attain plantable size so that it is often necessary to manipulate the watering and fertilization regime of the plantlets and seedlings in order to coordinate their growth.

Before field planting, both the plantlets and seedlings are gradually adapted to conditions outside the greenhouse. The succulent growth is hardened-off by first stopping fertilization and reducing watering. Next, the trees are transported outside in order to adapt to direct sunlight, natural photoperiod, and outdoor temperatures. After 2 to 4 weeks outside, trees are capable of surviving field planting.

The field tests are carefully site-prepared, hand-planted, and intensively managed. At the time of establishment, a soil analysis is conducted and any nutrient deficiencies are corrected. Additionally, approximately 50 g N in the form of ammonium nitrate is applied to every tree during the spring to enhance growth. Weeds in the plantings are controlled either by periodic mowing or with herbicides. Nantucket pine tipmoth (*Rhyacionia frustrana* Comst.) which often kills young loblolly pine shoot tips is controlled with Furadan applications.

Most of the field plantings contain paired row-plots of plantlets and seedlings from several half-sib families. The plantlets in a plot represent one clone produced from the cotyledons of a single embryo. The trees of the seedling plots are grown from seed of the half-sib family from which the plantlets were derived. Several of the field plantings contain clonal block plots of 16 or 25 plantlets compared to block plots of seedlings from the same half-sib family.

GREENHOUSE STUDIES

Several greenhouse studies have been conducted to quantify growth and developmental differences between plantlets and seedlings. In one study (13) the shoot dry weight production of plantlets and seedlings was compared over a 20-week period. During this period, 21 trees of each plant type were sampled at 2-week intervals. Figures 4a and 4b compare the shoot dry weight growth curves for plantlets and seedlings. After linearizing this relationship with a logarithmic transformation (Fig. 4b), both the slopes and intercepts differed significantly ($P < 0.01$) between the plantlet and seedling regression equations. Thus, the plantlets started the experiment at a smaller size (due to a 4-week growth lag in the mist bench before sampling began), yet accumulated shoot dry weight at a slightly faster rate than did the seedlings.

(a)

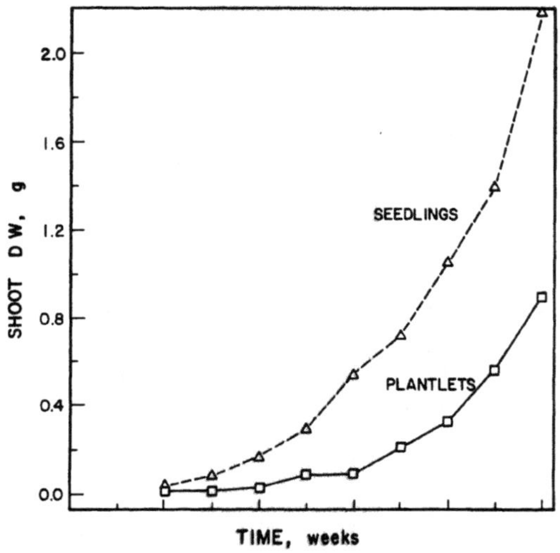

(b)

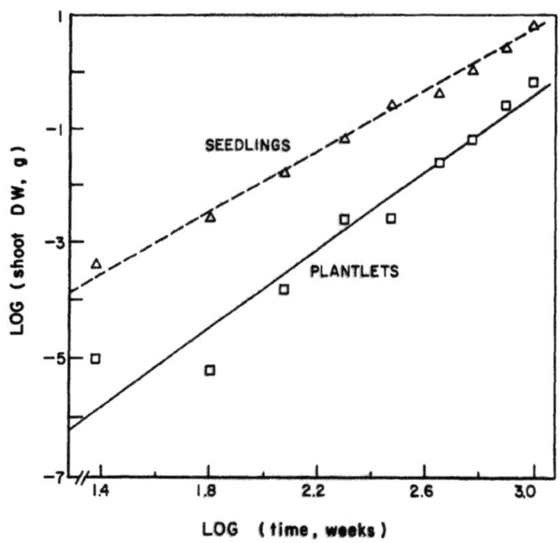

Fig. 4. Growth of shoot dry weights for plantlets and seedlings over a 20-week period. (a) Untransformed data. (b) Logarithmically transformed data (13).

Much work has emphasized morphological comparisons between plantlet and seedling root systems. Generally, plantlet root systems had fewer lateral roots which were distributed differently from those of seedlings. In one study (15), 20-week-old plantlets and seedlings averaged 10.3 and 18.3 dominant lateral roots per tree, respectively. The majority of the seedling lateral roots originated from the taproot in the upper third of the root system, while less than 20% originated from the bottom third (Fig. 5). In contrast, almost 50% of the plantlet lateral roots emerged from the main root in the bottom third of the root system while less than 15% originated in the upper third.

When both plant types were grown for 3 months under greenhouse conditions, but in large containers (946 cc milk cartons), total root dry weight did not differ significantly ($P < 0.05$) between plantlets and seedlings (14). However, there was a significant difference ($P < 0.05$) in the dry weight distribution of the root systems between the plant types. The plantlets and seedlings had 54.4 and 80.2% of their total root dry weight due to lateral roots, respectively.

One apparent problem with plantlet root systems is the inefficient uptake of nutrients. Plantlets are much less effective than seedlings at nitrogen and phosphorus uptake per unit root dry weight (13). However, when nutrient uptake was expressed per unit root surface area, plantlets and seedlings were equally effective. The differences between plant types were primarily related to root system morphology (fewer lateral roots, and thus, less root surface area for uptake) rather than physiological processes (less absorption per unit area).

Realizing these differences, a nursery study was conducted to investigate the effect of various root pruning treatments on root morphology, as well as to assess plantlet growth in the nursery (23). Conventional nursery practices were used to grow the trees

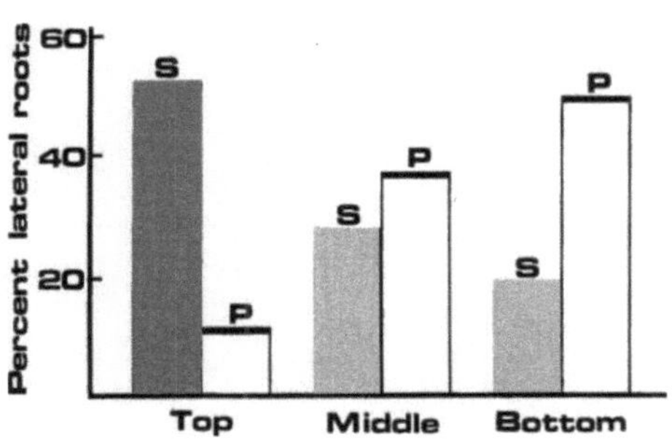

Fig. 5. Percentage of dominant lateral roots originating from the top, middle, and bottom third of the main root or taproot of plantlets (P) and seedlings (S) (15).

from May to October and pruning treatments were manually performed in August. The plantlet survival was excellent (97%). Plantlet height growth, although not as great as that of the nursery-grown seedlings, exceeded a control group of greenhouse-grown plantlets. Furthermore, the most severe root pruning treatment increased the similarity between the plantlet and seedling root system morphology. A portion of pruned and non-pruned plantlets and seedlings has been established in a field study to assess long-term effects of root pruning on plantlet and seedling performance.

FIELD PLANTINGS

The North Carolina State University Project on Tissue Culture has established 16 field plantings across the southeastern United States (Fig. 6). Over 3000 trees each of plantlets and seedlings have been planted representing over 25 half-sib families of loblolly pine.

The first series of 8 plantings were established in the summer and fall of 1981. Survival in these tests was excellent, exceeding 94% for both the plantlets and seedlings. First and second year results of these plantings are shown in Tab. 4. When comparing the growth patterns of the plantlets and seedlings in the field, a similar pattern to that observed in the greenhouse is apparent. In all 8 tests, first- and second-year heights of the seedlings signifi-

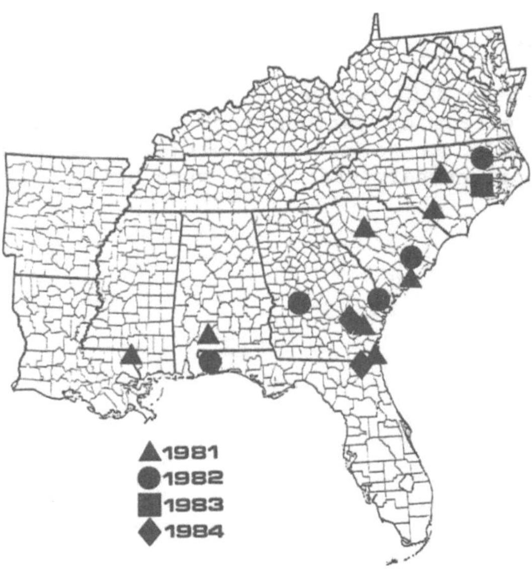

Fig. 6. Location of tissue culture field plantings of loblolly pine established by the Project on Tissue Culture at North Carolina State University (14).

cantly ($P < 0.05$) exceeded those of the plantlets. The height increase between the first and second year expressed as a percent of the first year's height was not significantly different between the plant types except in one test where the plantlet value exceeded that of the seedling value. This suggests that the plantlets experienced a period of adaptation to field conditions during which they lagged behind the seedlings in height growth. By the end of the second growing season, the plantlets had apparently adapted and grew at a rate similar to the seedlings. Most likely, the seedlings will remain taller than the plantlets in absolute height through rotation age. However, if the problem of the first- year lag in plantlet growth can be rectified, plantlet height growth should be similar to that of seedlings through rotation age. A similar relationship between plantlet and seedling growth is apparent for diameter and volume index (diameter squared times height). Measurements will continue in these tests to monitor long-term growth trends.

Fusiform rust (caused by the fungus Cronartium quercuum Miyabe ex Shirai F. sp. fusiforme (Cumm.) Burds. et Snow) is the most economically damaging disease of loblolly pine (11). Its primary symptom is gall formation on the stems and branches of infected trees. In the 1981 field plantings, the plantlets were less infected than the seedlings in all studies where rust infection was present (Tab. 4). Overall, the plantlets averaged 17.8% compared to 45.0% infection for the seedlings. In some areas of the southeastern United States where fusiform rust incidence is high, such reduction in infection would be of tremendous economic benefit and could more than offset slower growth rates. Although the reasons for these differences in infection are not known, some possible hypotheses follow:

1. Seedlings were taller than plantlets and, therefore, had more succulent tissue exposed for infection.

2. Plantlet phenology was different from seedlings so their new growth was unexposed during the 2- to 4-week period in the spring when basidiospore flight occurred.

3. New plantlet shoot growth was less succulent (more lignified) than seedling shoot growth.

4. Plantlets appeared morphologically more mature than seedlings and, therefore, more resistant.

More careful assessment of fusiform rust infection in the future, hopefully, will lead to rejection or acceptance of some of these hypotheses.

Tab. 4. First and second year results of the 1981 field plantings

		Stem Height (m)					Height Growth (%)				
		Year 1			Year 2		Year 0-1			Year 1-2	
Organization	Location	P		S	P		S	P		S	P		S
Federal Paper Board Co., Inc.	Lumberton, NC	-		-	1.31	*	1.86	-		-	-		-
Brunswick Pulp Land Co.	Jesup, GA	1.14	*	1.36	2.70	*	3.26	391.2		401.5	142.3		143.0
Westvaco Corp.	Summerville, SC	0.72	*	0.91	1.64	*	2.16	189.4		188.7	128.3		138.9
Scott Paper Co.	Monroeville, AL	0.68	*	0.91	1.81	*	2.28	174.0		191.1	170.8	*	156.2
N. C. State University	Raleigh, NC	0.46	*	0.62	0.92	*	1.18	67.6	*	98.5	89.1		91.7
ITT Rayonier, Inc.	Yulee, FL	0.51	*	0.61	1.42	*	1.82	109.1		94.5	179.9		197.5
Champion International Corp.	Newberry, SC	0.35	*	0.43	0.68	*	0.92	35.4		43.7	95.5		115.7
Crown Zellerbach Corp.	Bogalusa, LA	0.80	*	1.07	2.13	*	2.84	202.3	*	250.4	170.8		168.5
Weighted Mean[2]		0.63	*	0.82	1.52	*	1.99	151.0		169.2	138.5		144.9

[1] P = plantlet, S = seedling, * = significantly different at P $\leq$ .05 level.

[2] Each location mean was weighted by the number of observations in the test. Significant differences between the weighted means of plantlets and seedlings over locations were determined by a paired t-test.

Many differences between plantlet and seedling shoot morphology have been observed in the field plantings. Although the plantlets originate from embryonic material, their morphology appears more mature-like than seedlings. A detailed measurement of characteristics such as bud, needle, branch and crown size, number of branches and growth cycles, growth cycle lengths, and form quotient in one field study verified this observation (McKeand, unpub. data). The cause and permanence of this "maturation" phenomenon in the plantlets are unknown. It is suspected that the relative differences between plantlet and seedling morphology will lessen over time.

CONCLUSION

Loblolly pine propagation from embryonic explants has progressed to the point that about 3000 tissue-cultured plantlets are now in the field. Tissue culture procedures have shown growth regulator pulsing to be critical for the production of growth-competent shoots, and for high percentage rooting. Illumination was recognizably important in both shoot and root production. In both shoot and root growth, light was required. However, light requirements for cell divisions leading to shoots and roots differed. Pre-root cell divisions did not require light whereas pre-shoot divisions of the Project on Tissue Culture at North Carolina State University.

Stem Diameter (cm)				Diameter Growth (%)		Volume Index (D^2H)				Volume Growth (10^3%)		Fusiform Rust Infection (%)	
Year 1		Year 2		Year 1-2		Year 1		Year 2		Year 1-2		Year 2	
P	S	P	S	P	S	P	S	P	S	P	S	P	S
-	-	3.67 *	4.90	-	-	-	-	20.5 *	47.2	-	-	4.2	10.0
3.06	3.28	6.76 *	7.75	125.1	139.3	12.0	15.7	131.0 *	201.0	1.16	1.32	47.4	68.8
1.83 *	2.49	4.21 *	5.57	132.9 *	127.8	2.8 *	6.1	33.8 *	73.0	1.21	1.23	40.7 *	86.5
-	-	-	-	-	-	-	-	-	-	-	-	20.6	52.5
1.18 *	1.62	-	-	-	-	0.8 *	2.0	-	-	-	-	0.0	0.0
1.52 *	1.72	3.98 *	4.93	165.4	191.0	1.4 *	2.2	26.5 *	49.8	1.98 *	2.62	14.0 *	46.2
0.82 *	0.88	2.16 *	2.77	165.0	215.4	2.7 *	3.8	38.8 *	84.4	1.43	2.22	4.1 *	45.4
1.79 *	2.44	5.18 *	7.21	192.2	198.6	2.9 *	6.8	63.7 *	154.0	2.28	2.39	21.9	47.8
1.59 *	2.00	4.19 *	5.43	159.4	176.6	3.1 *	5.5	48.6 *	97.8	1.66	2.00	17.8 *	45.0

did. Root growth also was dependent on light quality and was enhanced by incandescent illumination. Recognition of some of the control features in loblolly pine organogenesis is a preliminary step towards a better overall understanding of pine propagation.

Early research with tissue culture plantlets of loblolly pine in greenhouse and field trials has identified differences in growth, morphology, and possibly physiology between plantlets and seedlings. Differences in size between plantlets and seedlings appear to be due to an initial lag period perhaps related to root morphology when plantlets are adapting to their new environment. Hopefully, treatments such as root pruning or alterations in laboratory techniques can correct this problem. Causes of other differences between plantlets and seedlings, such as greater fusiform rust resistance and expression of mature-like morphological characteristics, are being explored and may aid in developing a better understanding of the control of these processes.

REFERENCES

1. Abo El-Nil, M.M. (1982) Method for asexual reproduction of coniferous trees. U.S. Patent No. 4,353,184.

2. Aitken, J., K.J. Horgan, and T.A. Thorpe (1981) Influence of explant selection on the shoot-forming capacity of juvenile tissue of Pinus radiata. Can. J. For. Res. 11:112-117.
3. Amerson, H.V., S.E. McKeand, and R.L. Mott (1981) Tissue culture and greenhouse practices for the production of loblolly pine plantlets. In Proc. 16th South. For. Tree Impr. Conf., pp. 168-173.
4. Ball, E.A. (1950) Differentiation in a callus culture of Sequoia sempervirens. Growth 14:295-325.
5. Brown, C.L., and R.H. Lawrence (1968) Culture of pine callus on a defined medium. For. Sci. 14:62-64.
6. Cheng, T.-Y. (1977) Factors affecting adventitious bud formation of cotyledon culture of Douglas fir. Plant Sci. Lett. 9:179-187.
7. David, A. (1982) In vitro propagation of gymnosperms. In Tissue Culture in Forestry, J.M. Bonga and D.J. Durzan, eds. Martinus Nijhoff/Dr. W. Junk Publishers, The Hague, pp 72-104.
8. David, A., and H. David (1977) Manifestations de diverses potentialites organogenes d'organes ou de fragments d'organes de Pin maritime (Pinus pinaster Sol.) en culture in vitro. C.R. Acad. Sci. Paris 284:627-630.
9. Durzan, D.J., and R.A. Campbell (1974) Prospects for the mass production of improved stock of forest trees by cell and tissue culture. Can. J. For. Res. 4:151-174.
10. Gautheret, R.J. (1937) Nouvelles recherches sur la culture de tissu cambial. C.R. Acad. Sci. Paris 205:572-573.
11. Holley, D.L., and M.A. Veal (1977) Economic impact of fusiform rust. In Management of Fusiform Rust in Southern Pines, R.J. Dinus and R.A. Schmidt, eds. University of Florida, Gainesville, pp. 39-50.
12. McKeand, S.E. (1981) Loblolly pine tissue culture: Present and future uses in southern forestry. School of For. Res. Tech. Rep. No. 64, N.C. State University, 50 pp.
13. McKeand, S.E., and H.L. Allen (1984) Nutritional and root development factors affecting growth of tissue culture plantlets of loblolly pine. Physiol. Plant 61:523-528.
14. McKeand, S.E., and L.J. Frampton, Jr. (1984) Performance of tissue culture plantlets of loblolly pine in vivo. In Proc. Intl. Symp. of Recent Adv. in For. Biotechnology, Traverse City, Missouri (in press).
15. McKeand, S.E., and L.A. Wisniewski (1982) Root morphology of loblolly pine tissue culture plantlets. In Proc. 7th No. Am. Biol. Workshop, B.A. Thielges, ed., pp. 214-219.
16. Mehra-Palta, A., R.H. Smeltzer, and R.L. Mott (1977) Hormonal control of induced organogenesis: Experiments with excised plant parts of loblolly pine. Tappi 61(1):37-40.
17. Mott, R.L. (1981) Trees. In Cloning Agricultural Plants Via In Vitro Techniques, B.V. Conger, ed. CRC Press, Boca Raton, Florida, pp. 217-254.
18. Mott, R.L., and H.V. Amerson (1981) A tissue culture process for the clonal production of loblolly pine plantlets. North Carolina Ag. Res. Ser. Tech. Bull. #271, 14 pp.

19. Patel, K.R., and T.A. Thorpe (1984) In vitro differentiation of plantlets from embryonic explants of lodgepole pine (Pinus contorta Dougl. ex. Loud.). Plant Cell Tissue Organ Culture 3:131-142.
20. Sommer, H.E., and C.L. Brown (1979) Applications of tissue culture to forest tree improvement. In Plant Cell and Tissue Culture: Principles and Applications, W.R. Sharp, P.O. Larsen, and V. Raghaven, eds. Ohio State University Press, Columbus, pp. 461-491.
21. Sommer, H.E., and H.Y. Wetzstein (1982) Application of tissue culture to forest tree improvement. Proc. 8th Long Ashton Symposium (in press).
22. von Arnold, S. (1982) Factors influencing formation, development and rooting of adventitious shoots from embryos of Picea abies (L.) Karst. Plant Sci. Lett. 27:275-287.
23. Wisniewski, L.A., S.E. McKeand, and R.E. Brooks (1983) Growth of tissue culture plantlets of loblolly pine in a nursery and greenhouse. Proc. 17th So. For. Tree Improv. Conf., University of Georgia, Athens, pp. 186-193.

COUNTERING THE EFFECTS OF TROPICAL DEFORESTATION WITH MODERN TECHNOLOGY

Ariel E. Lugo

Institute of Tropical Forestry
Southern Forest Experiment Station
U.S. Department of Agriculture Forest Service
P.O. Box AQ
Rio Piedras, Puerto Rico 00928

INTRODUCTION

Through history, humans have mistreated forests by clearing extensive areas without giving attention to the regeneration of new tree crops. The vast tropical lands of Central America were once deforested to support the civilizations that existed before the European discovery of America. Today, ecologists studying forests in Panama or Costa Rica find charcoal, ruins, and human artifacts in the soil profiles of, what appear to be, virgin forests. The dramatic return of forests occurred not because of human management, but because humans abandoned the lands.

Today, the destruction of forests by humans occurs indirectly. This new threat is acid rain, an insidious stress that could change the economic importance, structure, species composition, and productivity of vast areas of forest lands. Bormann (3) outlined the pattern of decline of New England forests resulting from acid rain (Tab. 1). He discussed evidence that indicates that New England forests are at the threshold of stage III, implying the beginning of severe damage to the ecosystem, loss of productivity, and less forest control over the biogeochemistry of the region.

So far, I have only mentioned some of the negative impacts that humans have had over forests. There are many positive effects as well that are implemented by forest managers. With the development of new technologies, such as tissue culture, the ability of humans to mitigate global damage to forests has increased considerably. In this article, I address the problem of forest destruction, briefly

Tab. 1. Stages of decline of forest ecosystems under pollution stress (3).

Stage	Level of pollution	Severity of impact
0	insignificant	pristine
I	low level	relatively unaffected
IIA	inimical to sensitive organisms	possible changes in competitive ability of sensitive species
IIB	increased pollution stress	resistant species substitute for sensitive ones
IIIA	severe pollution stress	large scale changes in original ecosystem
IIIB	very heavy pollution stress	completely degraded ecosystem

discuss some of its serious consequences, and suggest how techniques of tissue culture could be used in concert with other silvicultural techniques to mitigate the human impact.

The tropics contain about one half of the world's forests, store a significant portion of the biosphere's biomass, and supply a considerable portion of the global demand for forest products. The main product of these forests is biomass energy for millions of people who have no energy sources other than wood. Tropical forests harbor the majority of genetic resources of the world and it is feared that significant losses of tropical forests will lead to the massive extinction of species. In addition, significant changes in tropical biota could be accompanied by changes in the earth's climate as well. Of the 2 main threats to forests (acid rain and deforestation), tropical deforestation is the most serious and will be the focus of this article. It is unlikely that acid rain will soon become an important issue in these lands (assuming that acid rain clouds from intensively developed areas don't cause extensive damage to tropical forests).

DEFORESTATION IN THE TROPICS

The rate and consequences of deforestation in the tropics are controversial issues because very little information is available; what is available is interpreted in different ways by different people, and the subject itself is an emotional one because it presents a dilemma to our civilization (5). On the one hand, almost 2 billion people are in need of food, shelter, and the means of survival. They can partially mitigate these needs through clearing forest land. On the other hand, satisfaction of these needs involves possible destruction of conditions that support life on the planet (by changing the habitat and climate and reducing the diversity of genetic resources) without a guarantee that the needs of the needy will be satisfied sustainably. This dilemma motivates reactions that range from those who suggest that there is no problem that technology cannot solve or that its impacts cannot be tolerated by natural systems (12), to the other extreme, that predicts the loss of most tropical forests by the year 2000 and the potential extinction of over one million species (1).

The best available assessment of the rates of tropical deforestation (8) suggests that deforestation in the tropics is indeed a formidable problem facing the world. Arguments to the contrary are not supported by fact or by the condition of millions of hectares of degraded land that are products of uncontrolled human activity. Worse yet, the 1980 rate of clearing of 11.3 million ha/yr is predicted to increase in the near future. However, the data also show that lands supporting secondary forest (240 million ha) are increasing in area, which suggests some forest recovery after abandonment (8).

Current rates of deforestation, even if extrapolated into the future, will not destroy most tropical forests by the year 2000. Countries with the largest areas of tropical forests (e.g., Brazil, Zaire, and Indonesia) exhibit low rates of deforestation, and their rates are not likely to increase significantly in the next few decades (8). In countries with small areas of tropical forests, however, rates of deforestation can be very high and are already at critical stages (e.g., Nepal, Costa Rica, Ivory Coast, and Haiti). Small countries with serious deforestation problems outnumber large ones with small rates. However, the large countries, because of the size of their forest resource, exert control on the global rate of forest loss. There should be concern for the loss of forests in small countries and the human suffering this entails; however, conditions in one country cannot be extrapolated to another nor to the global scale without taking extreme care. In most cases, the extrapolation would be erroneous.

EFFECTS OF TROPICAL DEFORESTATION

One major effect of tropical deforestation is the loss of genetic material in the form of species extinctions. This is a problem that now concerns most biologists and should also concern all people. The important question on this issue is the relationship between rate of deforestation and the rate of species extinction.

If the rate of deforestation is not as high as assumed in Barney (1), who made predictions of massive extinctions (up to 1.8 million species) by the year 2000, the actual magnitude of species extinctions may be lower. But how much lower?

Our understanding of species extinctions is not adequate enough to tell what the relationship is between loss of forest areas and loss of species. This is an area of biology that requires considerable research attention. Experience in the island of Puerto Rico, where close to 85% of the land area was deforested at one time, has failed to show a direct relation between rates of species extinctions and loss of forests. To the best of our knowledge from available data, species extinctions of vertebrates and spermatophytes have been less than 10% (Wadsworth, pers. comm.). But there is no evidence to suggest that this is typical of islands or of continents.

Tropical deforestation causes increased input of CO_2 to the atmosphere and this in turn could affect the heat balance of the earth (6). Other scientists worry about the impact of deforestation on the hydrological cycle (11). Both of these areas of concern are undergoing extensive research.

THE NEED FOR SOLUTIONS

Irrespective of scientific understanding, we know from experience that the current intensity of deforestation in the tropics is not good for the long-term welfare of humans. Techniques for mitigating human impact on the biosphere are needed. Management of forests through silvicultural treatment of natural stands or establishment of tree plantations are the 2 main tools available to accomplish this goal and to provide for the future needs of humanity. Both of these techniques are fairly well developed for temperate regions but not for tropical environments. The complexity of tropical forests, remoteness and variability of tropical habitats, absence of trained foresters, lack of funds, prevalence of forest exploitation over management activities, and a poor information base are some of the factors that have retarded the development of tropical forestry. Now that forestry problems in the tropics have been recognized to have global impacts on climate, survival of species, and on social problems, it may be possible to use modern technological tools to accelerate the management of tropical forest lands (10).

A ROLE FOR TISSUE CULTURE

One technological development, tissue culture, can be used to help tropical foresters catch up with their temperate region coun-

Tab. 2. Land use significance of forest plantation production projected to AD 2000 (from Wadsworth, Ref. 13).

% of consumption from plantation	Timber production plantations	Timber production native	Other forests	Deforestation
	----------------	million km^2	----------------	
12*	0.21	6.00	0.60	9.20
50	0.86	3.40	2.87	8.87
100	1.71	0.00	5.84	8.84

* Current demand.

terparts. It is clear that this technology has important implications for tropical forestry (2). Through tissue culture techniques, it may be possible to regenerate tree species whose seeds have short periods of viability, a characteristic that is typical of many tropical primary forest trees and which prevents their use in plantation forestry. Furthermore, propagation and conservation of endangered genetic material could also be accomplished through tissue culture techniques. However, these applications will require considerable research effort to make them useful on a large-scale basis and under the usually stressful tropical conditions. A more immediate application of tissue culture techniques with global implications would be achieved in the propagation of genetically superior trees for plantation purposes.

The area of tropical tree plantations is only about 0.6% of the area of tropical forests (8). Only one ha is planted for every 10 that are deforested in the world. However, because of their high yields per unit area, these plantations can supply 12% of the projected world timber demand in the year 2000 if current plantation programs are implemented. Wadsworth (13) has calculated that 91% of the available natural forest lands will have to be used to satisfy the rest of the demands on tropical woods projected for the year 2000. This means that very few areas of natural forests (9% of the total available today) would be left for reserves and uses that are noncompatible with timber harvesting. However, if plantation establishment rates could be improved through tissue culture propagation techniques, considerable gains could be made in the fundamental problems of deforestation and species extinctions. Significant areas of forest lands can be freed for other uses, and the deforestation rate would be reduced measurably (Tab. 2). As a side benefit, new plantations would remove CO_2 from the atmosphere and help mitigate atmospheric CO_2 enrichment caused by burning of forests and fossil fuels.

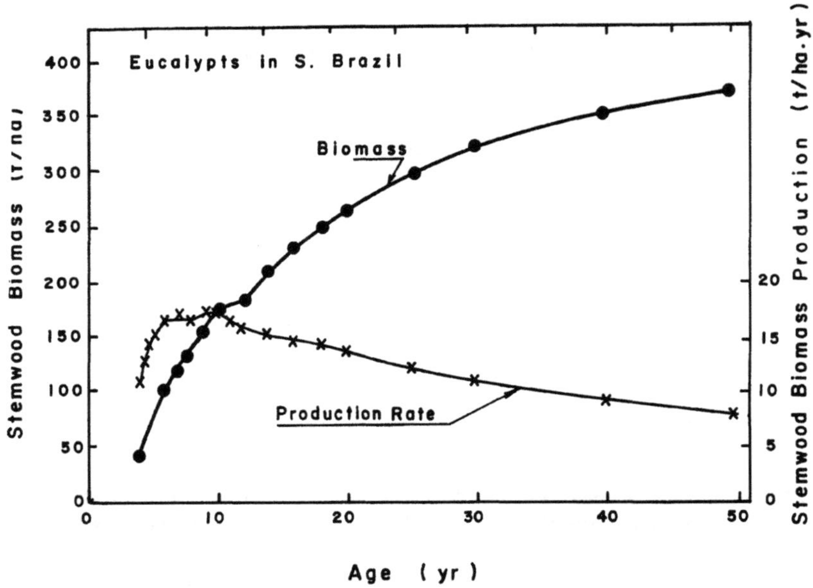

Fig. 1. Biomass dynamics of Eucalyptus plantations in southern Brazil.

If future wood demand projections hold, it is inevitable that higher wood yields will be needed from tropical lands. The best alternatives today are tree plantations or intensive management of secondary forests. Management of secondary forests is currently limited by lack of knowledge, but plantation culture techniques are better developed. In Brazil, vegetative regeneration of Eucalyptus is being used in combination with other management techniques (7) to increase stemwood yields to up to 70 m^3/ha and 5-27 t/ha yr (Fig. 1). This operation may be a harbinger of what can be done if the techniques are applied more extensively with the proper safeguards. However, high yields cannot be expected in all tropical localities. The maximum yield of the land has limits that are set by climate, soil, and genetic constraints. Though management can remove many of the constraints, there will always be productivity limits and differences among sites (4,9). Costs of management and new technology also need to be considered when projecting higher yields.

CONCLUSION

In summary, tissue culture techniques could provide impetus to massive tree planting in the tropics if techniques to propagate superior tropical broadleaf species could be developed. Such a breakthrough would have global implications because it would counter the current global trend of forest loss. Tropical tree plantations produce 10 times more wood per unit area and time than unmanaged natural forests and produce about 2 to 3 times more wood volume than temperate tree plantations. Thus, plantation establishment and management is a powerful tool for conserving natural forests and endangered genetic material, while providing alternatives to satisfy human needs. Research and development of the needed management techniques represent the most relevant subsidy that the more advanced countries can contribute towards the development of tropical countries.

REFERENCES

1. Barney, G.O. (1980) The Global 2000 Report to the President Vol. 2, Council on Environmental Quality and U.S. Department of State, Washington, D.C.
2. Bonga, J.M., and D.J. Durzan, eds. (1982) Tissue Culture in Forestry, Martinus Nijhoff/Dr. W. Junk Publishers, The Hague, 420 pp.
3. Bormann, F.H. (1982) The effects of air pollution on the New England landscape. Ambio 11(6):338-346.
4. Brown, S., and A.E. Lugo (1982) The storage and production of organic matter in tropical forests and their role in the global carbon cycle. Biotropica 14(3):161-187.
5. Brown, S., and A.E. Lugo (1985) Deforestation of tropical rain forests. In Environmental Science: A Framework for Decision Making, D.D. Chiras, ed. Benjamin Cummings, Menlo Park, California.
6. Clark, W.C. (1982) Carbon Dioxide Review: 1982. Clarendon Press, Oxford, 469 pp.
7. Ikemori, Y.K. (1976) Resultados preliminares sobre enraizamiento de estacas de Eucalyptus spp. Aracruz. Informativo Tecnico No. 2, Abril, 3 pp.
8. Lanly, J.P. (1982) Tropical forest resources. FAO Forestry Paper No. 30. FAO, Rome, Italy.
9. Lugo, A.E., R. Schmidt, and S. Brown (1981) Preliminary estimates of storage and production of stemwood and organic matter in tropical tree plantations. In IUFRO/MAB/Forest Service Symposium: Wood Production in the Neotropics via Plantations, J.L. Whitmore, ed. U.S. Man and the Biosphere Program, Washington, D.C., pp. 8-17.
10. Office of Technology Assessment (1984) Technologies to Sustain Tropical Forest Resources. OTA-F-214, Washington, D.C.
11. Salati, E., and P.B. Vose (1984) Amazon basin: A system in equilibrium. Science 225:129.
12. Simon, J.L. (1981) The Ultimate Resource. Princeton University Press, Princeton, New Jersey, 415 pp.
13. Wadsworth, F.H. (1983) Production of usable wood from tropical forests. In Tropical Rain Forest Ecosystems, A. Structure and Function, F.B. Golley, ed. Elsevier Scientific Publishing Co., Amsterdam, The Netherlands, pp. 279-288.

POSTER ABSTRACTS

(See Table of Contents for Alphabetical Listing)

SOMATIC EMBRYOGENESIS IN COMMON BERMUDAGRASS

B.J. Ahn, F.H. Huang, and J.W. King

Department of Agronomy, Horticulture, and Forestry
University of Arkansas
Fayetteville, Arkansas 72701

A culture method capable of plant regeneration is necessary to utilize tissue and cell culture techniques in bermudagrass research [Cynodon dactylon (L.) Pers.], but no plant regeneration has been reported in the literature. The objectives of this research were to: a) establish an in vitro culture system for callus induction and plant regeneration of common bermudagrass, and b) examine the mode of regeneration. Explants from young leaves were cultured on Murashige and Skoog (MS) and B_5 media supplemented with 2,4-dichlorophenoxyacetic acid (2,4-D) (2.0, 4.0 mg/l) and kinetin (0.2 mg/l). Immature inflorescences at different developmental stages (0.5, 1.0, 1.5, 2.0, and 2.5 mm long) were grown on MS and N_6 media (2.0 mg/l 2,4-D; 2 or 6% sucrose). The most leaf callus growth occurred on MS containing 2.0 mg/l 2,4-D without kinetin in the dark. However, calli induced from immature inflorescences grew better on N_6 than on MS. A higher sucrose concentration (6%) affected the compactness of embryogenic callus and frequency of embryogenic callus compared with a lower concentration (3%). Calli originating from leaf or node tissues were soft and white. These nonembryogenic calli consisted of large, tubular cells. Upon transfer to hormone-free medium, they readily produced roots without shoot formation. From 19 inflorescences less than 0.5 cm long, 16 embryogenic calli (84%) were produced, but no embryogenic callus was observed from those over 1.5 cm. Embryogenic callus on N_6 medium (2 mg/l 2,4-D; 6% sucrose) spontaneously formed somatic embryos on its surface. Somatic embryos germinated, and from each embryogenic callus numerous plantlets were produced and grown to mature plants in soil. Scanning electron microscopy further confirmed bermudagrass regeneration through embryogenesis. Two grams of embryogenic callus were put into 250-ml flasks containing 60 ml of N_6 medium (1 mg/l 2,4-D; 6% sucrose). The culture was agitated at 100 rpm on a gyratory shaker and subcultured with fresh medium every 2 weeks. A single-cell suspension with a density of 10^5 cells/ml was established. The suspension formed numerous compact calli when cultured in agar medium. These culture methods make it possible to apply cell culture techniques for modification of bermudagrass and other perennial grasses at the cellular level.

SOMATIC EMBRYOGENESIS IN CALLUS AND CELL SUSPENSION CULTURES IN THREE SPECIES OF MEDICAGO

Atanas Atanassov and Mariana Vlachova

Institute of Genetics
Bulgarian Academy of Sciences
P.O. Box 96
Sofia 1113, Bulgaria

Somatic embryogenesis is a widespread phenomenon in alfalfa callus and cell cultures. However, many problems have to be solved before this technique can be routinely applied to breeding and agriculture. In this study we tried to clarify the significance of several factors related to the improvement of somatic embryogenesis in *Medicago sativa* L. cvs. Rangelander, 74RS, and Furez; *M. varia* cv. 868 Rambler; and *M. falcata* cvs. 39, 47, and 49. To simplify the examination, only B5 modified media (B5-II) was used in the course of this study: macro- and microsalts, vitamins [Gamborg et al. (1968) *Exp. Cell Res.* 50:151-158], casein hydrolisate (500 mg/l), myoinositol (500 mg/l), FeEDTA [Murashige and Skoog (1962) *Physiol. Plant* 15:473-497], sucrose (30 g/l), L-glutathione (10 mg/l), 2,4-dichlorophenoxyacetic acid (2,4-D) (1 mg/l), kinetin (0, 2 mg/l), adenine (1 mg/l), and bactoagar (8 g/l).

Differences occurred among the investigated genotypes when cotyledons and hypocotyls were used as explant material from each cultivar. Only *M. sativa* L. cv. Rangelander was distinguished by 100% embryogenic potential. The other cultivars varied from 50 to 60% for *M. sativa* L. cv. 74RS and *M. falcata* cv. 47, 25 to 30% for *M. falcata* cvs. 39 and 49, and 6% for *M. sativa* cv. Furez. Recurrent selection was performed among the secondary regenerants as well as R_1 seeds produced by them. The regeneration capacity was increased at least twice in cultures re_established from the regenerants; once again, large variability was observed in the cultures arising from seeds. Callus cultures from the regenerants were used to investigate the significance of increased concentrations of: (a) 10 times 2,4-D, (b) 40 times adenine, and (c) 5 times L-glutathione on the base of B5-II media. Compared with the control B5-II, the developmental processes of embryo induction and, later, of plant formation were stimulated in *M. sativa* L. cvs. Rangelander and 74RS in the case of A media, in *M. falcata* cvs. 39, 47, and 49 in the case of B media, and in *M. varia* cv. 868 Rambler in the case of C media. Numerous, visually recognized embryos and embryo-like structures were formed on these media. For whole shoot development, such cultures needed to be transferred to B5 basal media plus 0, 2 mg/l benzylaminopurine (BA) or B5 plus 2 g/l yeast extract. The shoots were rooted easily on B5 basal media. Experiments were carried out on *M. falcata* cv. 47_1, a highly embryogenic cell line, to synchronize the processes of embryogenesis. It was observed that B5 base media plus 6 mM L-proline and 402 mg/l $(NH_4)_2SO_4$ can effectively promote the formation of embryos and the normalization of their growth over that of the control media B5-II. It can be concluded that a system for promoting secondary embryogenesis and development of whole plants from them in callus cultures of 6 cultivars in 3 species of genus *Medicago* has been established.

MICROPROPAGATION OF ENDANGERED NATIVE

RHODODENDRON CHAPMANII GRAY

Lee R. Barnes

Department of Ornamental Horticulture
University of Florida
Gainesville, Florida 32611

Rhododendron chapmanii Gray (Chapman's rhododendron) is a highly desirable ornamental shrub which is threatened with extinction due to limited numbers (less than 3000 plants surveyed in the wild) and distribution, habitat alteration and destruction, and illegal collection from the wild. Micropropagation using shoot-tips has produced hundreds of rooted plantlets with minimal impact on existing populations.

Decapitated shoot-tips with 5 nodes were surface sterilized, and then leaves were removed and explants were placed horizontally into the agar. Greatest multiplication occurred on Woody Plant Media amended with 50 µM (10 ppm) dimethylallylamino purine (2iP), 80 mg/l adenine sulfate, 100 mg/l myo-inositol, 100 mg/l NaH_2PO_4, 3% sucrose, and pH adjusted to 5.2.

Initial multiplication was 5.8-fold resulting in shoot-clumps from each node in 6 to 8 weeks. Clumps were divided roughly into 1-cm^2 clumps and subcultured to fresh multiplication media. Additional 6- to 8-week subcultures resulted in an average of 7.8-fold increase for 3 cycles, but declined to 5.0-fold by the fifth subculture.

Rooting occurred in 4 to 6 weeks after shoots greater than 15 mm were placed in cell-packs containing 1:1:1 screened peat:perlite:vermiculite and covered with clear plastic or a translucent flat. Use of multiple shoots (2 to 3 shoots per explant) resulted in greater survival and more uniform plantlets.

OPTIMIZATION OF ELECTROFUSION FOR THE FORMATION AND CULTURE OF PROTOPLAST HETEROKARYONS

George W. Bates

Department of Biological Sciences
Florida State University
Tallahassee, Florida 32306

Electrofusion has been demonstrated with a wide variety of plant protoplasts. Nonetheless, it has not been shown that the fusion products are viable, nor has the technique been refined to make it a practical alternative to fusion by PEG. Protoplasts derived from suspension cultures of *Nicotiana plumbaginifolia* have been fused with *N. tabacum* mesophyll protoplasts. By fusing mesophyll protoplasts with suspension culture protoplasts, hybrid cells can be visually identified in both living and fixed material. This system has been used to determine the optimal conditions for electrofusion and for development of a protocol for the isolation and culture of protoplast heterokaryons.

The production of hybrids by electrofusion has been optimized by examining the effects of various treatments using a fusion index (based on the number of nuclei per cell). The best results were obtained when the protoplasts were aligned in an AC field of 75 V/cm and fused using 2 DC pulses of 1000 V/cm, 50 µsec long. High protoplast densities were also beneficial. Comparison of a number of different types of fusion chambers (both flow-through and nonflow-through) has lead to the development of a chamber which is both simple to use and highly efficient. This chamber, which holds 0.3 to 0.5 ml, produced overall fusion rates as high as 40% (20% hybrids).

Following fusion the protoplasts were cultured by removing them from the fusion chamber and resuspending them in microdrops. Successful culture of the fusion products was achieved using a modification of Maliga's K3 medium (*Z. Pflanzenphysiol.* 78:353-355) at half strength. Development was never observed using the undiluted medium. Electrofusion was synchronous and rapid, however the suspension/mesophyll cell hybrids took about 1 hr to round up into a single spherical body. Wall regeneration was observed in 12 to 24 hr based on the reappearance of nonspherical cells in the culture. After 3 days in culture the hybrid cells became difficult to distinguish from the unfused mesophyll cells. Therefore, 24 hr after fusion the hybrid cells were manually isolated and cultured either in microdrops or in Cuprak dishes. When cultured in a conditioned medium the hybrids began to divide by day 3 to 6, and grew quickly into callus. Plant regeneration is underway.

THE CULTURE OF *ERAGROSTIS TEF* IN VITRO

Seifu M. Belay

Department of Botany
Howard University
Washington, D.C. 20059

A tissue culture system was used to determine the nutritional and hormonal requirements for the culture of *Eragrostis tef* in vitro since conventional methods of breeding gave little or no results. Calli were initiated from the nodal stem segments on Heller's medium supplemented with 2,4-dichlorophenoxyacetic acid (10 mg/l). The calli thus formed at the nodes were isolated and subcultured at 4- to 6-week intervals on either the initiating medium or on a medium containing an additional amino acid complex. Single roots, multiple roots, and plantlets were thus induced from the second and third subculture calli on Heller's medium without growth substances. Another significant finding was that shoots with roots, shoots with inflorescence only, and shoots with callus can be initiated directly from the nodal stem segment by the manipulation of the components of the media. Plants that developed directly from the nodal stem segments in vitro were isolated and grown in a greenhouse environment. A multiple-stemmed plant was formed that produced viable seeds. This technology could set *E. tef* breeders free from the historical constraints of working with normal sexual cycles, maturation, and growth characteristics.

THE EFFECTS OF PUTRESCINE ON SOMATIC EMBRYOGENESIS OF DAUCUS CAROTA AS EXAMINED BY TWO-DIMENSIONAL ELECTROPHORESIS

Peter M. Bradley, Fawzi El-Fiki, and Kenneth L. Giles

Department of Biology and Biotechnology
Worcester Polytechnic Institute
Worcester, Massachusetts 01609

Plant developmental processes can be studied using tissue culture models, e.g., the production of somatic embryos using cell suspension cultures. In this study, a putrescine/arginine treatment was used to synchronize the embryogenic process, and two-dimensional electrophoresis was used to compare the translational profiles at different stages of embryo formation [P.M. Bradley (1984) Ph.D. Thesis, Worcester Polytechnic Institute].

Wild carrot cells were cultured in B5 medium containing 2,4-dichlorophenoxyacetic acid (2,4-D). On transfer to medium lacking 2,4-D, embryogenesis occurred asynchronously to give mature embryos (control cultures). When cells were treated with 2,4-D plus 0.03 μM putrescine, embryogenesis proceeded only to the globular stage. No further development was obtained when the globular embryos were transferred to medium containing L-arginine. However, when they were passed to medium lacking L-arginine, mature embryos were obtained in a more rapid and better synchronized manner than with the control cultures [P.M. Bradley et al. (1984) Plant Sci. Lett. 34:397-401].

At each stage, cells were harvested and ground with sand in cold buffer to extract cellular protein. The protein was precipitated with cold acetone and subjected to electrophoresis by the NEpHGE method (first dimension). Electrophoretic separation in the second dimension was in 10 to 20% polyacrylamide-SDS linear gradient gels. The appearance or disappearance of several polypeptides at different stages of embryogenesis was seen in the gels following electrophoresis and staining the gels with silver salts. This approach could be used to dissect the embryogenic process. Future immunocytochemical work might identify specific proteins on the gels, and it may be possible to relate important biochemical control points to the electrophoretic data.

Polyamines (e.g., putrescine) and their associated enzymes are important in the control of growth and development in plants. L-arginine is a precursor for putrescine, and polyamines can block the conversion of methionine to ethylene. This suggests a control system in which ethylene, arginine, and polyamines play an interrelated role in the control of somatic embryogenesis.

Polyamine treatments may be useful for the long-term storage of cells or embryos. The putrescine/arginine treatment may lead to a synchronized embryogenic system for further biochemical studies, or to provide embryos for large-scale propagation purposes.

RAPID PROPAGATION OF BOSTON FERNS BY TISSUE CULTURE

James D. Caponetti and Thomas E. Byrne

Department of Botany
The University of Tennessee
Knoxville, Tennessee 37996-1100

Boston Fern and several of its cultivars are popular, ornamental, home and office plants. During the 1970s, about 2 million Boston Ferns were sold each year in the United States. Most of the production at that time was by one-for-one runner tip multiplication in the greenhouse. This method of propagation yielded low numbers of plantlets per unit time. Boston Ferns can be propagated rapidly in large numbers by shoot multiplication in sterile tissue culture. At the present time, the most common method of shoot multiplication of Boston Ferns in several nurseries is directly from shoot tissues. This method has been refined in Fishtail Fern [M.J. Beck and J.D. Caponetti (1983) Amer. J. Bot. 70:1-7].

Runner tips are placed in culture on Murashige Fern Multiplication Medium (MFMM). Four weeks later about 3 shoots develop from buds. These are separated and placed back in culture singly on MFMM supplemented with kinetin and naphthaleneacetic acid (NAA). The most practical shoot production was achieved on MFMM containing kinetin between 5×10^{-6} and 5×10^{-5} M, and NAA at 10^{-7} M. About 4 weeks later, 6 to 10 sizeable shoots develop. The separated shoots can be further multiplied in large flasks and the excess shoots can be placed in the greenhouse under mist propagation for further growth and root development to saleable size. Given the proper work force and space, several thousand shoots can be produced each week. In order to reduce the handling of cultures and thus help to reduce costs, 2 other tissue culture propagation methods for Boston Ferns are under investigation in our laboratory. One is multiple shoots from shoot fragments and the other is multiple shoots from shoot callus.

For multiple shoots from shoot fragments in Boston and Fishtail Ferns, runner tips are allowed to develop numerous shoots and runners which fill the tube in about 8 weeks. The whole mass is then reduced to small fragments in a blender with liquid MFMM. Small aliquots of the fragment suspension are transferred to large flasks containing fresh MFMM solidified with agar. Numerous shoots are produced in about 4 to 6 weeks, and then transferred to the greenhouse for further development.

Stolon tips from Boston Fern and 2 of its cultivars, 'scotti' and 'dwarf boston', can be induced to form callus on MFMM modified by the addition of 2,4-dichlorophenoxyacetic acid (2,4-D). Callusing occurs in apical and lateral buds of stolon tips in 4 to 6 weeks. Basal medium with 3% sucrose and 0.5 mg/1 2,4-D produces the greatest increase in fresh weight of subcultured callus in all cultivars. Organogenesis into shoots and roots from callus occurs on basal medium with combinations of 5×10^{-7} M kinetin plus 5×10^{-7} M NAA, 10^{-6} M kinetin plus 10^{-5} M NAA, and 10^{-6} M kinetin plus 5×10^{-7} M NAA in 12 weeks. Attempts to induce organogenesis with indoleacetic acid (IAA) and dimethylallylamino purine (2iP) produce only roots at all hormone levels tested.

Viable suspension cultures are produced from stolon callus of all cultivars employing a liquid basal medium containing 0.5 mg/1 2,4-D. Callus clones from single cells in suspension cultures are obtained by the Bergmann Technique [L. Bergmann (1960) J. Gen. Physiol. 43:841-851]. The clones are transferred to organogenic medium and form complete new plants.

IN VITRO CULTURE AND SOMATIC EMBRYOGENESIS IN *CUCUMIS* SPECIES

B.J. Disney,[1] N.D. Camper,[1] and G. Fassuliotis[2]

[1]Department of Plant Pathology and Physiology
Clemson University
Clemson, South Carolina 29631

[2]U.S. Department of Agriculture
Agricultural Research Service
Vegetable Breeding Lab
Charleston, South Carolina 29407

Cucumber (*Cucumis sativus*) and cantaloupe (*Cucumis melo*) have been reported to form somatic embryos in vitro. The objective of this study was to establish a protocol for the in vitro culture of several *Cucumis* spp. and for the induction of embryogenic callus and suspension cultures. In addition to cucumber and cantaloupe, *C. metuliferus* and *C. anguria* were included in this study. Leaf discs excised from the first leaf of young plants were placed on Murashige and Skoog (MS) medium supplemented with naphthaleneacetic acid (NAA) and benzylaminopurine (BA) at 1 mg/l. Cultures were incubated at 30 ± 2°C with a 16-hr photoperiod. At 2 to 3 weeks, callus which proliferated from these explants was placed on media containing a range of NAA and BA concentrations (10^{-7} to 10^{-3} M). Optimum callus production and variations in callus morphology, color, and texture were determined. Dense meristematic zones developed under certain regimes. Cucumber seeds were germinated aseptically and various explants investigated for callus production and regenerative potential on the callus induction medium. Cotyledon explants regenerated directly with low frequency. Suspension cultures initiated from 5-week-old, root-derived callus developed abnormal embryoids and plantlets after 3 months without a transfer.

EMBRYOGENESIS AND PLANT REGENERATION IN TISSUE CULTURES DERIVED FROM MATURE AND IMMATURE SOMATIC TISSUES OF RICE (*ORYZA SATIVA*) AND WHEAT (*TRITICUM AESTIVUM*)

Thomas A. Dykes and Murray W. Nabors

Department of Botany
Colorado State University
Fort Collins, Colorado 80523

Immature embryos at various stages of development, whole excised mature embryos, detached scutella of mature embryos, whole seeds, and shoot apical tissue of rice and wheat were cultured on various callus-inducing media. In rice, embryogenic callus was formed from the scutellum of immature embryos at all stages of development examined, as well as from the scutellum of physiologically mature embryos. Rice embryogenic callus produced well-defined somatic embryos and exhibited a high capacity for regeneration. The regeneration potential of embryogenic callus derived from the scutellum of mature embryos did not differ significantly from that of embryogenic callus derived from immature scutella. In wheat, callus produced from the scutellum of immature embryos decreased with increasing size of the embryo. No callus was formed from physiologically mature wheat scutella. Initiation of embryogenic callus in both rice and wheat was more dependent on the source of explant material than on the initiation medium. The ability of mature rice scutella to produce embryogenic callus is a demonstration in the Gramineae that some highly differentiated cells, that have ceased to divide in vivo, can remain competent to form somatic embryos and subsequently regenerate plants in vitro.

SOMATIC EMBRYOGENESIS FROM COMMERCIALLY

IMPORTANT MAIZE INBREDS

J.W. Fahey, J.N. Reed, T.L. Readdy,
and G.M. Pace

Crop Science Laboratory
Allied Corporation
P.O. Box 6
Solvay, New York 13209

Preliminary studies conducted in the summer of 1983 suggested that somatic embryogenesis could be achieved from certain field-grown, commercially important inbred lines of maize, as well as from crosses of these lines with A188. Three of these genotypes (B73, Mo17, and LH38) were subsequently selected for further evaluation in 1984, using greenhouse-grown donor plants and up to 8 different media. Immature embryos (ie's) were excised at 10 to 17 days post pollination and handled according to procedures developed by C.E. Green [C.E. Green and C.A. Rhodes (1982) Plant regeneration in tissue culture of maize. In Maize for Biological Research, W. Sheridan, ed. Plant Molecular Biology Assn., Charlottesville, Virginia].

Over 1500 ie's of B73 and 2900 ie's of LH38 were plated, but due to poor pollen shed, only 400 ie's were plated from Mo17. Most of the ie's plated produced callus. Overall response, reported as a percentage of the ie's plated that developed embryogenic callus, was 2.1%, 1.6%, and 26% for LH38, B73, and Mo17, respectively. Best response on a given medium for each of these genotypes was 9.2% (LH38), 4.4% (B73), and 100% (Mo17). Medium for the former 2 genotypes was adapted from Potrykus et al. [I. Potrykus et al. (1979) Theor. and Appl. Genet. 54:209-214], and for the latter, from Yu-pei medium [M.A. Rapela (1984) Maize Genet. Cooperative Newsletter 58:106-108]. Plantlets continue to be regenerated from selected treatments.

The anther culture literature indicates that temperature shock can have a dramatic effect upon embryogenesis. Heat shock experiments were therefore performed, but have thus far yielded equivocal results. Some crosses of inbred lines with A188 were evaluated for response to heat shock (2 hr at 40°C). Heat shock produced a dramatic increase in the proliferation of embryogenic callus and increased plantlet production from these genotypes. In contrast, quantitative experiments performed with BMSxA188 showed that treatment for 2 hr at 40°C actually caused a reduced number of regenerating calli (18%) compared to nonheat shock controls (36%). In subsequent studies, greenhouse-grown LH38 also failed to respond to heat shock. Treated ie's in that experiment produced the same level of response as the controls (ca. 10%).

Other parameters examined for their effects on production of embryogenic callus included plant age, embryo age (days post pollination), self vs. sib pollination, ear ranking (1st, 2nd, or 3rd ear), and transfer of nonresponding callus to responsive media.

PLANT REGENERATION FROM EMBRYOS OF *CUCUMIS METULIFERUS* CROSS-POLLINATED WITH *CUCUMIS ANGURIA*

George Fassuliotis and B.V. Nelson

U.S. Vegetable Laboratory
2875 Savannah Highway
Charleston, South Carolina 29407

The African-horned cucumber (*Cucumis metuliferus* Naud.) and the West India gherkin (*C. anguria* L.), both carrying resistance to root-knot nematodes (*Meloidogyne* spp.), were cross-pollinated successfully with *C. metuliferus* as the female parent. Embryo development was abnormal and arrested in the "rabbit-ear" stage. Occasionally, mature embryos were recovered, but they did not germinate on nutrient media tested. Presumptive hybrid plants were cultured on a basal medium containing Murashige and Skoog (MS) salts, vitamins, m-inositol (100 mg/l), casein hydrolysate (1 g/l), and sucrose (30 g/l), supplemented with indole-3-acetic acid (IAA) (5 and 10 μM) and N^6-benzyladenine (BA) (5 and 10 μM), and solidified with purified agar (7 g/l). Cultures were incubated at 25°C in the dark for 7 to 14 days, transferred to 16 hr of low light (50 fc) for 5 days, and then transferred to high light (750 fc). Embryoids and leaflets differentiated on the hybrid embryo cotyledons on media containing IAA/BA (5/10, and 10/5 μM). Leaflets from 10/5 μM IAA/BA were converted to whole plants by 2 successive subcultures on 1/3 MS and 1% sucrose. The hybrid plants were abnormal and contained flattened stems with short internodes surrounded by multiple small leaves. They were chlorophyll-deficient and did not grow to maturity after transplanting to a soil mix. Further work is in progress to increase the efficiency of converting hybrid embryos to plants via plant regeneration of embryos in culture.

CHARACTERIZATION OF CHLORATE RESISTANT MUTANTS OF *ARABIDOPSIS THALIANA* IN TISSUE CULTURES

K.A. Feldmann and R.L. Scholl

Department of Genetics
The Ohio State University
1735 Neil Avenue
Columbus, Ohio 43210

Mutagen-treated populations of *Arabidopsis thaliana* were screened for chlorate resistance to obtain new lines suitable for the study of nitrate metabolism and for use as markers in tissue culture studies. Plants and cultured tissues of the resistant mutant lines were characterized biochemically and physiologically.

Seeds from the race WS of *A. thaliana* were treated with 40mM EMS for 8 hours. The M_2 seeds were collected and tested for resistance to chlorate. Fourteen independently isolated mutants [8 putative structural or regulatory (sr) mutants and 6 putative uptake mutants) were selected from a total of 75 chlorate resistant lines for complete genetic and physiological characterization. These have been characterized for growth, nitrate and chlorate uptake, nitrate reductase (NRA), nitrite reductase (NiRA), and glutamine synthetase activity (GSA) on 3 nitrogen sources (NH_4NO_3, NH_4, and NO_3).

Enzyme activity comparisons suggest that reduced NRA is the cause of chlorate resistance in a number of mutants. The relative NRA of mutants in tissue cultures is similar to that in whole plants. In tissue culture 7 mutants had less than 30% of the NRA of WS. One mutant had less than 10% and another less than 3% of WS activity, both in whole plants and tissue culture. This places these in the same category as one line previously characterized as a nitrate reductase-less mutant of *A. thaliana*.

The NR-deficient mutants all possessed elevated NiRA, and grew either poorly or not at all as whole plants or in tissue culture on medium containing nitrate as the sole nitrogen source. The comparative growth rates of different lines on different nitrogen sources in whole plants and tissue culture is discussed.

A second group of chlorate-resistant mutants possessed normal NRA but had altered nitrate and/or chlorate uptake. The NRA of the putative uptake mutants was greater than 50% of WS in all lines. Two of these mutants took up less NO_3 and ClO_3 than WS. They appear to be phenotypically similar to a previously reported uptake line. Another mutant took up nitrate normally but excluded chlorate. The comparative kinetics of uptake in the 2 types are being examined in both whole plants and tissue culture.

These mutants are being further characterized in callus and plants. Relative chlorate resistance in tissue culture, inheritance and linkage relationships, and the biochemical nature of the lesions are being examined.

RAPID IN VITRO PROPAGATION OF VIRUS-INDEXED FREESIA

M.J. Foxe, A.E. Logan, and F.W. Zettler

Department of Plant Pathology
University of Florida
Gainesville, Florida 32611

An efficient tissue culture method has been developed for the production of more than 50,000 virus-free gladiolus transplants within 30 weeks [A.E. Logan and F.W. Zettler (1984) Acta Horticulturae (in press)]. This method has been used, with modifications, to successfully establish a rapid in vitro propagation system for virus-indexed freesia (*Freesia refracta*). Using *Freesia refracta* cvs. Ballerina and Rose Marie, apical shoot-tips (0.5 to 0.7 mm) were established (Stage I) on an agar medium containing Murashige and Skoog salts, 2.0 mg/l kinetin, and 0.1 mg/l naphthaleneacetic acid (NAA). Plantlets were transferred to a similar medium containing 4.0 mg/l kinetin and no auxin 6 to 7 weeks later for shoot proliferation (Stage II). Plantlets were subcultured on Stage II medium every 4 to 5 weeks as needed and yielded approximately 12 axillary shoots per culture. Roots developed readily in this medium suggesting that it may not be necessary to transfer the shoots to a specific root-inducing medium as is the case with gladioli. Virus indexing was accomplished using several methods, including electron microscopy and enzyme-linked immunosorbent assay.

STABILITY OF NUCLEAR DNA CONTENT IN LONG-TERM CALLUS CULTURES OF LOBLOLLY PINE

Chandrasekaran I. Franklin, Ralph L. Mott, and Jerome P. Miksche

Department of Botany
North Carolina State University
Raleigh, North Carolina 27650

Callus cultures from embryo and stem explants of loblolly pine (*Pinus taeda* L.), maintained in culture on 3 different basal media for more than one year, were screened for their nuclear DNA content using feulgen cytophotometry.

The nuclear DNA content corresponding to the 2c condition was predominant (about 80% of nuclei examined), and ploidy levels higher than 4c were totally absent in all the calli studied. No significant differences in ploidy levels were observed among calli derived from stem or embryo, or from those grown on different media. These results indicate a stable 2c condition in long-term loblolly pine calli, irrespective of tissue source or the media on which they are grown.

IN VITRO CULTURE OF *CAMELLIA SINENSIS*

Carl H. Frisch and N.D. Camper

Department of Plant Pathology and Physiology
Clemson University
Clemson, South Carolina 29631

Camellia sinensis (tea) has played an important role in the development of world trade and politics. However, compared to other woody crops such as coffee and citrus, little in vitro technology has been applied to tea. Research reported to date showed slow callus growth from tea explants and contamination problems which reduce the number of usable aseptic cultures. The objective of this study was to establish an appropriate explant and in vitro culture protocol for tea.

Disinfestation of the plant tissue was carried out with 7.5% sodium hypochlorite and 7.5% $CaCl_2$. Explants (divots) were cut from green-wood stem internodes excluding as much xylem as possible. The salt base media [Murashige and Skoog (MS) and B5] were compared. Cultures were grown in the dark and transferred at the end of the fifth week. Observations were recorded at the end of the tenth week. No differences were observed between the 2 media; MS modified with sucrose (30 g/l), inositol (100 mg/l), and thiamine-HCl (1.3 mg/l) was selected for the subsequent experiments.

With a constant kinetin concentration (10^{-5} M), synthetic auxins 2,4-dichlorophenoxyacetic acid (2,4-D), 2,4,5-trichlorophenoxyacetic acid (2,4,5-T), naphthaleneacetic acid (NAA), and 4-amino-3,5,6-trichloropicolinic acid [Tordon (Picloram)] were compared individually and in a series of concentrations (0,10^{-8} to 10^{-4} M). Concentrations which induced the most callus were: 2,4-D (10^{-7} M), 2,4,5-T (10^{-8} and 10^{-7} M), NAA (10^{-7}, 10^{-6}, and 10^{-5} M) and Tordon (10^{-7} M). Analysis of growth at the end of the tenth week showed that Tordon (10^{-7} M) resulted in the greatest callus development. Using the modified MS salt base, all ratio combinations of 2,4,5-T (0,10^{-9} to 10^{-5} M) and kinetin (0,10^{-8} to 10^{-4} M) were tested. The best callus growth was obtained with 10^{-7} M 2,4,5-T over all levels of kinetin tested. Kinetin at 10^{-5} M induced the most callus growth with all combinations of 2,4,5-T. Consequently, the most productive ratio was 2,4,5-T (10^{-7} M)/kinetin (10^{-5} M). On this ratio after 12 weeks, roots differentiated from callus growing on divots isolated from stem tissue collected in the spring. The divot technique could be applied to further in vitro research of tea or other woody plants.

TISSUE CULTURE OF *PINUS ELDARICA* (MED.)

Heather G. Herrera and Gregory C. Phillips

Department of Horticulture
and Plant Genetic Engineering Laboratory
New Mexico State University
Las Cruces, New Mexico 88003-3530

Afghan pine [*Pinus eldarica* (Med.)] was introduced to the United States in the mid-1960s from the Near East (Russia to Pakistan), and is of particular interest in the arid southwestern United States. This pine is very vigorous and tolerant of drought and alkaline soils. It is useful for Christmas tree production, ornamental landscaping, and windbreaks, with potential for fuelwood and pulpwood production. Exploitation is hindered by limited seed availability. Tissue culture techniques offer alternatives for rapid propagation of this important pine.

Hypocotyls, intact cotyledons/shoot-tips (excised from 1- to 2-month-old seedlings), apical shoot-tips, and needle fascicles (excised from 6-month-old seedlings and mature trees) were used to initiate tissue cultures. All were incubated at 25°C with a 16-hr photoperiod. Five contrasting basal nutrient media were evaluated for supporting growth of these cultures. The best basal medium was a modified, half-strength Murashige and Skoog. Three carbon sources were evaluated, glucose and sucrose proving to be superior to fructose over 4 monthly passages. Direct shoot development and multiplication were observed in 46% of the 1- to 2-month-old cotyledon/shoot-tip explants after one month. This response was maintained for at least 7 monthly passages. Only 14% of the cultures had shoots at the end of the seventh month, with as many as 14 and as few as 2 shoots per explanted line. Direct shooting from mature tree explants was observed much less frequently, but was achieved from a needle fascicle collected during a spring growth flush. Combinations of indole-3-butyric acid (IBA) at 0.05 to1.0 mg/l with kinetin at 0.1 to 1.0 mg/l elicited the best shooting responses. Other combinations, including 4-amino-3,5,6-trichloropicolinic acid (picloram), 1-naphthaleneacetic acid (NAA), and/or 6-benzylaminopurine (BA), proved less useful for shooting. Callus cultures were obtained by using NAA or IBA with BA at 0.5 to 1.0 mg/l each. Suboptimal balances of growth regulators in a shoot multiplication system also encouraged callus formation and suppressed bud development. Transfer to hormone-free medium stimulated bud development. Unorganized callus regenerated shoots upon transfer to shooting-type and/or hormone-free media. Cultures maintained as long as 14 months as unorganized callus regenerated shoots from 12% of the inocula, and another 24% formed bud-like structures. Callus derived from dormant mature tree tissue has produced bud-like structures.

Additional research needs to evaluate rooting treatments and establishment of tissue culture-derived plants. Juvenile explants establish in vitro cultures readily and demonstrate potential for micropropagation, which, if developed for mature tree explants, would extend the technique. Plants can be regenerated from callus, which further enhances micropropagation potential and permits cellular approaches to genetic improvement.

SELECTION AND MICROPROPAGATION OF *EUCALYPTUS* CLONING CANDIDATES

Gary P. Howland,[1] George Meskimen,[2]* and Milton J. Constantin[3]

[1]Clonal Products, Inc.
1056 Captiva Point
Lakeland, Florida 33801

[2]Forest Research Consultant
4459 Riverside Dr., S.E.
Ft. Myers, Florida 33905

[3]Phyton Technologies, Inc.
7327 Oak Ridge Highway
Knoxville, Tennessee 37931

Eucalyptus species demonstrating spectacularly rapid growth, adaptability to diverse sites, genetic plasticity, and coppice regeneration are prime candidates for afforestation of vast lands on short development schedules. An operational eucalyptus technology exists in south Florida, based upon commercial planting which began in 1972. This work was developed by the U.S. Forest Service researchers working cooperatively with state, industrial, and private interests. Over a period of 10 years, 8.8 million seedlings of *Eucalyptus grandis* Hill ex. Maiden have been planted on 6,475 hectares in southwest Florida. Every outplanted seedling was grown from genetically improved seed collected from local seed orchards.

In 1982, workers at the U.S. Forest Service laboratory at Lehigh Acres selected 451 cloning candidates scattered over 40 sites in 17 counties in central and southwest Florida. Desirable cloning candidates demonstrate superior growth and form, ability to coppice

---

* Eucalyptus Tree Breeder with the U.S. Forest Service at Ft. Myers, Florida through September 1984.

from the stump after harvest, and resiliency following freeze damage. Ramets were produced by girdling these select ortets to induce juvenile, basal sprouts which were sectioned into cuttings and rooted in mist beds. In addition to *E. grandis*, clonal sections of *E. robusta*, *E. camaldulensis*, *E. rudis*, *E. tereticornis*, and *E. saligna* x *tereticornis* are also under evaluation. A total of 244 clones are under test at one or more of 5 locations planted in central and southwest Florida. The additional 207 selections that were girdled failed to produce rooted cuttings for field testing.

It is clear to us that only cloning can bring commercial eucalyptus forestry to central Florida within the next decade. The requisite combinations of cold tolerance, growth rate, and site adaptation are too rare and elusive in seedling populations to support operational planting.

Following the severe freeze of Christmas 1983, ortets produced juvenile, epicormic sprouts. These sprouts have provided explants for the establishment of tissue cultures. Many of the sprouts had matured past the optimal juvenile physiology by the time we collected them (May 10; 138 days after the freeze). Still, two-thirds of the ortets provided explants that were capable of development in culture.

Nodal explants usually developed preventitious sprouts within a week following introduction into culture. Transfer of these in vitro sprouts to an appropriate medium resulted in rapid shoot multiplication. Work on defining the clone-specific culture conditions is continuing. Multiplication rates of 10-fold per month have been obtained. In the absence of outside funding for this work, only 10 clones have been cultured, but hundreds of worthy cloning candidates have been identified, while at least a 1,000 more are there for the finding in commercial plantations. Ramets of captured clones will be available for further field evaluations and pilot projects.

IN VITRO REGENERATION OF PLANTLETS FROM CULTURED TISSUES OF SCOTS PINE (*PINUS SYLVESTRIS* L.)

Feng H. Huang and Sonia Tsai

Department of Horticulture and Forestry
University of Arkansas
Fayetteville, Arkansas 72701

After germination and 4 weeks of culture on a modified Murashige and Skoog (MS) medium, apical slices from Scots pine seedlings, which included the base of the cotyledonary whorl subtended by a stub of hypocotyl, were established in culture. Adventitious buds developing on these apical slices were further multiplied at 1- to 2-month intervals. Two plantlets were regenerated when elongated adventitious shoots from these cultures were transferred to a rooting medium. Greater difficulty was encountered when ungerminated embryos were cultured; nevertheless, one plantlet was regenerated using this starting material. The 3 regenerated plantlets were transferred to vermiculite and maintained at high humidity for 2 months, during which time further shoot growth occurred.

EFFECT OF EXPLANT TYPE ON SHOOT PROLIFERATION AND PHYSICAL SUPPORT ON ROOT INITIATION FOR A RANGE OF HORTICULTURAL SPECIES

James F. Hutchinson

Department of Agriculture
Horticultural Research Institute
P.O. Box 174
Ferntree Gully 3156
Victoria, Australia

The use of shoot-tip culture is well-established for the rapid propagation of many species. Proliferating shoot cultures generally fall into one of 2 types: those that form a mass of shoots from the base, e.g., gerbera and venus fly trap, and those with a branched structure, e.g., apple, cherry, and rose. The former are subcultured by dividing the shoot clumps, and the latter by removing the shoot-tips and often discarding the remainder.

Detailed studies with proliferating shoot cultures of the apple cultivar 'Northern Spy' have shown that shoot-tips are not the best type of explant to use for routine maintenance of cultures [Hutchinson (1984) Scientia Hortic. 22:347–358]. Additional research has been done using proliferating shoot cultures of the cherry rootstock F12/1, the pear rootstock D-6, and the hybrid tea rose Fraser McClay. With F12/1, nodal explants produced twice as many shoots and basal mass explants produced 3 times as many shoots as shoot-tip explants, whereas with D-6, basal mass explants produced more than twice as many shoots as shoot-tips. However, with Fraser McClay, neither nodal or basal mass explants were as good as shoot-tips. While basal mass explants are, with some species, a very productive source of shoots, their use should be treated with caution until more is known of the origin of shoots.

Root initiation is usually achieved by placing shoot-tips on an agar-gelled medium. While agar medium is easy to prepare, agar is variable and different brands can result in different responses. With 'Northern Spy,' F12/1, and D-6 shoot-tips, a range of alternative physical supports such as coarse sand, perlite, and liquid rotated medium were compared with agar. Percentage root initiation was similar with all physical supports, but the number and length of roots were greater with agar alternatives. While longer roots may not be an advantage in the survival of plants once transferred to potting medium, the presence of a greater number may be.

The use of alternatives to shoot-tips for shoot proliferation has the advantage that all shoot-tips can be used for root initiation and other explant types can be used for the routine maintenance of cultures. Alternatives to agar for root initiation overcome the variation in agar quality and allow for some control over the number and length of roots.

ORGAN FORMATION FROM CALLUS CULTURES OF BANANAS AND PLANTAINS

R.L. Jarret, R.E. Litz, and J. Fisher*

Institute of Food and Agricultural Sciences
University of Florida
Tropical Research and Education Center
18905 S.W. 280th Street
Homestead, Florida 33031

Spherical swellings which resembled globular embryos developed on the surface and within leaf base explants of banana (Musa acuminata Colla. cvs. Dwarf Cavendish and Valery), plantain (Musa x paradisiaca L. cvs. Mysore and Apple), and bluggoe-type cooking banana (Musa x paradisiaca L. cv. Chato) when cultured without illumination for 30 to 60 days on a basal medium containing Murashige and Skoog (MS) salts, i-inositol (100 mg/l), thiamine-HCl (0.4 mg/l), sucrose (30 g/l), supplemented with 0.1% (w/v) activated charcoal, various concentrations of 2-methoxy-3,6-dichlorobenzoic acid (dicamba), and solidified with 0.3% (w/v) agar. Inclusion of kinetin or N^6-benzyladenine (BA) in addition to dicamba during culture initiation suppressed the formation of these structures. Transfer of leaf base explants to hormone-free basal medium or to MS with a lowered concentration of dicamba resulted in the rapid enlargement and elongation of the globular masses. Small pieces of callus which were shaken free after transfer to liquid medium continued to divide, and eventually organized into discrete structures resembling those which arose from the globular swellings. In both instances the structures closely resembled those described elsewhere as "somatic embryos." However, shoot formation from these structures was never observed. Root elongation occurred. Histological examination revealed these structures to be unipolar and without typical embryo anatomy. Swellings in the basal area, presumed to be nascent shoot axes, were identified as secondary root axes.

---

* Fairchild Tropical Garden, Miami, Florida 33158

SEASONAL VARIATION IN CALLUS PROLIFERATION FROM EXPLANTS OF MATURE *PINUS STROBUS* TREES

Karan Kaul*

CRS Plant and Soil Science Research
Kentucky State University
Frankfort, Kentucky 40601

In vitro propagation from explants of mature trees must be achieved before plant tissue culture technique can become a viable method for mass propagation of superior forest trees. One possible route for in vitro mass propagation is differentiation of plantlets from callus cultures. As the first step towards achieving this goal, experiments were done to establish callus cultures from shoots of 15- to 18-year-old *Pinus strobus* (white pine) trees. The study was started in the spring of 1982 just after the end of dormancy of winter buds. Young shoots were collected early in the morning and were surface sterilized by treatment with 2.625% sodium hypochlorite for 15 min, followed by 70% ethanol for 2 min and 2.625% sodium hypochlorite for an additional 5 min. The surface-sterilized shoots were thoroughly washed with sterilized double-distilled water and cut into approximately 5 mm long segments, which were put on nutrient media. Cultures were incubated at 26 ± 1°C under 450-500 $\mu W/cm^2$ cool white fluorescent illumination. Day length was 16 hr. Shoot samples were collected at 2-week intervals during May and June, and at monthly intervals during the rest of the year. Best callus proliferation occurred from segments of shoots collected in early spring. A modified Murashige and Skoog medium supplemented with 0.2 mg/l α-naphthaleneacetic acid (NAA) and 2 mg/l N^6-benzyladenine (BA) was the best for callus proliferation. The callus could be maintained on this medium through several subcultures. The extent of callus proliferation varied with the age of the shoots. Surface contamination was a major problem in shoots collected in late spring or later. Surface contamination could be controlled to a large extent by heat treatment of explants during the process of surface sterilization.

* Supported by USDA/CSRS Grant No. KYX1282000008.

PLANT REGENERATION FROM TISSUE CULTURES OF AMARANTHUS SPECIES

Robert D. Locy and Lawrence C. Fisher

NPI
417 Wakara Way
Salt Lake City, Utah 84108

As one of the few genera of dicotyledonous plants which use the C_4 pathway of photosynthetic carbon fixation, Amaranthus may be interesting both for its future potential as a crop plant and for studying the genetic regulation of the C_4 photosynthetic pathway. Thus, we are attempting to develop systems for whole plant regeneration from tissue and cell cultures, and from protoplasts of Amaranthus spp. Amaranthus callus cultures can be initiated from hypocotyl segments prepared from aseptically germinated seeds by placing such segments on media containing Murashige and Skoog (MS) salts and vitamins, 3 or 10 mg/l 2,4-dichlorophenoxyacetic acid (2,4-D), and 0.05 mg/l kinetin. MS media containing indoleacetic acid (IAA) or naphthaleneacetic acid (NAA) at concentrations up to 10 mg/l are ineffective at producing a proliferating callus in the presence or absence of N^6-benzyladenine (BA) or kinetin at concentrations up to 10 mg/l. The callus produced on cytokinin-containing media forms progressively more brown pigment and grows more slowly as the cytokinin concentration increases.

Plantlet regeneration can be obtained by transferring callus cultures grown as described above to media containing MS salts and vitamins, 3 mg/l kinetin, and 0.3 mg/l IAA. Green, leafy structures appear on the cultures within 2 transfers on this media. The leafy structures appear to be organized into shoots, some of which are rooted. However, the formation of intact rooted shoots and whole plants from these structures has not yet been demonstrated. The ability to undergo morphogenesis is dependent on the genotype of the original plant. Whether the mechanism for plantlet regeneration involves somatic embryogenesis or adventitious shoot formation is currently under investigation.

ORGANOGENESIS AND EMBRYOGENESIS

FROM MAIZE INBRED LINE B73. K.

S. Lowe, N.P. Everett, D.B. Taylor,
and K.E. Paterson

Stauffer Chemical Company
1200 S. 47th Street
Richmond, California 94804

Although the maize inbred line B73 is the leading parent for hybrid corn production in the United States, there are no reports of regeneration from established cultures of this inbred. Organogenic tissue, embryogenic callus, and embryogenic cell suspensions have now been obtained from embryo-derived tissue cultures.

Immature B73 embryos were cultured on Murashige and Skoog (MS) medium with 0.5 mg/l 2,4-dichlorophenoxyacetic acid (2,4-D) and 12% sucrose. A compact white tissue proliferated from the scutella of cultured embryos. A slow-growing, green, compact morphogenetic tissue was selected from this white compact tissue on a 2% sucrose medium. This tissue was termed organogenic since the shoot meristems were not directly associated with root meristems. Shoot meristems were usually naked (not enclosed by sheathing leaves) and never associated with the scutella or coleoptiles characteristic of cereal embryos. This tissue has been in culture for over 2 years and is still capable of producing plants.

After one year in culture, embryogenic callus was visually selected from the organogenic tissue. Unlike the organogenic tissue, this callus was visibly undifferentiated, friable, highly mucilaginous, and characterized by a relatively fast growth rate. The secretion of mucilage was stimulated by 2,4-D, and its composition was found to be different from root slime. In the absence of 2,4-D, this callus organized into discrete globular structures which, in some cases, germinated to form plants. Embryogenic callus was termed embryogenic for the following reasons:

a. Plants developed directly from discrete globular structures.

b. Germinating globules were morphologically similar to late coleoptile-stage zygotic embryos.

c. The somatic embryos were discontinuous with the underlying callus.

d. Somatic embryos were morphologically similar to zygotic embryos matured in vitro.

Despite the morphological similarities, somatic embryos were not anatomically equivalent to zygotic embryos. Embryogenic cell suspensions were easily established from the embryogenic callus and have retained their totipotency for over one year.

To further characterize the ontogeny of plants arising from culture, the organogenic and embryogenic cultures were analyzed for the presence of tissue-specific isozymes. In general, embryogenic cultures expressed the same isozymes as zygotic embryos, but organogenic cultures gave more complex results.

The anatomical and biochemical differences between the 2 modes of regeneration cannot be the result of culture conditions or genotype-induced abnormalities, since both types of cultures are derived from the same inbred and are maintained and regenerated on identical media.

SOMATIC EMBRYOGENESIS FROM MATURE WHEAT EMBRYOS

Christy MacKinnon and Murray W. Nabors

Department of Botany
Colorado State University
Fort Collins, Colorado 80523

Cereal calli which are most responsive to regeneration attempts have been those derived from immature embryos, inflorescences, or leaves. The basis of these regenerative responses appears to be somatic embryogenesis. There is little evidence in the literature of "mature" tissue (e.g., whole caryopses, isolated mature embryos, and expanded leaves) yielding significant numbers of embryoids. This report details the consistent formation in vitro of significant amounts of somatic embryos from either whole caryopses (seeds) or isolated mature embryos of Triticum aestivum (cv. Glennson) cultured on a high (5 mg/1) 2,4-dichlorophenoxyacetic acid (2,4-D) concentration.

Embryoids increase substantially after the first 6 weeks of culture; after 3 months of subculture, numbers decrease. However, selective transfer can be used to maintain embryoids for 10 months or more. Significantly, embryoids retain their "competence" for regeneration/germination after 10 months in vitro on high auxin concentrations. The presence of kinetin in initiation and/or maintenance media does not appear to significantly influence the incidence of embryoid formation, nor does it appear to increase plantlet formation. Experiments conducted to date demonstrate that after an average of 8 months of in vitro culture, an estimated average of 65 plants/gram of embryogenic callus can be obtained.

ASEXUAL PROPAGATION OF COFFEE THROUGH SHOOT-TIP CULTURE

Meena S. Moses and June E. Sullivan

NPI
417 Wakara Way
Salt Lake City, Utah 84108

An in vitro propagation method for coffee using shoot-tip culture has been developed. Shoot tips, either terminal or axillary, were initiated on an agar-solidified medium containing half-strength Murashige and Skoog (MS) salts, and the following: i-inositol (100 mg/l), thiamine-HCl (0.4 mg/l), pyridoxine-HCl (0.5 mg/l), nicotinic acid (0.5 mg/l), glycine (2.0 mg/l), sucrose (30 g/l), naphthaleneacetic acid (NAA) (0.2 mg/l), and N^6-benzyladenine (BA) (3.0 mg/l). Shoot tips grew and developed 2 to 4 leaves in about 3 weeks. The small shoots could be subcultured on the same medium to obtain a multiplication rate of 1.5 to 2.0 shoots every 4 weeks. To optimize in vitro multiplication, the effects of various levels of BA, NAA, MS nitrates, MS FeEDTA, and activated charcoal have been evaluated. Attempts are also being made to root the shoots in vitro.

NITROGEN AND GROWTH REGULATOR EFFECTS ON SHOOT AND ROOT GROWTH OF SOYBEAN IN VITRO

Ralph L. Mott,[1] John M. Cordts,[2] and Ann M. Larson*

[1]Department of Botany
North Carolina State University
Raleigh, North Carolina 27695

[2]U.S. Department of Agriculture
Appalachian Fruit Research Station
Kearneysville, West Virginia 25430

Effects of different nitrogen sources and growth regulator concentrations on explants from soybean seed embryos [Glycine max (L.) Merr. cv. Braxton] were studied to identify reasons for the general lack of regeneration of soybean plants from callus. The nitrogen source supplied in the agar medium was critical to shoot growth. Ammonium stunted shoot growth unless high relative concentrations of nitrate were also supplied. However, stunting of shoots was reversible upon transfer to a medium more promotive to growth, i.e., low ammonium, high nitrate. Actively growing shoots were also adversely affected by a high ammonium concentration as indicated by cessation of apex growth and severe leaf senescence. A test of 4 diverse genera showed that ammonium/nitrate inhibition of shoot growth was neither limited to soybean, nor common to all legumes. Depressed pH of the medium due to high ammonium/nitrate ratios could not account for decreased shoot growth. While glutamine, substituted for ammonium, did not depress shoot growth, nitrate was still required for growth to proceed. Root initiation was prolific on all nitrogen combinations tested, but root growth was reduced by ammonium and promoted by nitrate. Glutamine had little effect on root initiation or growth.

The cytokinin N^6-benzyladenine (BA) inhibited both shoot and root growth at 0.5 μM or greater concentrations on all media. The

* Associate Professor of Biology, Sangamon State University, Springfield, Illinois 62708

auxin naphthaleneacetic acid (NAA) promoted both shoot growth and root initiation, although root elongation was depressed with increasing auxin concentrations. High concentrations of NAA ($\geq$ 5 μM) suppressed root initiation and interfered with shoot growth as callus formation was enhanced.

PLANT REGENERATION FROM EMBRYOGENIC SUSPENSION-DERIVED PROTOPLASTS OF SANDALWOOD (SANTALUM ALBUM)

P. Ozias-Akins,[1] P.S. Rao,[2] and O. Schieder*

[1]Department of Botany
University of Florida
Gainesville, Florida 32611

[2]Bhabha Atomic Research Centre
Plant Morphogenesis and Tissue Culture Section
Bio-organic Division
Bombay 400 085, India

Protoplasts of sandalwood (*Santalum album* L.), an economically important tree in India and Southeast Asia, divided to produce embryogenic callus from which plants could be regenerated. Suspension cultures which provided a convenient source for protoplasts were established with callus originating from cultured shoot segments from 20-year-old trees. Suspensions maintained by subculture every 4 to 5 days in a modified Murashige and Skoog (MS) medium containing 0.5 or 1.0 mg/l 2,4-dichlorophenoxyacetic acid (2,4-D) produced embryoids when plated in stationary liquid media or on agar media with 0.1 to 2.0 mg/l benzylaminopurine (BA). Plants could be obtained from secondary embryos. Cells grown in suspension afforded a good protoplast yield with an enzyme solution consisting of 1% Cellulase RS, 1% Macrozyme R-10, 0.5% Driselase, and 0.55 M mannitol. Protoplasts cultured in V47 medium [H. Binding (1974) *Z. Pflanzenphysiol.* 74:327] containing 3 mM 2-(N-morpholino)ethanesulfonic acid, 0.5 mg/l 2,4-D, and 0.5 mg/l BA, and adjusted with mannitol to 750 mOs/kg H_2O, divided within 3 to 4 days. Multicellular colonies present after 2 weeks continued to develop when the osmalality was gradually reduced by dilution every 5 to 7 days. Numerous somatic embryos were formed after plating the protoplast-

* Max-Planck-Institut für Züchtungsforschung, 5000 Köln 30, Federal Republic of Germany

derived colonies on MS agar medium supplemented with indole-3-acetic acid and BA. Complete plants could be obtained from these embryoids. Regeneration of plants from protoplasts previously has been accomplished in only one other woody plant, *Citrus*.

REGENERATION OF BROCCOLI FROM BUDS AND LEAVES

Peter Petracek and Carl E. Sams

Department of Plant and Soil Science
The University of Tennessee
Knoxville, Tennessee 37901-1071

Broccoli leaves and buds were used as explants on Schenk-Hildebrandt media with 8.0 mg/l indoleacetic acid (IAA) and 0.026 mg/l kinetin. After 30 days on the media, plantlets forming from the explants were removed and planted in Jiffy pots. Fifteen days later they were potted in a soil mixture. The regenerated plants were greenhouse-grown to maturity in 130 days compared with a 100-day maturation time of seeded broccoli. Plantlet formation from explants was 90% from buds and 40% from leaves; plantlet survival rate was 80% from buds and 70% from leaves. Histological sectioning shows vascular interconnections of the explant to newly formed roots and shoots. Fifteen percent of all regenerated plants grown to maturity showed growth atypical for this cultivar. Although only a small number of plants has been evaluated thus far, the preliminary observation of variability in regenerated plants indicates that commercial application of tissue culture as a means of plant regeneration for broccoli may be limited.

HISTOLOGY OF STRUCTURES FROM A SOYBEAN (GLYCINE MAX) SOMATIC EMBRYOGENESIS PROTOCOL

Paulette Pierson, Susan Koehler, and Martha Wright

Monsanto Agricultural Products Company
800 N. Lindbergh Avenue
St. Louis, Missouri 63167

Globular and heart-shaped soybean (*Glycine max*) embryoids were obtained from the Phillips and Collins somatic embryogenesis procedure (*Plant Cell Tissue Organ Culture* 1:123-129). These structures, in addition to comparable stage zygotic embryos, were fixed, sectioned, and then stained with safranin-fast green. Although the somatic structures appear similar to zygotic embryos in gross morphology, the somatic structures are generally 2 to 5 times larger than comparable stage zygotic embryos. When thin sections of the somatic structures and the zygotic embryos were compared, the somatic structures lacked both a true epidermis and an internal organization capable of producing a shoot meristem.

IN VITRO REGENERATION CAPACITY OF CORN AND TEOSINTE GENOTYPES

L.M. Prioli,[1] W.J. Silva,[1] P. Arruda,[1] and M.R. Sondahl[2]

[1]Department of Genetics and Evolution
IB, UNICAMP
Campinas, SP 13100
Brasil

[2]DNA Plant Technology Corporation
2611 Branch Pike
Cinnaminson, New Jersey 08077

Germplasm sources from the Maydea Tribe are being utilized for maize improvement. Several species of this Tribe have been introduced into a research program and are being exploited through conventional breeding, tissue culture, and techniques of molecular biology, with the aim of gaining access to new corn genotypes. Teosinte species, as well as isogenic lines and commercial hybrids of *Zea mays*, are being used for tissue culture screening for regeneration capacity. Among teosinte materials, *Zea diploperennis* seems very attractive due to its perennial habit, whereas the annual *Zea mexicana* is the closest relative to domesticated corn. Interspecific hybrids from corn and teosinte species are also being utilized in this program.

Shoot apices, immature inflorescences, immature embryos, and anthers have been used for callus induction. Callus tissues were cultivated on Murashige and Skoog (MS) medium supplemented with pyridoxine (5 μM), thiamine-HCl (15 μM), nicotinic acid (15 μM), inositol (550 μM), sucrose (3%, 6%, 12%), agar (0.7%), and various concentrations of 2,4-dichlorophenoxyacetic acid (2,4-D). Primary cultures were kept in the dark at 28 ± 1°C and subcultured every 20 to 25 days. Secondary cultures were established with the same primary medium without 2,4-D in the presence of light.

In *Zea diploperennis*, optimum callus proliferation was achieved with 1.25 to 10 μM 2,4-D and 3% sucrose after 25 to 30 days of initial culture. Following 3 to 4 passages on primary medium, small callus fragments were transferred to secondary medium where plant regeneration was observed after 2 weeks. Plantlets (1 to 2 cm in height) were subcultured to half-strength MS, and finally transferred to vermiculite, then to the field. In this teosinte species, typical corn somatic embryogenesis was not observed as described for *Zea mays*.

Presently in *Zea mays*, 8 isogenic lines have been tested for regeneration using immature embryos as the explant source. One line (L 160) produced a high frequency of plant regeneration in 70 to 80% of cultivated callus fragments. In this L 160 line, the somatic embryos had a well-developed scutellum and a normal embryo axis similar to a zygotic embryo. Three other lines (L 937, L 903, and L 402) demonstrated intermediate to low regeneration in 10% of cultivated callus. The somatic embryos derived from these 3 lines frequently did not show normal development, demonstrating different culture requirements for these genotypes. The remaining 4 lines (L 1038, L RPS720, L 78, and L 1005) developed callus with watery and compact appearances, but without any regeneration.

The present data suggest that there is great potential for the use of corn ancestors for maize improvement through tissue culture technique. The differential regeneration capacities of 8 isogenic lines of *Zea mays* suggest that there are specific requirements for *in vitro* culture according to the genotype.

HIGH-FREQUENCY PLANT REGENERATION IN LONG-TERM TISSUE CULTURES OF RICE*

N.V. Raghava Ram[1] and M.W. Nabors[2]

[1]Sungene Technologies Corporation
3330 Hillview Avenue
Palo Alto, California 94304

[2]Department of Botany and Plant Pathology
Colorado State University
Fort Collins, Colorado 80523

Cereal tissues were considered recalcitrant in culture for plant regeneration. In few successful cases the cultures were initiated from immature embryos; however, in many cases, cultures did not maintain their regenerability during repeated subcultures. In the present investigation, we report high-frequency plant regeneration in long-term rice callus cultures initiated from seed.

The frequency of plant regeneration in rice callus cultures was markedly increased by a) initiation and preferential proliferation of embryogenic callus; b) optimization of medium for plant regeneration; c) optimization of callus to medium volume ratio; and d) utilization of regeneration medium previously conditioned by callus.

Rice seed callus consisted of 2 types of cells: long, tubular cells forming nonembryogenic (NE) callus, and small, isodiametric cells forming embryogenic (E) callus. The latter always regenerated large numbers of plants on suitable regeneration media. Kinetin at low concentrations (0.2 to 0.5 mg/1), in addition to 2,4-dichlorophenoxyacetic acid (2,4-D) (0.5 or 1.0 mg/1), best initiated and proliferated embryogenic callus. Tryptophan (50 to 100 mg/1) further enhanced embryogenic callus production in some cultivars. How-

* This work was done at Colorado State University and was supported by AID Contract No. AID/DSAN-C-0273.

ever, the effect of tryptophan could not be reproduced by using indoleacetic acid (IAA). Light and dark culture conditions significantly influenced the amount and the regenerability of embryogenic callus.

Regeneration media consisted of IAA as the sole auxin at 0 or 0.5 mg/l and benzylaminopurine (BA) at 0.4 to 0.5 mg/l. Embryogenic callus could be maintained for over 6 passages of 4 weeks each, on a medium containing 2,4-D and kinetin. At the end of each passage, embryogenic callus showed 2.5 to 4.0 times increase in its volume. Regeneration potential measured as plants produced per gram of callus on the best regeneration medium showed an increase during successive passages.

The frequency of plant regeneration on the best regeneration medium could be improved by optimizing the mass/volume ratio of callus to the medium to 6.5 mg/1.0 ml. Furthermore, an increase of 3.8 times in regeneration potential could be achieved by using the regeneration medium conditioned for 2 weeks by an optimal amount of callus. Experiments are in progress to determine the chemical nature of the conditioning factor, and to determine its species or cultivar specificity in cultures. This technique may well be applied to regenerate plants from recalcitrant cultures of other cereals.

Optimized culture conditions were successfully utilized to produce embryoids in high frequency in cell suspension cultures of rice.

IN VITRO CLONING OF *DALBERGIA LATIFOLIA* ROXB. (INDIAN ROSEWOOD)

K. Sankara Rao, G. Lakshmi Sita,*
and C.S. Vaidyanathan

Department of Biochemistry
Indian Institute of Science
Bangalore 560 012, India

Callus induction and regeneration of whole plants of the tree legume *Dalbergia latifolia* Roxb. (Indian Rosewood) was achieved in tissue culture under defined conditions. Callus was initiated from shoot and leaf segments and maintained on Murashige and Skoog (MS) basal medium that had both an auxin(s) and a cytokinin (kinetin). A compact and green callus that ultimately developed shoot buds was obtained on MS medium containing naphthaleneacetic acid (NAA) (3 mg/1) and benzylaminopurine (BA) (1 mg/1) in combination. Callus survived several subcultures. With an increased concentration of BA (3 to 5 mg/1) and a decreased concentration of NAA (0.5 mg/1) in the medium, 10 to 12 shoot buds appeared on the callus. Bud initiation, however, was effective with BA at 5 mg/1 and NAA at 0.5 mg/1. Buds developed well and elongated into shoots in BA at 3 mg/1 and NAA at 0.5 mg/1. Shoots were separated and rooted on half-strength MS medium with indolebutyric acid (1 mg/1).

Regeneration of plantlets from shoot and leaf callus cultures can be developed as a method of clonal propagation of elite Rosewood trees.

---

* Microbiology and Cell Biology Laboratory

IN VITRO CALLUS FORMATION AND PROPAGATION OF SUGARCANE

M. Sarwar

Tissue Culture Laboratory
Pakistan Agricultural Research Council
P.O. Box 1031
Islamabad, Pakistan

Cultures were established using the fully developed, top 4 buds from field-grown canes of *Saccharum officinarum* var. BL 4. During the initial culture, explant browning was controlled by presoaking 30- to 50-mm stem sets, having one bud each, for 4 to 5 hr in 5% Clorox. During subculture, the death of in vitro-produced plantlets because of browning was controlled by keeping the cultures in the dark for the first week. Maximum bud proliferation was obtained on the media having 1 to 5 μM benzylaminopurine or 20 to 40 μM kinetin. Up to 40 plantlets were obtained in 50 days. Only one culture of explanted stem pith produced callus on 1.5×10^{-5} M 2,4-dichlorophenoxyacetic acid medium.

VARIATION FOR SALT TOLERANCE IN TOBACCO CELL SUSPENSIONS

Sherry Rae Schnapp, Marla L. Binzel,
Paul M. Hasegawa, and Ray A. Bressan

Horticulture Department
Purdue University
W. Lafayette, Indiana 47906

Cells of tobacco, Nicotiana tabacum cv. Wisconsin 38, have been obtained, which are capable of active growth in media with 10 or 25 g/l NaCl (S-10 and S-25, respectively). The S-10 and S-25 cells, as well as a population of cells which had not been growing in salt (S-0), were evaluated for tolerance to salinity by examining their growth characteristics when inoculated into media with varying concentrations of NaCl. Tolerance of the cell populations to NaCl increased with the level of salt from which the cells were obtained. The enhanced tolerance of the S-10 cells was lost after one passage in the absence of salt, indicating that the tolerance of these cells was unstable. In contrast, the S-25 population retained a small but stable portion of its tolerance even after 40 recultures in the absence of salt. Analysis of the tolerance of individual clones isolated from the S-0 and S-10 populations indicated that substantial variation for salt tolerance exists within the cell populations, and that the S-10 population exhibited higher tolerances. The frequency distribution of tolerance for S-10 clones grown in the absence of salt was very similar to that of the S-0 clones. Increased tolerance of the S-10 cell population after exposure to salt appeared to be due to physiological adaptation since individual S-10 clones lost tolerance when grown in the absence of salt. The fact that the S-25 cells retained some degree of tolerance to salt suggests, that at high salt concentrations, it may be possible to identify both an unstable, adaptive component of tolerance, and a stable component which can withstand several recultures in the absence of salt.

SOMATIC EMBRYOGENESIS IN TOBACCO CALLUS

Sandra F. Simpson

FMC Corporation
Biotechnology Department
P.O. Box 8
Princeton, New Jersey 08540

Long-term callus cultures of Nicotiana tabacum cv. Wisconsin 38 were utilized to examine the auxin and cytokinin requirements, as well as the light requirement for somatic embryogenesis. Initial observations indicated that somatic embryos were formed in callus cultured on Murashige and Skoog (MS) medium supplemented with indoleacetic acid (IAA) (11.4 μM) and benzylaminopurine (BA) (22.1 μM) in continuous light of ca. 100 $\mu E/m^2/sec$ at 25°C. After 4 weeks of culture, these embryos were transferred to either MS without hormones, or to MS with IAA (11.4 μM) and BA (0.04 μM), the latter being the maintenance medium for the normally dark-grown callus. The somatic embryos were then placed in culture under a 16:8 light:dark cycle at 25°C and regenerated into whole plantlets. Several plantlets grew to maturity, flowered, and set seed. The genetic stability of this seed has not yet been tested. This basic experiment has been successfully repeated 4 times.

To determine the light requirement for somatic embryogenesis in the tobacco callus, callus was cultured under continuous light and a light:dark cycle of 16:8 at 25°C on the inductive medium (see above). Both light regimes were equally effective in stimulating somatic embryo formation, with globular embryos visible at 10 to 12 days of culture. In comparison, dark culture of the callus on the inductive medium resulted in very little callus growth, with a few somatic embryos formed on the lower side of the callus.

To examine the hormone concentration requirement for somatic embryogenesis in tobacco callus, the IAA concentration was held constant (11.4 μM) with the BA concentration varying from 0.22 to 22.1 μM. Tobacco callus cultured in continuous light on these medium variations formed somatic embryos only at the highest BA concentration tested. To determine whether this response was a ratio or an absolute concentration effect, the hormones were decreased in amount while maintaining the inductive ratio constant. Culturing callus on these medium variations in continuous light resulted in equivalent stimulation of somatic embryogenesis as compared with the original inductive medium (see above), with enhanced callus growth at the lowest hormone concentrations (IAA at 2.85 μM and BA at 5.55 μM). Thus, the ratio of auxin to cytokinin was inductive to somatic embryogenesis in tobacco callus, and light potentiated the callus response to these hormone ratios.

AGAR INDUCED VARIATIONS ON THE NUTRITIONAL COMPOSITION OF PEAR AND CRABAPPLE SHOOTS IN VITRO

Suman Singha,[1] Edwin C. Townsend,[1] and Gene H. Oberly[2]

[1]Division of Plant and Soil Sciences
West Virginia University
Morgantown, West Virginia 26506

[2]Department of Pomology
Cornell University
Ithaca, New York 14853

The objective of this investigation was to determine whether nutritional differences in explants on media solidified with 3 agar brands would explain agar-induced variations in proliferation and growth responses. Shoot-tips of 'Almey' crabapple and 'Seckel' pear were cultured on Murashige and Skoog (MS) salt mixture supplemented with 10 mg/l myo-inositol, 0.4 mg/l thiamine, 30 g/l sucrose, and 2 mg/l N^6-benzyladenine (BA). Media were solidified with either Bacto-agar, Phytagar, or T.C. agar at concentrations ranging from 0.3 to 1.2%. A liquid medium treatment was used to obtain a comparative benchmark for explant nutrient levels. Explant nutrient levels determined after 8 weeks were influenced both by agar brand and concentration. Although large differences in a number of elements occur both in agar brands and in explants cultured on media containing similar concentrations of these brands, variations in proliferation and growth cannot be explained based on differences in individual elements. From a nutritional standpoint, modification of the elemental composition of the basal medium may be one cause of the growth variations induced by different agar brands.

MORPHOGENETIC RESPONSE FROM ORCHARDGRASS PISTILS

D.D. Songstad and B.V. Conger

Department of Plant and Soil Science
The University of Tennessee
Knoxville, Tennessee 37901-1071

Unpollinated orchardgrass (Dactylis glomerata L.) pistils from an embryogenic genotype were cultured on a Schenk and Hildebrandt (SH) basal medium supplemented with 3% sucrose and 30 μM 3,6-dichloro-o-anisic acid (dicamba) in the dark at 25°C. The experimental design was a randomized complete block with 6 treatments (1- to 6-week culture periods) and 3 blocks where each experimental unit consisted of 15 pistils. The morphogenetic response was followed at weekly intervals for 6 weeks by stereoscopic and scanning electron microscopy. Root-like protrusions emerged basipetally from approximately 15% of the pistils cultured for one week; however, no plants were regenerated from these structures. Ovary and style regions became swollen after 2 weeks, and embryoids emerged directly from these areas during the third week of culture. These embryoids acted as secondary explants and initiated calli in 4-week-old cultures. During the fifth and sixth weeks, many embryoids were initiated from the newly formed calli. Three weeks after transfer of these cultures to dicamba-free medium, the embryogenic response was quantified by counting shoots. A significant F value ($\alpha = 0.05$) for the treatment effects was followed by mean separation through orthogonal contrasts. This revealed no significant difference between the 3- and 4-week treatments. However, significant F values were obtained when contrasting weeks 2 to 3 and 4 to 5. Plotting average shoot number vs. weeks in culture showed an overall increase in shoot number through the 6-week culture period, except for a decline during the fourth week. Such variations in mean shoot numbers, in conjunction with morphological observations, indicate that direct embryogenesis was followed by indirect embryogenesis from embryoid-derived callus.

DETECTION OF *CRONARTIUM QUERCUUM* F. SP. *FUSIFORME* IN *PINUS TAEDA* EMBRYOS USING THE ENZYME LINKED IMMUNOSORBENT ASSAY

Pauline C. Spaine, Henry V. Amerson,
and James W. Moyer*

Departments of Botany and Forestry
North Carolina State University
Raleigh, North Carolina 27650

The ELISA (Enzyme Linked Immunosorbent Assay) has been used for several years in the detection of animal and plant pathogens. An ELISA was developed to measure *Cronartium quercuum* f. sp. *fusiforme* infection of *Pinus taeda*.

The double antibody sandwich, indirect and competitive versions of the ELISA were compared. Comparison of the double antibody sandwich and indirect ELISA's for *C. quercuum* axenic cultures revealed the indirect assay to have greater sensitivity, with detection of the fungus possible down to 20 μg/ml. The use of several blocking agents, poly-L-lysine and Immulon I and II plates, were also tried in order to increase assay sensitivity.

It was shown in the indirect assay that there is competition for binding sites between healthy plant extracts and fungal mycelium on the polystyrene plate. Development of a competitive ELISA allowed plant material and fungal antigens to compete for specific antibody sites rather than nonspecific affinity of the polystyrene plate. The competitive ELISA has given a 10-fold increase in the sensitivity over the indirect method, with detection of axenic fungal cultures now possible to 2 μg/ml.

At present we can detect fungal antigens from in vitro-infected embryos when compared to uninoculated control embryos. We propose to use this ELISA for *C. quercuum* as an early disease resistance screening assay for susceptible and resistant genotypes of *P. taeda*.

---

* Associate Professor, Department of Plant Pathology.

CLONAL PROPAGATION OF CHILOPSIS LINEARIS (CAV.) THROUGH TISSUE CULTURE

David W. Still, John F. Hubstenberger,
Gregory C. Phillips, and Ronald F. Hooks

Department of Horticulture
and Plant Genetic Engineering Laboratory
New Mexico State University
Las Cruces, New Mexico 88003-3530

Desert willow (Chilopsis linearis Cav.) is a valuable tree native to the southwest United States that is used as an ornamental in landscaping, offering low maintenance and drought tolerance. Desert willow is also useful in soil stabilization and mine reclamation. Elite selections of desert willow are generally propagated asexually because it is not true-breeding. An efficient means of rapid clonal propagation through tissue culture could facilitate the use of elite selections.

Shoot-tips and cotyledons were excised from aseptically germinated seedlings, cultured on Schenk and Hildebrandt basal nutrient media with various combinations of indole-3-butyric acid (IBA) and benzylaminopurine (BA), and incubated at 25°C with a 16-hr photoperiod. Adventitious shoots were induced in the cotyledon cultures on all media tested at 85 to 90% frequency. Enhanced axillary branching occurred with shoots obtained from both explant sources at comparable frequencies, yielding a mean of 9.1 shoots per explant, with as many as 25 shoots at the end of the second monthly passage. The best media treatments included 0.03 to 0.05 mg/l IBA with 4 to 5 mg/l BA. Enhanced axillary branching was maintained through at least 8 passages with a mean multiplication rate of 3.7 every month. However, all cultures exhibited some degree of callusing at the basal end of the shoot masses. In a preliminary study, 20 to 33% of the shoots were rooted by treatment with IBA with or without BA, or with one-tenth strength mineral salts. Plants were established in the greenhouse. Apical and axillary shoot bud explants derived from mature trees responded with axillary branching at very low frequencies and with a maximum of only 3 shoots per explant. However, these explants produced callus at 100% frequency. Roots and bud-like structures have been observed in callus of up to one year in age, but, to date, no elongated shoots or plantlets have been regenerated from callus. Although further research is needed to define a system for explant tissue, our results clearly demonstrate that clonal propagation of desert willow through tissue culture is feasible using seedling-derived explants.

HIGHLY MORPHOGENIC CALLUS OF *ARACHIS PARAGUARIENSIS* FROM AGAR AND SUSPENSION CULTURES

Paul E. Still, María Inés Plata, and C.L. Niblett

Department of Plant Pathology
University of Florida
Gainesville, Florida 32611

Arachis paraguariensis (Chod. et Hassl.) is a wild species of peanut with 2n=20. Callus cultures derived from anthers of this species were highly morphogenic, producing shoots, leaves, and flowers at a high frequency. A much lower percentage regenerated into complete plants with roots. Anthers from flowers with no petal showing were used to initiate callus, using the salts and organics of N6 medium (Chu et al.) supplemented with casein hydrolysate, 500 mg/l; myo-inositol, 100 mg/l; sucrose, 20 g/l; L-proline, 3 g/l; 4-amino-3,5,6-trichloropicolinic acid (picloram), 0.2 mg/l; N^6-benzyladenine (BA), 0.25 mg/l; and agar, 6 g/l. Calli were maintained on the same N6 medium with picloram levels of 0.008 mg/l, 0.25 mg/l BA, and 60 g/l sucrose. Buds appeared on calli after 30 to 40 days. When these buds were transferred to Murashige and Skoog medium (MS) with 0.01 mg/l BA or with no hormones, shoots developed with leaves and flowers. Callus was used to initiate suspension cultures in N6 liquid medium that contained casein hydrolysate, 500 mg/l; myo-inositol, 100 mg/l; sucrose, 20 g/l; picloram, 0.008 mg/l; BA, 0.2 mg/l; and L-proline, 1.7 g/l. Suspension cultures were maintained by transferring clumps of tissue between 210 and 840 μm in size on a weekly basis. Shoots appeared in suspension cultures if clumps of cells larger than 840 μm were transferred to MS with no hormones. From an initial callus of 3 to 5 g, several hundred shoots were produced in a period of 2 months. Root formation occurred at a very low frequency in this system. One plant out of a total of 5 transferred to soil grew to maturity.

ANTHER CULTURE OF PAPAYA (*CARICA PAPAYA* L.)

C.Y. Su and H.S. Tsay

Department of Agronomy
Taiwan Agricultural Research Institute
Taichung, Taiwan 431, Republic of China

Papaya (*Carica papaya* L.) anthers containing microspores in tetrads to early-binucleate stages were successfully cultured in one-half strength Murashige and Skoog (MS) medium (full-strength of NaFeEDTA) supplemented with 2 mg/l naphthaleneacetic acid (NAA), 1 mg/l N^6-benzyladenine (BA), and 6% sucrose for callus formation. Highest frequencies of callus induction were obtained from anthers with uninucleate-mitosis stage microspores. Callus could be induced under either light or dark conditions although better results were obtained when anthers at uninucleate stage were cultured in darkness. Haploid plantlets and pollen-derived embryoids were obtained from anthers cultured at the uninucleate stage on 3% sucrose-containing MS medium without any growth regulator and low light intensity. Mass propagation was achieved when these embryoids were transferred to MS medium with 3% sucrose and no growth regulators. Some haploid plants were further developed from these embryoids. Examination of root tips of embryoid-derived plants showed the chromosome number of 9 (18 chromosomes for diploid papaya). It was thus concluded that the embryoids originated from pollen.

MORPHOLOGICAL ADAPTATION OF LEAVES OF STRAWBERRY PLANTS GROWN IN VITRO AFTER REMOVAL FROM CULTURE

Ellen G. Sutter,[1] Andrea Fabbri,[2]
and Sheryl Dunston[1]

[1]Department of Pomology
University of California
Davis, California 95616

[2]Instituto di Coltivazioni Arboree
Universita di Firenze
Firenze, Italy

Leaves from strawberry plants regenerated in vitro were examined by light and scanning electron microscopy to determine whether morphological changes occurred in leaves during the first 20 days after removal from culture, during which time cultured plants are extremely fragile. Strawberry plants regenerated in vitro from meristems were removed from culture and hardened off in a growth chamber for 20 days. Leaves examined included those present in vitro, leaves formed in vitro and present on the plant 10 and 20 days after removal from culture, leaves formed after removal from culture, and leaves from field-grown plants. Measurements of anatomical features were determined using digital image analysis. Leaves formed in vitro became somewhat thicker after removal from culture. Epidermal and palisade cells enlarged, but no additional layers of palisade cells were formed in these leaves. The percentage of mesophyll air space increased in leaves that were being acclimatized, but only after they had been out of culture for 20 days. Leaves formed after removal from culture had 2 layers of palisade cells and the same percentage of mesophyll air space as leaves formed in vitro 20 days after removal from culture. Leaves on field-grown plants were significantly thicker and had 3 layers or more of palisade cells. Stomates were clearly seen on the abaxial surface of leaves formed in vitro. On leaves formed after removal from culture, stomates were sunken and slightly fidden in the depressions formed at the junctions of the anticlinal walls of epidermal cells. Crystalline epicuticular wax was found on the surfaces of leaves from all stages, but the wax varied in structure and distribution among the different treatments. There was a lack of major change towards anatomical characteristics that conferred an adaptive benefit on cultured plants during the period immediately following removal from culture. This observation suggests that at the time plants are transferred from culture, survival depends as much on vigorous growth of new leaves which are adapted to greenhouse conditions as on the adaptation of leaves present on the plant at time of removal from culture.

CLONAL VARIABILITY AND FIELD PERFORMANCE

IN SWEET POTATOES PROPAGATED IN VITRO

K.M. Templeton-Somers and Wanda Collins

Department of Horticultural Science
North Carolina State University
Raleigh, North Carolina 27695-7609

Tissue culture techniques, particularly short-term culture procedures such as shoot-tip culture and regeneration from primary explants, have been proposed as methods for obtaining large numbers of plants identical to the plant used as an explant source. Sweet potatoes, although a clonally propagated crop, typically show large yield differences among plants at harvest, with variations that are significant even within rows that appear quite uniform and productive. This study was designed to compare the effects of propagation method, including 3 in vitro propagation methods (2 types of regeneration from leaf explants and lateral bud culture), nodal propagation, and slips and cuttings from bedded roots, on field performance and variability. The study was repeated a second year, and a carryover study, for which plants were obtained from the bedded roots of the first year's study, was also completed. Each hill was individually evaluated for yield, skin and flesh color mutation frequency, dry matter, and protein content.

The slips and cuttings from bedded roots produced the highest total yield and yield of marketable roots, while plants from an in vitro procedure in which shoots developed from adventitious roots on the leaf explant ranked significantly lower in yield than all other propagation methods. The frequency of roots with flesh color mutations was lower for the in vitro derived plants than for plants derived from bedded roots. This difference, as well as the yield differences, was insignificant in the carryover study. The percentage of roots with skin color mutations was also higher in plants from bedded roots, but this difference was maintained in the carryover study.

The origin of the explants was also considered for its effects on yield and mutation frequency. Although all the plants used as sources of explant material are the result of clonal propagation and intense selection for characteristics true to the variety, the explant origin was a significant factor in both yield and skin color mutation frequency.

Data from this study are also being used to determine the amount of variability within propagation procedures.

INDUCTION AND SELECTION OF ATRAZINE RESISTANCE VARIANTS FROM TOBACCO CELL CULTURES

Wen Chung Wang and G.B. Collins

Department of Agronomy
University of Kentucky
Lexington, Kentucky 40546

Photomixotrophic cell cultures of Nicotiana tabacum L. (var. Wisconsin 38) were obtained by repeated selection of cell cultures initially established from hypocotyl explants. Attempts to induce photomixotrophic cell cultures of N. tabacum L. var. Ky 14 were not successful. Calli derived from Wisconsin 38 hypocotyls regularly produced a greening response and were used for optimizing the growth conditions for photoautotrophic and photomixotrophic cultures. Two light intensities (67 and 7.6 µE/M^{2}sec) and 4 sucrose concentrations (0, 0.5, 1, and 2%) were tested for optimizing chlorophyll synthesis and callus growth rate. Calli growing on media containing 0.5% sucrose gave better chlorophyll synthesis and chlorophyll a/b ratios at high light intensity than did other treatments.

Photomixotrophic cell cultures of Wisconsin 38 were maintained by repeated selection of greening callus. Dosage response to atrazine concentrations of 10^{-6}, 10^{-5}, and 10^{-4} M was studied with photomixotrophic callus, with photomixotrophic cell suspension, with germinating seeds, and with 3-week-old seedlings. Response to atrazine was determined by measuring fresh weight, dry weight, and chlorophyll concentration. Dosage responses for the 4 plant materials revealed similar inhibition trends over the atrazine concentrations tested.

Eighteen green colonies were isolated from photomixotrophic suspension cells which were incubated on solid media containing 10^{-6}, 10^{-5}, or 10^{-4} M atrazine for 6 weeks. The selected green calli were grown on callus proliferation media and then transferred onto a regeneration media with different concentrations of atrazine to select regeneration-competent cells with tolerance to atrazine.

Three types of shoots have been regenerated: normal green, translucent, and albino. These shoots were placed on R1/2N rooting medium containing 10^{-6} M atrazine; tolerant green shoots rooted, whereas albino, translucent, and "susceptible" escape plants did not root. A total of 37 green plants have been isolated. Morphological abnormalities and chromosomal losses have been observed in some plants. Variants showed a low percentage of viable pollen, and variability in leaf shape and floral structure. Several methods to identify atrazine-resistant plants have been evaluated. These include an atrazine-binding assay, a floating leaf-disk method, a leaf-disk assay, a leaf-culture assay, and a leaf-fluorescence assay.

CALLUS MORPHOLOGY ASSOCIATED WITH SOMATIC EMBRYOGENESIS IN BARLEY

R.C. Weigel, Jr. and K.W. Hughes

Department of Botany
The University of Tennessee
Knoxville, Tennessee 37996-1100

Our experiments with Atlas 57 barley callus cultures indicate that somatic embryogenesis is preceded by the formation of callus with an unusual morphology.

Explants were taken from 9-day-old Atlas 57 barley plants which were grown under sterile conditions. The explants consisted of 1 cm long sections which included the apical meristem and adjacent stem tissue. These were transferred to basal medium (Murashige and Skoog salts, 5 mg/l thiamine-HCl, 30 g/l sucrose, 500 mg/l myo-inositol, and 0.8% agar) plus plant growth substances [10 μM indoleacetic acid (IAA), 15 μM 2,4-dichlorophenoxyacetic acid (2,4-D), and 1.5 μM dimethylallylamino purine (2iP)]. Plant growth substances were added by filter sterilization to basal medium plus agar which had been autoclaved for 15 min at 121°C, 1.46 kg/cm^2. Cultures were maintained at 24°C in a growth chamber with a 16-hr daylength provided by cool white fluorescent tubes. After 2 months the resulting callus cultures were transferred to basal medium without growth substances, were subcultured onto basal medium every 2 months for the next 6 months, and then were subcultured again on basal medium plus growth substances. Typical callus cultures in our experiments are clear to lightly colored and crystalline in appearance. Callus color density increased in one culture. After transfer of all cultures to basal medium plus plant growth substances for a 2-month period, a unique, lighter-colored callus formed from a portion of the darker callus. This unusual callus was white, opaque, and had a nodular, convoluted appearance, similar to literature descriptions of embryogenic callus. Plantlets with roots and shoots were produced after transfer of the embryogenic callus back to basal medium. The first plants were produced from the embryogenic callus when the cultures were about one-year-old. Subcultures from the embryogenic cultures are continuing to produce embryogenic callus and plantlets at the present time, about 2 years after the original cultures were initiated.

Similar results have been obtained in another experiment in which the callus culture had much less exposure to basal medium as opposed to plant growth substance medium. Embryogenic callus and plantlets continue to be produced about 1.5 years after initiation of the second experiment.

EMBRYO PROPAGATION BY DIRECT SOMATIC EMBRYOGENESIS AND MULTIPLE SHOOT FORMATION

E.G. Williams and G. Maheswaran

Plant Cell Biology Research Centre
School of Botany
University of Melbourne
Parkville
Victoria 3052, Australia

We are developing methods for rapid clonal propagation by direct somatic embryogenesis or direct multiple shoot formation in situ on immature zygotic emrbyos, with the following aims: a. to minimize culture-induced variation by avoiding cellular destabilization; b. to provide simple, rapid, space-efficient propagation techniques which can be applied as early in the life cycle as possible; c. to develop continuous embryogenic cultures with the potential for mechanized sterile propagation and harvesting.

Rapid production of primary somatic embryoids from immature embryos has been achieved for *Trifolium repens* (white clover), *T. pratense* (red clover), *T. resupinatum* (Persian or annual strawberry clover), *Medicago sativa* (alfalfa), *Lotus corniculatus* (birdsfoot trefoil), and *Brassica campestris* (oilseed rape) on the following media:

T. repens *T. pratense* *T. resupinatum* *M. sativa*	EC6 [Maheswaran and Williams (1984) *Ann. Bot.* 54: (in press)] with 1 g/l yeast extract (YE) plus 0.05 mg/l benzylaminopurine (BA).
L. corniculatus	EC6 (without YE) plus 1 to 2 mg/l BA
B. campestris	B5 [Gamborg et al. (1968) *Exp. Cell Res.* 50:151-158] plus 0.05 mg/l BA; *or* modified SH [Bhattacharya and Sen (1980) *Z. Pflanzenphysiol.* 99:357-365] plus 1 g/l YE, 0.022 mg/l 2,4-dichlorophenoxyacetic acid (2,4-D), and 0.00216 mg/l kinetin.

Secondary direct somatic embryogenesis from subcultured primary embryoids has also been achieved for T. repens and B. campestris on the following media:

T. repens	EC6 with 0.25 to 1 g/l YE plus 2 mg/l BA.
B. campestris	B5 with 2% sucrose, no hormones.

In both species secondary embryoids are extremely easily detached from the parent primary embryoid tissue, suggesting that mechanical harvesting of continuous embryogenic cultures may be a feasible development in these systems.

In addition, direct multiple shoot induction from immature embryos has been achieved for Lotus corniculatus, Vigna unguiculata (cowpea), and Lycopersicon peruvianum (wild tomato) on the following media:

L. corniculatus	EC6 or B&N [Williams and De Lautoure (1980) Bot. Gaz. 141:252-257], both with 1 g/l YE plus 0.05 mg/l BA.
V. unguiculata	B5 plus 0.005 mg/l indole-3-butyric acid (IBA).
L. peruvianum	EC6 with 1 g/l YE plus 1 to 2 mg/l BA.

Primary and secondary embryoids or multiple shoots of all species have been readily separated and rooted on the corresponding media without hormones before transfer to soil.

In T. repens the tissue origin of primary somatic embryoids has been examined by scanning electron microscopy and light microscopy of sectioned material. Embryoids arise superficially by proliferation of one or a few epidermal cells of the hypocotyl region of the original zygotic embryo.

Although the regenerants from these embryo propagation techniques have not yet been surveyed in detail for uniformity, by minimizing callus formation and cellular destabilization we hope to minimize tissue culture-induced variability. Apart from the obvious application in clonal propagation, direct multiplication of embryos may provide aseptic "seedling" clones for direct screening in host/pathogen or host/Rhizobium studies. Also, since embryoids appear to arise from single cells or from very small proembryonal complexes, direct somatic embryogenesis after mutagen treatment of immature embryos may improve recovery rates of whole-plant mutations and reduce the incidence of chimerism.

PROTOCLONES OF SORGHUM BICOLOR WITH UNUSUALLY HIGH MITOCHONDRIAL DNA VARIATION

A.J. Wilson, P.S. Chourey, and D.Z. Sharpe

U.S. Department of Agriculture
Agricultural Research Service
and University of Florida
Gainesville, Florida 32611

We are analyzing genomic variations at the molecular level in tissue-cultured cells of corn and sorghum with the ultimate aim of characterizing the parts of the genome that are stable or unstable in the tissue culture environment. The main focus of our effort is on tissue-cultured cells rather than regenerated plants since regenerated plants are lacking in most of the random variation due to the inherent developmental selection pressure in the plant regeneration process. We have obtained cell suspension cultures from 3 cultivars of sorghum. Protoplasts from these cultures are readily obtained and give rise to callus that can be returned to cell suspensions (Chourey and Sharpe, manuscript in preparation) which are designated here as protoclones. The analysis of 6 such independently derived, randomly selected protoclones from the cultivar NK 300 is reported in this study. Mitochondrial DNA (mtDNA) is prepared, restriction digested, and electrophoresed according to a newly developed rapid method for mtDNA isolation (Wilson and Chourey, manuscript submitted). In addition to comparing the UV fluorescing bands from stained agarose gels, we have also compared those segments which share homology with 4 EcoRl fragments analyzable by membrane hybridization techniques. The following observations are noteworthy.

1. A unique set of restriction fragments is displayed by each of the 6 protoclones, the parental cell suspension, a newer cell suspension, and NK 300 coleoptiles. Although a large number of restriction fragments are conserved in these samples, a certain proportion of the genome seems to be variable. Two protoclones show a complete loss of the largest EcoRl fragment.

2. A considerably larger amount of variation is uncovered by membrane hybridization analysis. Four different EcoR1 fragments were extracted from agarose gels of digested suspension culture mtDNA and used as hybridization probes. Although all 4 restriction fragments display variation, the largest EcoR1 hybridization fragment shows an exceptionally high capacity for loss and/or rearrangement (recombination?) which is similar to observations with *copia*-like elements in tissue-cultured cells of *Drosophila*. Further molecular characterization of this fragment is in progress to ascertain if it indeed shares the transposable DNA-like properties of *copia*. An alternate possibility is that these variations are directly correlated with the size of the DNA fragment under analysis, since the smaller EcoR1 fragments display less variation.

3. The basis of the tissue culture origin of this variability is at present unclear. The indirect evidence suggests that it is unlikely to be present in the original cell suspension culture. The parental cell suspension culture has been remarkably stable during the past 6 months of this investigation, and no alterations are seen at various stages of the cell cycle in the population. The possibility that this high level of variation is intrinsic to the protoplast culture process is being investigated.

FACTORS AFFECTING ISOLATION OF PROTOPLASTS FROM LEAVES OF GRAPE (*VITIS VINIFERA*)

Daniel C. Wright

Brooklyn Botanic Garden Research Center
712 Kitchawan Road
Ossining, New York 10562

A method of isolating grape mesophyll protoplasts was investigated to facilitate the use of genetic engineering techniques to improve this economically important woody plant species. Grape leaves were macerated under aseptic conditions using Cellulysin and Macerase in combination with a variety of different factors known to promote the isolation of protoplasts from herbaceous plants. The effects of several factors influencing protoplast isolation could be evaluated quickly by using leaf disks 1 cm in diameter and known volumes of maceration and wash media. The best yields of mesophyll protoplasts were obtained using medium-sized leaves of well-fertilized grapevines kept in the dark for 24 hr prior to maceration in 1% Cellulysin, 0.5% Macerase, 0.7 M mannitol, 5 ppm 2,4-dichlorophenoxyacetic acid, 0.1 ppm benzylaminopurine, and one-tenth strength Murashige and Skoog medium, then incubated at 22°C in the light (1,000 lux) for 24 hr. Over 5,000 protoplasts per cm^2 of leaf were produced using these conditions. This method of screening factors affecting protoplast isolation would be applicable to other species as well.

SOMACLONAL VARIATION IN TOBACCO AND SUGAR BEET BREEDING

Nedjalka Zagorska and Atanas Atanassov

Institute of Genetics
Bulgarian Academy of Sciences
P.O. Box 96
Sofia 1113, Bulgaria

Somaclonal variation is already a well proved and used phenomenon for introducing nonspecific variation into crop species, using standardized plant regeneration methods.

In the course of this study it was shown that the genetic variability occurred at different frequencies in tobacco and sugar beet tissue cultures. As a result of organogenesis in 5 cultivars of tobacco [Burley 21; Vranja 96, Nos. 4, 3942, and 3974 (from fourth to sixth subculture)] and 3 cultivars of sugar beet [diploid plurigerm line B032, diploid monogerm line 680, and tetraploid two-germ line 155 (first and second passages)], 1230 and 43 regenerants were obtained, respectively.

Many deviations, related to the morphology, habitus, shape and size of the leaves, flowers and reproductive organs, chromosome counts, and meiosis, were recovered in tobacco plants regenerated from culture. In R2-R4 generations, more than 100 new lines were tested. Finally, 10 stable breeding lines were selected which, when compared with the respective parent cultivars, were tolerant to *Perenospora tabacina*, *Thielavia basicola*, and TMV, and had higher yields and better technological qualities. In sugar beets, the shoot formation in somatic tissues of diploid lines B032 and 680 did not reveal any plants containing morphological or chromosomal deviations compared with the starting material. In contrast to these results, it was possible in tetraploid line No. 155, using callus cultures from flower buttons, to obtain 7 plants possessing a diploid chromosome number and monogerm seeds. R1 progenies of each plant maintained the new character and were morphologically identical. These results are of great interest for tobacco and sugar beet breeding programs.

ROSTER OF PARTICIPANTS AND SPEAKERS

ABO EL-NIL, MOSTAFA, Weyerhaueser Forestry Research Center, Centralia, WA
AHN, JOON, University of Arkansas, Fayetteville, AR
AMERSON, HENRY, North Carolina State University, Raleigh, NC
AMMIRATO, PHILIP V., Barnard College, New York, NY
ANDERSON, MARK M., Proctor and Gamble Co., Cincinnati, OH
ANDERSON, DAVID R., Clemson University, Clemson, SC
ANTHONY, JANICE, Texas A&M University, College Station, TX
ARCHER, VERNON G., Jackson, MS
ASOKAN, M.P., University of Florida, Homestead, FL
ATANASSOV, ATANAS IVANOV, Institute of Genetics, Sofia, BULGARIA
AYNSLEY, JOHN, International Plant Research Institute, San Carlos, CA
BAKER, CHARLEEN M., University of Florida, Homestead, FL
BALUCH, STEVE J., FFR Cooperative, W. Lafayette, IN
BARNES, LEE, University of Florida, Gainesville, FL
BATES, GEORGE W., Florida State University, Tallahasse, FL
BATES, RICK, West Virginia University, Morgantown, WV
BEATY, ROBERT, University of Tennessee, Knoxville, TN
BECWAR, MICHAEL, Purdue University, West Lafayette, IN
BENNETT, RANDY, The Hobbit's Glen Nursery, Pfafftown, NC
BENZION, GARY, University of Kentucky, Lexington, KY
BILKEY, PETER, George J. Ball, Inc., West Chicago, IL
BONGA, J.M., Maritimes Forest Research Center, Fredericton, CANADA
BRADLEY, PETER, Westboro, MA
BRAND, MARK H., Ohio State University, Columbus, OH
BRIGGS, BRUCE, Briggs Nursery, Olympia, WA
BRIGHTWELL, BLANCHE B., Monsanto Agricultural Products Corp., St. Louis, MO
BROOME, OLIVIA C., U.S. Dept. of Agriculture, ARS, Beltsville, MD
BUCKLEY, PAUL M., U.S. Dept. of Agriculture, ARS, Mississippi State, MS
BULLARD, RAY, University of Tennessee, Chattanooga, TN
BURKE, MARTHA ROE, Ceres 2000, Inc., Winter Haven, FL
CALIE, PAT, University of Tennessee, Knoxville, TN
CALLAHAN, MAUREEN L., Goodyear Tire and Rubber Co., Akron, OH
CAMERON, MARLENE, Michigan State University, East Lansing, MI
CAMPER, N. DWIGHT, Clemson University, Clemson, SC

CAPONETTI, JAMES D., University of Tennessee, Knoxville, TN
CARLSON, PETER S., Crop Genetics International, Dorsey, MD
CARNES, MICHAEL G., Monsanto Company, St. Louis, MO
CARSWELL, GLETA, Ciba-Geigy Corp., Research Triangle Park, NC
CASO, OSVALDO H., Centro de Ecolisiologia Vegetal, Buenos Aires, ARGENTINA
CHAMBERLIN, MARK A., North Carolina State University, Raleigh, NC
CHANG, YIN-FU, Ciba-Geigy Corp., Research Triangle Park, NC
CHING, ALIX, University of Arkansas, Fayetteville, AR
CHOU, TAU-SAN, George J. Ball, Inc., West Chicago, IL
CHOUREY, PREM S., University of Florida, Gainesville, FL
CHRISTIANSON, MICHAEL, Zoecon Corp., Palo Alto, CA
CHU, MEL C., University of Illinois, Urbana, IL
CHU, IRWIN Y.E., George J. Ball, Inc., West Chicago, IL
CLOSE, KELLY, Sungene Technologies, Palo Alto, CA
COHEN, MIKE, Agrogene Plant Science, Inc., Delray Beach, FL
COHN, CHARLES, Martin Marietta Corp., Baltimore, MD
CONGER, BOB V., University of Tennessee, Knoxville, TN
CONRAD, BETSY, Crop Genetics International, Dorsey, MD
CONSTANTIN, MILTON J., University of Tennessee, Knoxville, TN
COOK, JUDITH, Sungene Technologies, Inc., Palo Alto, CA
CORDITS, JOHN M., Appalachian Fruit Research Station, Kearneysville, WV
DAVIS, MELANIE E., Ohio State University, Columbus, OH
DERMODY, TOM, University of Tennessee, Knoxville, TN
DEWALD, STEVE, University of Florida, Gainesville, FL
DEWALD, MARIA G., University of Florida, Gainesville, FL
DERNAR, HELEN DOBER, Standard Oil of Ohio, Cleveland, OH
DISNEY, B.J., Clemson University, Clemson, SC
DONNAN, ALVAN, Botanical Resources, Forest City, FL
DUESING, JOHN H., Ciba-Geigy Biotechnology, Research Triangle Park, NC
DURE, LEON, III, University of Georgia, Athens, GA
DURZAN, DON J., University of California, Davis, CA
DYKES, THOMAS A., Colorado State University, Fort Collins, CO
EARLE, ELIZABETH D., Cornell University, Ithaca, NY
EDGEWORTH, DEBORAH, Western Nurseries, Inc., Hopkinton, MA
EIZENGG, GEORGIA C., University of Kentucky, Lexington, KY
EL-BAKRY, AHMED, University of Kentucky, Lexington, KY
ESSER, PETER, NUNC, Roskilde, DENMARK
EVANS, GLENN F., Eli Lilly and Co., Greenfield, IN
FAHEY, JED W., Allied Corp., Savay, NY
FASSULIOTIS, GEORGE, U.S. Dept. of Agriculture, ARS, Vegetable Lab, Charleston, SC
FEJES, ERZSEBET, Advanced Genetic Sciences, Manhattan, KS
FELDMANN, KENNETH A., Ohio State University, Columbus, OH
FERCHAK, JOHN, Morris Arboretum, Philadelphia, PA
FIELDS, MARY, Ursinus College, Collegeville, PA
FIGLIOLA, SUZANNE S., Clemson University, Clemson, SC
FINER, JOHN J., Ciba-Geigy Biotechnology, Research Triangle Park, NC
FINK, CHARLES, Ohio State University, Columbus, OH
FORDHAM, INGRID, U.S. Dept. of Agriculture, ARS, Beltsville, MD

FOXE, MICHAEL J., University of Florida, Gainesville, FL
FRAMOND, ANNICK DE, Ciba-Geigy Biotechnology, Research Triangle Park, NC
FRANCISCO, NICHOLAS, Blodgett Gardens and Nursery, Orlando, FL
FRANKLIN, CHANDRA, North Carolina State University, Raleigh, NC
FRISCH, CARL H., Clemson University, Clemson, SC
GIESMAN, LARRY A., Northern Kentucky University, Highland Heights, KY
GLAVS, KENT R., Mellon Institute, Pittsburgh, PA
GORDON, PHIL, Yale University, New Haven, CT
GRAVES, DUANE, University of Tennessee, Knoxville, TN
GRAVES, CLINTON, H., Mississippi State University, Mississippi State, MS
GRAY, DENNIS J., University of Tennessee, Knoxville, TN
GREEN, C. ED, Molecular Genetics, Inc., Minnetonka, MN
GRESSHOFF, PETER M., Australian National University, Canberra, AUSTRALIA
GUPTA, PREM P., University of Florida, Gainesville, FL
HAISSIG, BRUCE E., U.S. Dept. of Agriculture-Forest Service, Rhinelander, WI
HALDEMAN, JANICE H., Erskine College
HARMS, CHRISTIAN T., Ciba-Geigy Biotechnology, Research Triangle Park, NC
HARRIS, RITA, Monsanto Agricultural Products Corp., St. Louis, MO
HARTMAN, ROBERT, Hartman's Bromeliads, Palmdale, FL
HENDERSON, WAYNE, Corn Products, Inc., Summit-Argo, IL
HENKE, RANDOLPH, University of Tennessee, Knoxville, TN
HENNEN, GARY, Oglesby Plant Labs, Hollywood, FL
HERMAN, PATRICIA L., Michigan State University, East Lansing, MI
HERMAN, EDWIN B., Agricell Report, Shrub Oak, NY
HERRERA, HEATHER G., New Mexico State University, Las Cruces, NM
HIATT, EMMETT E., Clemson University, Clemson, SC
HOLLAENDER, ALEXANDER, Council for Research Planning in Biological Sciences, Inc., Washington, D.C.
HOWLAND, GARY P., Clonal Products, Inc., Lakeland, CA
HUANG, FENG HOU, University of Arkansas, Fayetteville, AR
HUANG, LEAF, University of Tennessee, Knoxville, TN
HUGHES, KAREN W., University of Tennessee, Knoxville, TN
HUTCHINSON, JAMES F., Dept. of Agriculture, Victoria, AUSTRALIA
HYNDMAN, SCOTT E., Purdue University, West Lafayette, IN
BANNER, WARREN A., III, Phyton Technologies, Knoxville, TN
FENNELL, THOMAS A., III, Orchid Jungle Labs, Homestead, FL
IMBRIE, KATY, Purdue University, West Lafayette, IN
IZHAR, SHAMAY, The Volcani Center, Bet Dagan, ISRAEL
JARRET, ROBERT L., University of Florida, Homestead, FL
JAYNE, SUSAN M., Ciba-Geigy Biotechnology, Research Triangle Park, NC
JONES, JEANNE BARHILL, HERS, Dallas, TX
KAPP, FRED, Bessemer State Technical College, Bessemer, AL
KARLSSON, SYLVIA, University of Florida, Gainesville, FL

KAUL, KARAN, Kentucky State University, Frankfort, KY
KEESE, RENEE J., Clemson University, Clemson, SC
KERNS, HOLLY ANN, University of Illinois, Urbana, IL
KING, SANDRA, University of Delaware, Newark, DE
KINZY, TERRI GOSS, Standard Oil of Ohio, Cleveland, OH
KITTO, SHERRY, University of Delaware, Newark, DE
KLOC-BAUCHAN, FRANCINE, U.S. Dept. of Agriculture, ARS, Beltsville, MD
KNOBLETT, JOYCE, University of California, Davis, CA
KODRZYCKI, ROBERT, North Carolina State University, Raleigh, NC
KOERTING, LOLA E., Brooklyn Botanic Garden, Ossining, NY
KORBAN, SAFI, University of Illinois, Champaign, IL
KUHR, STEPHEN L., University of Nebraska, Lincoln, NE
KURTZ, SHARON, Harris Moran Seed Co., Inc., Rochester, NY
KUYKENDALL, HOLLI, Clemson University, Clemson, SC
LAIRD, DAVID W., Mississippi State University, Mississippi State, MS
LASPERBAUER, M.J., U.S. Dept. of Agriculture, ARS, Florence, SC
LAZZERI, PAUL A., University of Kentucky, Lexington, KY
LEE, NI, University of Georgia, Athens, GA
LENGEN, MARGARET, Uniroyal Chemical, Naugatuck, CT
LEVIN, ROBERT G., Miloda Ltd., Mobile Post Ashrat, ISRAEL
LINEBERGER, R. DANIEL, Ohio State University, Columbus, OH
LITVAY, JOHN, Institute of Paper Chemistry, Appleton, WI
LITZ, RICHARD E., University of Florida, Homestead, FL
LLOYD, GREG, Agrogene Plant Sciences, Inc., Delray Beach, FL
LOCKARD, JOYCE M., 271 Malabu Drive, Lexington, KY
LOCY, ROBERT D., NPI, Salt Lake City, UT
LOVE, STEPHEN L., Clemson University, Clemson, SC
LOWE, KEITH S., Stauffer Chemical Co., Richmond, CA
LUGO, ARIEL E., U.S. Dept. of Agriculture, Rio Piedras, PUERTO RICO
LUJAN, ANITA, University of Tennessee, Knoxville, TN
LUTZ, JOSEPH D., Agrigenetics Corp., Boulder, CO
MACE, SANFORD, Walt Disney World, The Land, Lake Buena Vista, FL
MACKINNON, CHRISTY, Colorado State University, Fort Collins, CO
MADISON, RITA A., The Dow Chemical Co., Midland, MI
MALCOLM, SANDRA K., Eli Lilly and Co., Indianapolis, IN
MALIGA, PAL, Advanced Genetic Sciences, Inc., Manhattan, KS
MALINICH, TIMOTHY J., Portland, OR
MARCOTRIGIANO, MICHAEL, University of Massachusetts, Amherst, MA
MARDEN, LINDA, University of Tennessee, Knoxville, TN
MATTHEWS, CHARLES, Clemson University, Clemson, SC
MCKENTLY, TEXIE, University of Florida, Orlando, FL
MENCZEL, LASZLO, Advanced Genetic Sciences, Manhattan, KS
MERKLE, SCOTT A., University of Georgia, Athens, GA
MEYERS, JAMES R., University of Kentucky, Lexington, KY
MICHLER, CHARLES H., Ohio State University, Columbus, OH
MIKSCHE, JEROME P., North Carolina State University, Raleigh, NC
MONTAGNO, THOMAS J., Ohio State University, Columbus, OH
MOSES, MENA S., NPI, Salt Lake City, UT
MOTT, RALPH L., North Carolina State University, Raleigh, NC

MULLIN, BETH, University of Tennessee, Knoxville, TN
MYERS, PAMELA J., NPI, Salt Lake City, UT
NEGROTTO, DAVID, University of Tennessee, Knoxville, TN
NELSON, BARBARA, U.S. Dept. of Agriculture, ARS, Vegetable Lab, Charleston, SC
NIBLETT, CHUCK, University of Florida, Gainesville, FL
NOVITZKY, ROSEANN T., Ciba-Geigy Biotechnology, Research Triangle Park, NC
OWEN, HENRY, Virginia Polytechnic Institute & S.U., Blacksburg, VA
OZIAS-AKINS, Peggy, University of Florida, Gainesville, FL
PASCHALL, JEANETTE, ITT Corp., New York, NY
PATTERSON, GORDON, Hershey Foods Corp., Hershey, PA
PEDERSON, GARY A., U.S. Dept. of Agriculture, ARS, Mississippi State, MS
PERANI, LAURA ANN, Calgene, Inc., Davis, CA
PETERSON, D.J., Pfizer Central Research, Groton, CT
PETERSON, MICHAEL A., W-L Research, Inc., Highland, MD
PETRACEK, PETER, University of Tennessee, Knoxville, TN
PIERSON, PAULETTE, Monsanto Agricultural Products Co., St. Louis, MO
PRIOLI, LAUDENIR, State University of Campinas, BRAZIL
PULLMAN, GERALD, Weyerhaeuser Co., Tacoma, WA
RAM, RAGHAVA, Sungene Technologies Corp., Palo Alto, CA
RANGASNAMY, NAGMANI, Maritimes Forest Research Center, Fredericton, CANADA
RAO, K. SANKARA, Indian Institute of Science, Bangalore, INDIA
REGER, BONNIE J., U.S. Dept. of Agriculture, ARS, Russell Research Center, Athens, GA
REPORTER, MINOCHER, Battelle, C.F. Kettering Research Labs, Yellow Springs, OH
RESCIGNO, CAROLE MARIE, Ceres 2000, Inc., Winter Haven, FL
REYNOLDS, JOHN F., The Upjohn Co., Kalamazoo, MI
RHODES, BILLY B., Edisto Experimental Station, Blackville, SC
RILEY, MELISSA B., Clemson University, Clemson, SC
ROBERTS, JEAN L., Lilly Research Labs, Greenfield, IN
ROMAINE, CHARLES PETER, Pennsylvania State University, University Park, PA
ROSTEN, ANDY, Del Monte Corp., San Leandro, CA
ROTH, L. EVANS, University of Tennessee, Knoxville, TN
RUPERT, EARLENE, Clemson University, Clemson, SC
SAGAWA, YONEO, University of Hawaii, Honolulu, HI
SCHALFENBERG, CARL, Yoder Brothers, Inc., Alva, FL
SCHLARBAUM, SCOTT E., University of Tennessee, Knoxville, TN
SCHNAPP, SHERRY RAE, Purdue University, West Lafayette, IN
SCHROLL, SANDY, Purdue University, Lafayette, IN
SCHWARZ, OTTO J., University of Tennessee, Knoxville, TN
SECKINGER, GARY, ARCO Seeds, Brooks, OR
SHAHIN, ELIAS A., ARCO Plant Cell Research Institute, Dublin, CA
SHARPE, DIANA Z., University of Florida, Gainesville, FL
SIMPSON, SANDRA F., FMC Corp., Princeton, NJ
SINGHA, SUMAN, West Virginia University, Morgantown, WV
SLAY, RAYMOND M., Purdue University, West Lafayette, IN

SMITH, GARRY A., Colorado State University, Fort Collins, CO
SOCHA, NANY D., Dow Chemical Co., Midland, MI
SOMMER, HARRY E., University of Georgia, Athens, GA
SONDAHL, MARO R., DNA Plant Technology Corp., Cinnaminson, NJ
SONGSTED, DAVE, University of Tennessee, Knoxville, TN
SPAINE, PAULINE, North Carolina State University, Raleigh, NC
STACY, GARY, University of Tennessee, Knoxville, TN
STAMP, ANNE-MARIE, North Carolina State University, Raleigh, NC
STANDARDI, ALVARD, U.S. Dept. of Agriculture, ARS, Beltsville, MD
STELZER, HENRY, Purdue University, West Lafayette, IN
STILL, PAUL, University of Florida, Gainesville, FL
STILL, DAVID, New Mexico State University, Las Cruces, NM
STRODE, RANDALL E., Agri-Starts, Inc., Longwood, FL
STUART, DAVID, Plant Genetics, Inc., Davis, CA
STYER, D.J., DNA Plant Technology Corp., Cinnaminson, NJ
SULLIVAN, June E., NPI, Salt Lake City, UT
SULLIVAN, KELLEN, Galena, OH
SUTTER, ELLEN, University of California, Davis, CA
SWEDLUND, BRAD, NPI, Salt Lake City, UT
TABAEIZADEH, ZOHREH, University of Florida, Gainesville, FL
TARDIF, RICHARD R., Phyton Technologies, Knoxville, TN
TECHMAN, THOMAS M., Techman Nursery, Yalaha, FL
TEMPLETON-SOMERS, KAREN, North Carolina State University, Raleigh, NC
THOMPSON, DAVID G., International Paper Co., Tuxedo Park, NY
TRIGIANO, ROBERT, University of Tennessee, Knoxville, TN
TRINITY, PHILIP, O.M. Scott and Sons, Marysville, OH
TSAY, HSIN-SHENG, Taiwan Agricultural Research Institute, Taichung, TAIWAN
TULECKE, WALT, University of California, Davis, CA
ULRICH, TOM, NPI, Salt Lake City, UT
UMBECK, PAUL F., Cetus Madison Corp., Madison, WI
VASIL, INDRA, University of Florida, Gainesville, FL
VASIL, VIMLA, University of Florida, Gainesville, FL
VEILLEUX, RICHARD E., Virginia Polytechnic Institute and S.U., Blacksburg, VA
VIDRA, ALICE, Agrogene Plant Science, Inc., Delray Beach, FL
WALKER, KEITH, Plant Genetics, Inc., Davis, CA
WANG, CHEN-YEN, University of Tennessee, Knoxville, TN
WANG, YI-CHANG, Purdue University, West Lafayette, IN
WANG, WEN CHUNG, University of Kentucky, Lexington, KY
WANSTREET, AUDREY, O.M. Scott and Sons, Marysville, OH
WEIGEL, RUSSEL, University of Tennessee, Knoxville, TN
WESTBROOK, ROBERT, Oklahoma University, Norman, OK
WETHERELL, D.F., University of Connecticut, Storrs, CT
WETZSTEIN, HAZEL, University of Georgia, Athens, GA
WILKINS, MUFF, Greenville, SC
WILLIAMS, ELIZABETH G., University of Melbourne, Parkville, AUSTRALIA
WILLIAMS, W. PAUL, U.S. Dept. of Agriculture, ARS, Crop Science Research Lab, Mississippi State, MS

WILSON, CLAIRE M., Council for Research Planning in Biological Sciences, Inc., Washington, D.C.
WOLFF, DAVID, Virginia Polytechnic Institute and S.U., Blacksburg, VA
WONG, JIM, Agrigenetics Corp., Boulder, CO
WRIGHT, DANIEL C., Brooklyn Botanic Garden, Ossining, NY
WRIGHT, MARTHA, Monsanto Agricultural Products Co., St. Louis, MO
YODER, DAVE, Gilroy Foods, Inc., Gilroy, CA
ZANKOWSKI, PAUL M., University of California, Davis, CA
ZAPATA, FRANCISCO J., International Rice Institute, Manila, PHILIPPINES
ZIEG, ROGER, George J. Ball, Inc., West Chicago, IL
ZIMMERMAN, ELIZABETH S., University of Minnesota, Minneapolis, MN
ZIMMERMAN, RICHARD H., U.S. Dept. of Agriculture, ARS, Beltsville, MD
ZYGMONT, NANCY JEAN, University of Tennessee, Knoxville, TN

INDEX

MIX
Papier aus verantwortungsvollen Quellen
Paper from responsible sources
FSC® C105338

If you have any concerns about our products,
you can contact us on
ProductSafety@springernature.com

In case Publisher is established outside the EU,
the EU authorized representative is:
Springer Nature Customer Service Center GmbH
Europaplatz 3, 69115 Heidelberg, Germany

Printed by Libri Plureos GmbH
in Hamburg, Germany